# The Universe Unfolding

# The Universe Unfolding

Edited by
**Sir Hermann Bondi**
and
**Miranda Weston-Smith**

CLARENDON PRESS • OXFORD
1998

*Oxford University Press, Great Clarendon Street, Oxford OX2 6DP*

*Oxford New York*
*Athens Auckland Bangkok Bogota Bombay Buenos Aires*
*Calcutta Cape Town Dar es Salaam Delhi Florence Hong Kong*
*Istanbul Karachi Kuala Lumpur Madras Madrid Melbourne*
*Mexico City Nairobi Paris Singapore Taipei Tokyo Toronto Warsaw*

*and associated companies in*
*Berlin Ibadan*

*Oxford is a trade mark of Oxford University Press*

*Published in the United States*
*by Oxford University Press Inc., New York*



*The chapters by Lyttleton, Chandrasekhar, Rees, Cowling, Wolfendale, Kendall, King-Hele, McCrea, Fowler, Atiyah, Radhakrishnan, Dyson, Longair, and Sciama have previously appeared as articles in the* Quarterly Journal of the Royal Astronomical Society, *and are reproduced (some in a modified form) by permission of Blackwell Science Ltd.*



*A catalogue record for this book is available from the British Library*

*Library of Congress Cataloging in Publication Data*
*The universe unfolding / edited by Hermann Bondi and Miranda Weston-Smith.*
*Collection of twenty Milne lectures, which started in 1977.*
*1. Astronomy. 2. Astrophysics. 3. Cosmology. I. Bondi, Hermann. II. Weston-Smith, Miranda.*
*QB51.U54 1997*
*520—dc21 97–14899 CIP*

*ISBN 0 19 851188 4*

*Typeset by Footnote Graphics, Warminster, Wilts*
*Printed in Great Britain by Bookcraft Ltd., Midsomer Norton, Avon*

# Foreword

## How the sponsorship began—the view from IBM

IBM is proud to have sponsored the series of twenty Milne Lectures, which started in 1977. Every year IBM receives a great number of requests for sponsorship, so that, to be a winner, an application must stand out from many other contenders. In the proposals from The Milne Society two features aroused our interest. There was an extremely distinguished list of academic supporters committed to the project, and in Milne's granddaughter, Miranda Weston-Smith, there was a youthful campaigner of unusual talents.

Computers are a vital tool in astronomy, so that IBM was glad to create a link with their academic users. IBM also noted that, in funding a lecture series on a specialist subject, the audience would be primarily young people, the students of Oxford University. Not only were they potential recruits to IBM, but in the future they were likely to be decision makers in science.

The Milne Society initially approached Nigel Corbally Stourton, who held certain purse strings at IBM, and he handed the matter over to me, as I was the Manager of Education and Scientific Relations. I became particularly enthusiastic since the proposals struck a personal chord. As a boy, I had been inspired by (Sir) Fred Hoyle's radio talks on 'The Origin of the Universe' and I decided to study science instead of classics. None other than Sir Fred Hoyle was to be the inaugural Milne Lecturer. After a meeting with Sir Hermann Bondi, who had helped me with a previous project, I felt deeply committed towards the Milne Lecture and successfully convinced IBM to fund the series.

Sir Fred Hoyle's Milne Lecture was memorable on at least two counts. In expounding his latest theory on comets and the origin of life, he ruffled not a few feathers of the astronomical establishment. Additionally, the event drew a huge audience, estimated at about five hundred people, who had to be accommodated in an overflow room, which was supplied with closed circuit television. Since that spectacular start the Milne Lecture has gone from strength to strength, and flourished for two decades, with a canon of speakers that reads like a *Who's Who* in astronomy.

October 1996 Brian Kington

# Contents

# Milne lecturers

**Sir Fred Hoyle** Formerly Plumian Professor of Astronomy in the University of Cambridge and formerly Director of the Institute of Theoretical Astronomy

**Raymond Lyttleton** Latterly Professor of Theoretical Astronomy in the University of Cambridge

**Subrahmanyan Chandrasekhar** Latterly Emeritus Professor, Laboratory for Astrophysics and Space Research in the University of Chicago; Nobel Prizewinner for Physics, 1983

**Sir Martin Rees** Astronomer Royal, and Royal Society Research Professor in the University of Cambridge

**Thomas Cowling** Latterly Emeritus Professor of Mathematics in the University of Leeds

**Sir Arnold Wolfendale** President, Institute of Physics, and Emeritus Professor of Physics in the University of Durham

**David Kendall** Emeritus Professor of Mathematical Statistics in the University of Cambridge

**Desmond King-Hele** Formerly Deputy Chief Scientific Officer, Space Department, Royal Aircraft Establishment, Farnborough

**Sir William McCrea** Emeritus Professor of Astronomy in the University of Sussex

**William Fowler** Latterly Emeritus Professor of Physics in the California Institute of Technology; Nobel Prizewinner for Physics, 1983

**Sir Michael Atiyah** Formerly Director of the Isaac Newton Institute for Mathematical Sciences in the University of Cambridge and Master of Trinity College, Cambridge

**V. Radhakrishnan** Formerly Director of the Raman Research Institute, Bangalore

**Thomas Gold** Emeritus Professor of Astronomy in Cornell University

**Dennis Sciama** Professor of Astrophysics in the International School of Advanced Studies, Trieste

**Freeman Dyson** Emeritus Professor of Physics in the Institute for Advanced Study, Princeton

**Malcolm Longair** Jacksonian Professor of Natural Philosophy in the University of Cambridge

**Joseph Taylor** James McDonnell Distinguished Professor of Physics in Princeton University; Nobel Prizewinner for Physics, 1993

**Robert Kirshner** Professor of Astronomy in Harvard University

**John Mather** Senior Astrophysicist in NASA Goddard Space Flight Centre

**Sir Roger Penrose** Rouse Ball Professor of Mathematics in the University of Oxford

# Introduction

There have long been many reasons for all who have an interest in astronomy to remember E. A. Milne with appreciation and gratitude. He contributed mightily (and often controversially) to several topics in the subject. Three are of particular note. First, his pioneering treatment of radiative transfer in stellar atmospheres, and thus of the formation of spectral lines, still excites admiration. He ventured to study the subject before the quantum theory had reached a stage at which it could be applied unambiguously to material at such a high temperature. In spite of this difficulty, he carried out investigations of lasting value on such matters as limb darkening and the ability to make deductions on the composition and temperature profile of stellar atmospheres from their observed spectra. A fuller description of Milne's central contribution to this subject will be found in Chandrasekhar's Milne Lecture in this volume.

A second topic on which he turned the searchlight of his insight was the structure of the stars. This was a field that Eddington had just established brilliantly, but he had rushed over a number of difficulties in rather a cavalier manner. Our understanding has been greatly helped by Milne's penetrating criticisms, which served to identify and clarify these problems. In the end, many of Eddington's guesses turned out to be fruitful, but our understanding owes much to Milne.

Lastly one must mention his keen attack on the problems then posed in cosmology by Hubble's discoveries. Not only did Milne manage to construct a remarkable theory of the universe using only special (and never general) relativity, he also showed, with McCrea, that some of the most important equations of the subject could be established on a purely Newtonian basis.

In all this work he had ideas of astounding originality. What I recall with particular joy is his introduction of the radar definition of distances (in terms of echo-time) many years before knowledge of radar became public. For it is this definition that enables the hoary and fundamentally non-relativistic 'rigid measuring rod' to be banished from physics, making it a far more logical subject. On a more personal note, I entered the subject of astronomy 'after hours', while working full-time on naval radars during the war. To find this close link between my daily work and evening interest was a pleasure. A decade and a half later I used Milne's notion of radar distances to put the foundation and presentation of special relativity on a far sounder basis.

I also greatly admire his brilliant idea that different types of clock may measure different times. Alas, this has not so far been found useful in science.

This volume gives us all yet another reason to honour his memory. For this is indeed a unique collection of twenty lectures given by outstanding astronomers over a period of twenty years. These Milne Lectures, that have been given annually in Oxford in the period 1977–1996, were intended to be on topics fructified by him, and this remit has been interpreted liberally. Indeed this is essential, as all these subjects have been in a state of rapid development since Milne's untimely death in 1950. Appropriately, the final lecture was given in the year of the centenary of his birth. Equally appropriately, the Lectures have been linked with his old Oxford college, Wadham.

What makes this collection of lectures so remarkable is that each gives a personal and authoritative view of a topic in a widely intelligible form. This really is 'astronomy for pleasure'! For once in such a collection the differences in outlook, attitude and style of the various authors actively add to the enjoyment of the reader, who will receive a beautifully rounded picture of what goes into the making of science.

One of my strongest dislikes is for the widely purveyed picture of scientists as objective and cool, working and thinking in a thoroughly impersonal manner. This absurd view is unhappily held by many who could enjoy science and contribute to it. It cannot be dispelled by just teaching more science for examinations (which may make it look even more dull and soulless), but only by seeing it in the making. Preferably such a picture should be conveyed by its active creators, presenting their corners of the subject in a thoroughly personal manner and in a widely intelligible form, an aim achieved by this volume of the Milne Lectures in an exemplary manner.

Finally I thank all those who made the Milne Lectures actually happen. The drive and determination of Milne's granddaughter, Miranda Weston-Smith, turned her concept into reality. Moreover the organisation of it all was carried through by her. The farsightedness and generosity of IBM (UK) Ltd made possible the vital financial support, but in addition IBM helped greatly through their constant involvement. Most of the Lectures were published one by one in the *Quarterly Journal of the Royal Astronomical Society*. I also wish to mention that the late Professor R. A. Lyttleton chaired the Milne Society through the majority of these years.

*Cambridge*
April 1996

Sir Hermann Bondi
President, the Milne Society

# Comets: a matter of life and death*

Sir Fred Hoyle

## 1 A definition of life

The atoms present in living systems are no different from similar atoms in non-living material. An atom of carbon in our bodies has the same individual physical properties as a carbon atom in a flake of soot. Yet the cooperative properties possessed by the arrangements of atoms in living matter are astonishingly different from those in inanimate material. You could store equal quantities of carbon dioxide and free hydrogen in a bell jar in the laboratory for an eternity and that is the way they would stay. But introduce a special kind of bacterium into the bell jar and the gases will go in short order into methane and water. The bacteria in question are of a special kind which in recent years have become known as archaebacteria. They form a special kingdom, apparently without microbiological connections to other bacteria or to the larger so-called eukaryotic cells of which ordinary plants and animals are built.

Defining the nature of life is one of those questions which becomes harder and harder, the more you look into it. Instinct tells us that a snail is radically different from a stone. But why is it different?

Let us start an attempt to answer this question by noticing that the issue of which assembly of molecules is most stable (the proportions of their constituent atoms being specified) depends on the temperature. At laboratory temperature the most stable form for a suitable mixture of hydrogen, oxygen and carbon is methane and water. But at the temperature of a wood fire the most stable form is carbon dioxide and hydrogen. Add to this the fact that mixtures of atoms do not necessarily reach their most stable forms under normal conditions. At higher temperatures like the log fire they usually do,

*The lecture was concerned with work done jointly with Professor N. C. Wickramasinghe who has kindly assisted me in recovering the lecture from my surviving notes of it. Some confirmatory later observational results are mentioned towards the end.

but at laboratory temperatures they may not. Start from methane and water in the laboratory and heat the mixture. Given adequate time, it will go to hydrogen and carbon dioxide. Now cool the mixture. It will not return to methane and water, no matter how slowly you cool it. Unless archaebacteria happen to be present.

The most stable forms for mixtures with atoms of hydrogen, carbon, nitrogen and oxygen, the commonest atoms in living material, behave in exceedingly complex ways at laboratory or lower temperatures. But the most stable forms are generally not attained by inanimate mixtures. They are attained, however, or nearly attained, when living organisms are present. It is this property of being able to reach the stable forms of mixtures at temperatures characteristically found on the Earth (300 Kelvin) that best defines the nature of life.

The mixtures on which living systems operate in this way are usually derived at higher temperatures. It is a general property that, as mixtures change their most stable forms with decreasing temperature, energy is released, not absorbed. Thus the ability of life forms to reach equilibrium states with lowering temperature provides them with energy sources. It is on such sources that life in its simplest forms depends.

A classification of the various sciences can be made in terms of the magnitude of the energy transitions which they involve. The biggest steps are those found in particle physics, running to thousands of millions of electron volts (eV). Accumulating basic information about such steps is difficult, and consequently expensive. Most of the basic data on which theories in particle physics are based could be written on three sheets of paper—data which has cost billions of dollars to obtain. In contrast, basic data at energy steps of a few eV, obtained in the 19th and early years of this century, cost sums measured only in thousands of dollars. This was the data of atomic physics that led in its highest theoretical form to the development of quantum mechanics. Because of its history, the habit of thought in physics is to relate subtlety directly to energy: the larger the energy step, the greater the measure of subtlety. Biology challenges this point of view. Biology says, conversely, that the lower the energy step, the greater the measure of subtlety. It is perhaps because of this inversion of attitude that physics and biology have become so sharply separated in our educational system.

The chemical bonds between atoms that have to be changed in reactions at low temperatures in order to achieve the most stable states are pretty much the same as in atomic physics, energy steps of a few electron volts. But whereas state-changes in atomic physics are achieved by radiation units, quanta, with energies that are the same as those of the changes in question (quanta of a few eV), in biology the state-changes are achieved with quanta of much lower energy, typically of about 1/40 of an electron volt. This is done by exceedingly subtle accumulations of energy, by pumping through sequences of metastable states. An analogy might be to surmount a high wall step-by-step up the many rungs of a ladder. Or one might think of charging an electric battery over a lengthy period and of then discharging it in short order.

In one important respect, biology also takes advantage of quanta with energies of about three electron volts, in the process of photosynthesis whereby carbon dioxide and water are reduced to oxygen and sugars, a similar result to the operation of archaebacteria but very different in its detailed operation. The operation is by no means completed through the higher energy quanta. Their absorption serves as an energy source which gives rise to a train of reactions of the more usual lower energy type.

The substances that control the small energy steps of biology are proteins. Proteins consist characteristically of linked chains of amino acids, of which 20 different kinds dominate the situation in biology. Only an exceedingly small fraction of the possible chains of amino acids are biologically relevant—just how small this fraction is will form the main topic of the next section. Also characteristically, a biologically important protein (enzyme) will have a number of amino acids in its chain, ranging from about 100 on the low side to about 1000 on the high side, with 300 as a fair average.

Although it is useful for diagrammatic purposes to think of a protein as a linear chain, enzymes actually take up amazingly complicated shapes in three dimensions, especially when suspended in water. The primary structure is a helix. Water is repellent to a fraction of the 20 amino acids and these, wherever they are in the chain, form a central region in the presence of water, so as to become shielded from the water by the others. This leads to a hugely complicated shape which is then given stability by chemical linkages, as for instance between the sulphur atoms that are present in just one of the 20 amino acids, methionine. Such linkages are like the spars used in buildings to give strength to a man-made structure. Notice that although these so-called disulphide bonds occur between amino acids that are adjacent in space, such neighbours are not usually neighbours in the original chain. They have been brought together by the manner in which the original chain has been folded by its water-repellant members. The extreme complexity of the situation is illustrated by a particular example in Fig. 1. Determining a structure like that shown in Fig. 1 is a difficult job for the experimentalist. So, not unnaturally, the experimentalist chooses the relatively simplest cases to study. Thus Fig. 1 is towards the simpler end of the class of enzymes.

Enzymes do not have simple surfaces. On the outside they are irregular with one specially important cavity, the so-called active site. The shape of this cavity is crucial to making chemical reactions 'go' that would not 'go' under inanimate conditions, like those reactions which promote the con-

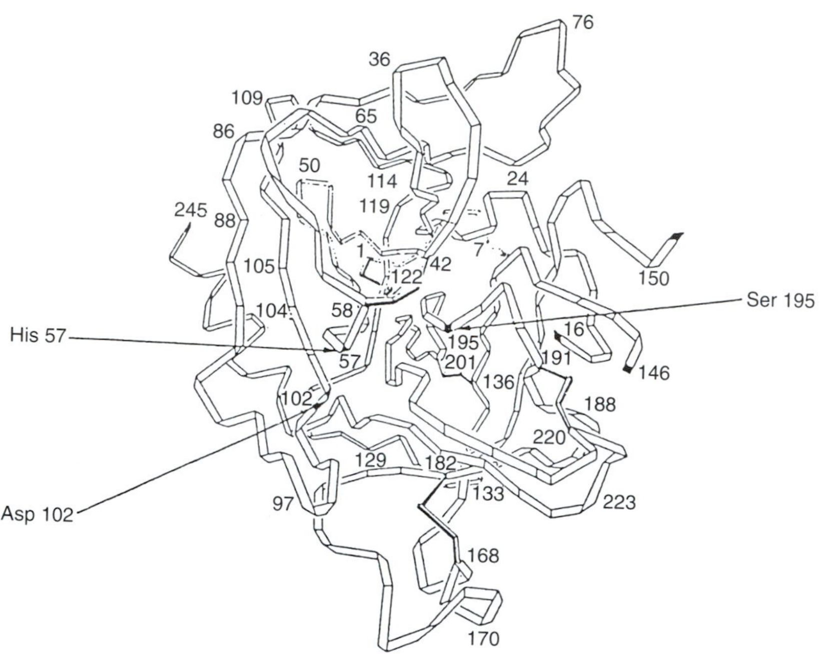


**Fig. 1** The structure of the enzyme α-chymotrypsin. (Adapted from *Enzymes*, 3rd edn (Boyer, P. D. ed.), Vol. 3, p. 194.

version of carbon dioxide and hydrogen to methane and water in the case of the archaebacteria. What happens for a particular reaction is that the chemicals involved fit with startling precision into the cavity of the relevant enzyme, not just as pieces of a jigsaw fit, but in a specially reactive orientation with respect to each other. Moreover, the chemicals are jostled so as to promote the reaction by the amino acids with which they are in contact, the amino acids forming the active site. The jostling is not random. It is organised in the sense of the ladder-over-a-wall analogy. When the reaction is completed, with the reacting chemicals having changed their shapes, they no longer fit the enzyme cavity as before. Consequently they break away from the cavity, freeing it to promote the same reaction yet again—and again and again, in the manner of a catalyst. A catalyst is defined in chemistry as a substance which promotes a chemical reaction without itself being changed. Enzymes are catalysts analogous to man-made catalysts, but they are millions of times more effective.

Since an enzyme depends for its operation on a hugely precise matching of its shape to that of chemicals in a particular reaction, and since chemicals in different reactions have different shapes, enzymes are highly specific. Each promotes one particular reaction but not others, which is why living systems need many enzymes. The simplest living system needs 2000 or more enzymes, each matched to a particular reaction among the complex network required to sustain the system. In a complex system like ourselves, upwards of 100 000 highly specific amino acid chains are probably required to sustain the human network, although a precise count is hardly possible because of the network's immense complexity.

A living system has need of many copies of each of its enzymes. A literal accurate copying, amino acid by amino acid, of a structure like Fig. 1 would be so difficult as to be hardly feasible. Just as we ourselves copy buildings from blueprints rather than by copying brick-by-brick or stone-by-stone, so copies in living systems are obtained from a blueprint. The blueprint is carried by four characteristic markers (nucleotides) read in blocks of three (codons) on the now-famous double-helix structure of DNA. The reading process is also vastly complicated. It is done mostly by the enzymes themselves. The first step is to construct an intermediate sequence of blueprints (the various forms of RNA). It is a case of the master blueprint of DNA producing through enzymic activity (not through its own activity—by itself DNA is very inactive) blueprint A, which then produces blueprint B, which produces blueprint C, . . . , until ultimately a considerably simplified and fragmented form is used to construct the enzyme in question. The raw materials for constructing the enzyme are separated amino acids which have to be linked together in the order prescribed by the eventual blueprint. A similar logic is used in constructing a man-made building. The architect's drawings are more complicated than those which are issued to individual workmen. But the human situation is simpler than the biological situation by a huge margin.

If one thinks there was a time before which life did not exist, a conundrum arises in understanding its origin. Which came first, the blueprint for an enzyme or the enzyme itself? If one says DNA came first, the problem is that DNA is inactive. If one says the enzymes came first, enzymes apparently cannot copy themselves. The favoured answer among biologists is to say that an intermediate blueprint came first, a blueprint expressed by RNA not by DNA. In recent years, RNA has been shown to possess a limited degree of activity of its own, although whether the activity is sufficiently diverse as to be capable of maintaining a replicative system remains a question. The problem is one already hinted at above. The bond strengths, whether in RNA or proteins, are in the region of 4 eV, much too strong to be broken thermally. Thus a failure to find a working system at the first joining of atoms stops there. Without enzymes to break the bonds a second trial cannot be made, except by flooding the material with so much energy that everything is smashed back into the constituent atoms. But such extreme violence cannot lead anywhere, since floods of energy would also destroy anything useful that might arise. There is but one way out of this logical impasse, in my opinion, which is to make trials, not repeatedly on a limited sample of material as in Darwin's 'Warm little pond', but to make just one trial on a breathtakingly large number of samples. Just how large the number that would be needed before anything interesting happened will be the topic of the next section.

## 2 The information content of life

With the invention of computers in the 1940s the idea of measuring the information content of a message was born, and a mathematical theory of how this might be done emerged, to widespread applause from the scientific community. However, one might be pardoned for not joining too vigorously in the applause, because the applicability of the mathematics seemed too restricted in its scope to be of much interest. What one would really like to be able to do would be to give a logical numerate meaning to the difference in the information content in the following two messages, supposed to reach the German Chancellory in Berlin on 1 June 1944:

*Message 1*
This morning the British Prime Minister, Winston Churchill, ate bacon and eggs for breakfast. Yesterday he smoked eleven cigars and sniffed brandy throughout the day. It is anticipated he will do the same on the 6th of the month.

*Message 2*
Early on the 6th, the Allies will attempt to land very large forces on the Normandy beaches, from St Germaine in the west to Ouistreham in the east. There will be no landing in the Pas de Calais.

The mathematical theory of information does not attempt to grapple with cases like these. Yet it is situations like these that are most important. Similar but still more awkward problems arise when the information content of life is at issue. Were a refined theory available for estimating the information content of DNA it would, in our opinion, be immediately apparent from its overwhelming content that life could never have arisen on a minuscule planet like on Earth. It would be seen that, to match the information content of even the simplest cell, nothing less than the resources of the entire Universe are needed. This is an opinion that can be backed up by making a shot at estimating the information content, noticing that if on reasonable grounds the answer turns out as vast beyond all precedent, it does not matter in its implications just how vast it really is, because one huge number would have the same implications as another.

For every enzyme needed to make a chemical reaction 'go' in the large complex of reactions that maintains a living cell, a number can be estimated in the following way. Take first the total number of proteins that can be constructed by assembling at random the 20 biologically significant amino acids in chains of the same length as the enzyme in question, a length typically of some 300 amino acids. For such a length this number is unequivocal. It is about $10^{390}$, i.e. 1 followed by 390 zeros. Next, divide by the number of possibilities in this set that serve to make the particular chemical reaction 'go' at an adequate speed to sustain the cell, a number $f$ on the average, say. Do this for every enzyme, 2000 in the case of a simple cell, 100000 for a complex organism like ourselves. The result for the information content is then:

$$(10^{390}/f)^{2000} \quad \text{simple cell}$$
$$(10^{390}/f)^{100\,000} \quad \text{complex organism}$$

The situation is still unequivocal. Scope for argument arises only when we come to estimate the likely average value of $f$. We saw in the previous section that an enzyme has to possess exceedingly specific properties in relation to the reaction which it catalyses. It has to curl up into a three-dimensional structure with a surface cavity that provides a precise and special fit to the shape of the reacting chemicals. Moreover, the amino acids forming the cavity, the active site, have to be capable of jostling the reacting chemicals in a highly organised way. These properties depend crucially, not only on particular amino acids which form the active site, but on the positioning of the water-repellant amino acids which play a critical role in deciding the three-dimensional structure. Another necessary property not mentioned in the preceding section is that an enzyme must be controllable. It must be capable of being switched on and switched off by chemical agents controlling the behaviour of a cell. Uncontrolled behaviour is what happens with cancers and this is to be avoided. Clearly all these drastic and precise requirements will not permit $f$ to be unduly large, nothing like as large as the number $10^{390}$ appearing in the above formulae.

An extreme position would be to say that all these special requirements demand that the chain of amino acids be unique for each enzyme, requiring $f = 1$. This appears to be close to the truth in some cases. The protein histone-4 is found in both plants and animals and it has essentially the same amino-acid structure in every organism. Little or no variants have been permitted throughout biological evolution. Human DNA has some thirty distinct genes coding for histone-4. Variants are found among the thirty but they are all of the kind that lead to the same chain of amino acids (same-sense mutations). Other proteins are not as restrictive as histone-4, however.

But every enzyme that has been examined in detail has been found to vary among plants and animals only to a moderate degree. Summing up what has been found as fairly as possible, about one-third of the amino acids in a typical enzyme are obligate, which is to say a particular amino acid must occupy each of about 100 positions in a chain of 300. The remaining 200 positions are by no means free choices. Each of them can be occupied by three or four among the bag of 20 amino acids, not by any member of the bag. Arguing thus leads to $f = 4^{200} = 10^{120}$ (to sufficient accuracy) and $(10^{390}/f) = 10^{270}$, giving the following for the information content:

| | |
|---|---|
| $10^{540\,000}$ | simple cell |
| $10^{27\,000\,000}$ | complex organism |

These are not 'astronomical numbers', the description used popularly for large numbers. They are hugely greater than astronomical numbers, the largest of which is obtained by dividing the distances of the most remote galaxies, $10^{28}$ centimetres, by the scale of an atomic nucleus. This yields the number $10^{40}$,

$$10\,000\,000\,000\,000\,000\,000\,000\,000\,000\,000\,000\,000\,000$$

when written out in full, certainly a big number, but nothing to compare with the above numbers, which can be considered by thinking how long one would need to write them out in full, and how much paper would be used up in the process. Reckoning you could write a zero in every second, it would take only 40 seconds to write out $10^{40}$. But it would take nearly 2 years working 12 hours a day to write out $10^{27\,000\,000}$, and it would use up some 10,000 sheets of paper.

Evidently then, we are dealing with *superastronomical* numbers on a grand scale. Moreover, when one ponders over the unequivocal expression $(10^{390}/f)^{100\,000}$, it is clear that no reasonable choice for $f$ can possibly lead to anything other than a hugely superastronomical number. Cavilling over the value of $f$ will not lead to anything different. One superastronomical number is the same as any other in its significance, for it means that if we are to understand anything of the nature and origin of life we must search the universe for other superastronomical numbers. Only when we can match the superastronomical number from biology with a superastronomical number from cosmology can we expect to arrive at an insight into biology.

## 3 Immense numbers from cosmology

In this section we shall search for corresponding superastronomical numbers from cosmology. We begin by noting that, with the exception of hydrogen, all elements originate in stars, especially in supernovae. Thus stars provide the feedstock of life, just as they provide the inanimate materials of everyday life; the iron in the steel bodywork of a car, for example.

The distribution of the elements is moderately uniform throughout our galaxy, and is believed to be much the same in most other galaxies. There is thus an approximately uniform distribution of the abundances of the elements throughout the universe. This cosmic distribution mirrors quite well the distribution of the life-forming elements, except that hydrogen is much more abundant cosmically than it is in living material. Carbon, nitrogen and oxygen are about ten times more abundant both cosmically and in life than the next group consisting of sodium, magnesium, silicon, phosphorus, sulphur, chlorine, potassium, calcium and iron, while the latter are about a thousand times more abundant than the trace elements. If one had to pick out an exception it would be phosphorus, which is some ten times more abundant in life than it is cosmically.

The complexity of the network of chemical reactions which define the nature of life depends crucially for its remarkable versatility on the properties of the carbon atom. Thus, in estimating the quantities of potential life-forming material in various places within the universe, it is sufficient to specify the quantity of carbon, since the other elements follow along with the carbon in generally the required proportions. How these estimates go for a number of locales is shown in Table 1.

It is seen that superastronomical numbers appear in the second part of the

**Table 1** Estimates of the quantity of carbon in various places in the universe

| **Place** | **Amount of carbonaceous material (g)** |
|---|---|
| Earth | $10^{23}$ |
| Outer regions of solar system (Uranus, Neptune, comets) | $10^{30}$ |
| Molecular cloud (e.g. Orion Nebula) | $10^{35}$ |
| Interstellar material through our galaxy | $10^{40}$ |
| All detectable galaxies | $10^{50}$ |
| Limit for interrelated quantities of material in big-bang cosmology: | |
| Interval in Hubble times (about $12 \times 10^9$ years) | Quantity of interrelated carbonaceous material in steady-state cosmology (g) |
| 1 | $\sim 10^{50}$ |
| 10 | $10^{63}$ |
| 100 | $10^{180}$ |
| 1000 | $10^{1350}$ |
| 1 000 000 | $10^{1\,300\,000}$ |
| 100 000 000 | $10^{130\,000\,000}$ |

table, but not in the first part. The meaning of the quantities in the second part is that if one starts with a chemical message (as for instance DNA is a chemical message) at a particular place at a particular time, and if the message can be copied, then after the time intervals in the first column the message will have been spread by copying through the quantities of material in the second column. In the extreme case of the last line of the table, after a hundred million Hubble times (about $1.2 \times 10^{18}$ years) the message will be spread through $10^{130\,000\,000}$ grams of material, a number that is in a class which matches the biological superastronomical numbers of the preceding section. This suggests that life might be produced in a time interval of $10^{18}$ years provided the cosmology is steady-state.

Table 1 gives scope for a great deal of discussion. Here we shall simply indicate how the vast quantities of carbonaceous material in the second part of the table have been calculated. Biological cells typically have sizes of the order of one ten-thousandth of a centimetre, which happens to be just the size at which small particles are effectively repelled by the pressure of light, picking up speeds in the galaxy from starlight of several hundred kilometres per second. This is sufficient to spread a biological message everywhere through a galaxy in a time even less than a single Hubble time. It is indeed sufficient, just about, to spread the message from our galaxy to another, but only between neighbours. A still more powerful mode of spreading turns on the properties of iron as it is expelled from a supernova.

When metallic vapours are slowly cooled in the laboratory, condensation eventually occurs, not into more or less spherical globules, but into threads of 'whiskers'. Diameters of whiskers are typically about a millimetre, giving the very large ratio of about 100 000 : 1 for the length to diameter. Such metallic particles are extremely strongly repelled by radiation in the far infrared region of the spectrum, and since molecular clouds in galaxies emit radiation strongly in the far infrared, whiskers can be repelled from galaxies into extragalactic space at speeds upward of ten thousand kilometres per second, when distant galaxies can be reached from the galaxy of their origin in only a single Hubble time. About a million galaxies can be reached in this way. Whiskers from galaxies mix on this scale, the products of a million galaxies together, thereby producing a very uniform distribution for iron whiskers in extragalactic space.

Of course iron carries no biological message in itself. But contiguous particles in a near vacuum have a marked tendency to stick together. A carbonaceous particle carrying a biological message could quite well stick to an iron whisker, hitch-hiking a lift across extragalactic space. One is reminded of the story of how the birds, after quarrelling as to who among them should be King, decided that it should be the one that in a trial was able to fly highest. Each kind fell back in the trial, leaving the eagle eventually to soar above the others. Yet even the eagle at last reached the height of exhaustion. When it did so, the wren, which had so far travelled unnoticed on the eagle's back, took off and with an effort attained a few feet more. So it came about that the wren became the King of birds.

The expansion of the universe does the rest of the spreading of the message. After reaching some million galaxies in the first Hubble time, the expansion increases, by the exponential factor, the radius of the cosmological region containing the message for every succeeding Hubble time. After a million Hubble times the radius of the region therefore increases by $e^{1\,000\,000}$, and the amount of material in the region increases by $e^{3\,000\,000}$. In the latter connection it will be recalled that the essential difference between big-bang cosmology and steady-state cosmology is that the universe does not empty as it expands in the steady-state case. It is this critical property of steady-state cosmology that leads to the vast quantities of material in the second part of Table 1, quantities that match biological requirements.

There is no reason why the standard qualitative picture in biology of the origin and evolution of life should not be given expression in this way. But it must be given expression in a cosmological setting, and the cosmology must be steady-state. These are startling conclusions on which a great deal of evidence can be brought to bear. But for the present let us conclude by mentioning another way of arriving at significant superastronomical numbers. Start with a single living cell, say a bacterium. A typical doubling time by binary fission for a bacterium supplied with appropriate nutrients would be two or three hours. Continuing to supply materials, the initial bacterium would generate some $2^{40}$ bacteria in 4 days, yielding a culture of the size of a pinhead. Continue for a further 4 days and the culture, now containing $2^{80}$ bacteria, would have the size of a village pond. Another 4 days and the resulting $2^{120}$ bacteria would have the scale of the Pacific Ocean. Yet another 4 days and the $2^{160}$ bacteria would in quantity be comparable to a molecular cloud like the Orion Nebula, and another 4 days, bringing the total time interval to only 20 days, and the scale in quantity would be that of a million or more galaxies. In a year there would be some $2^{3650}$ bacteria and in a thousand years the total would be $2^{3\,650\,000}$ bacteria. Thus biology yields superastronomical numbers as well as depending on them.

Nutrients could not be continuously supplied, it might be objected. Yet, cosmically speaking, the situation is nearer to a continuous supply than one might at first think. Formaldehyde ($COH_2$) is built as a weakly-bound molecule from the two commonest molecules in the universe, carbon monoxide ($CO$) and hydrogen ($H_2$). Although formaldehyde is not itself a substance of surpassing interest, take five or six formaldehyde molecules, swop atoms a little from one to another and join them appropriately, and you have all the sugars, the driving foodstuff of biology. Eliminate a water molecule between sugars and you have all the carbohydrates. Join sugars through nitrogen atoms and you have materials like the shells and claws of prawns and lobsters. A continuous supply is pretty well what one really does have, in fact. It is rather here on the Earth where supply is limited, not in the universe at large.

Let us now draw together what can be said from the above considerations. At first sight it might seem from Table 1 that, whereas the steady-state theory is readily consistent with the existence of life, the big-bang theory is not. But this is to overstate the situation. The numbers show that once a replicative system emerges in the steady-state theory, a replicative system of any kind, it will spread throughout volumes of space and quantities of material that increase exponentially with time. Life need not have arisen all in one go. There could be a sequence of steps A, B, C, . . . , with accumulating associations AB, ABC, ABCD, . . . , one step being piled on another in an evolutionary process, again provided the reproductibility criterion is satisfied at all stages. Then in the steady-state theory the probability of life arising is of the order of the *sum* of the probabilities for each of A, B, C, . . . taken separately. For big-bang cosmology, on the other hand, because of the limited time scale, which prevents spreading exponentially to superastronomical numbers, the probability of life arising is of the order of the *product* for each of A, B, C, . . . taken separately.

Although the probability of life arising in big-bang cosmology is therefore superastronomically small, it is possible to defeat even a superastronomical improbability in an open cosmology. Somewhere among the *infinite* amount of material in an open cosmology, even a superastronomical improbability will occur. The difference between steady-state and big-bang is that when life arises in the former case it will be found to be spread throughout a superastronomical quantity of material. In the big-bang case, however, the appearance of life would be essentially a point affair, not spread throughout any large quantity of material. To proceed further it is evidently necessary, therefore, to take a look at the extent to which life appears to be spread throughout the visible universe. A thoroughgoing discussion of this question would unfortunately go beyond present-day knowledge, but some progress towards it can be made, as will be shown in the final section. To conclude the present discussion it is worth noting the extent to which the combination of big-bang cosmology with the point-appearance of life is a creationist position, with both the origin of the universe and of life being two acts of special creation.

## 4 Comets as the distributors of life

It would be a natural consequence of life being a cosmological development of the universe, that the basic components of life should have been present in our galaxy at the time of its origin. The basic components would be viruses, single-celled eukaryotic cells or even genetic fragments down to the level of a single gene. It seems helpful to think of the origin of such components as an extragalactic or cosmological phenomenon, while the assembly of the components into the genetic structures leading to multi-celled organisms can be seen as a phenomenon taking place within our galaxy itself. The immediate question then is whether there is evidence to support the position that small particles of biological origin are widespread throughout the galaxy, and this is the question now to be considered.

The particles responsible for the extinction of visible starlight are of the order of 0.5 microns in their typical dimensions, and this agrees with the general scale of the biological components just mentioned. Moreover, the amount of the extinction requires interstellar grains to be mostly composed of the CNO group of elements, since elements heavier than these such as Mg, Al, S, are not present in the interstellar medium in adequate quantity. This is a strong indication that the bulk of the grains must be of an organic nature, although it does not in itself prove that they must be organics of biological origin. Yet biology is by far the most efficient producer of organic material, as the above discussion of the potentially explosive character of bacterial growth made clear.

Although the plot of the visual extinction of starlight as a function of wavelength appears to be a simple curve, it has proved remarkably difficult to fit it accurately with the effects of grain models, despite many free parameters being used in the models, explicitly the freedoms to adjust compositions and size distributions. Throughout the 1960s it gradually became apparent that the trouble lay with the real part of the refractive index of the material of the particles, which would minimally be about 1.35 and more typically around 1.55. Yet lowering the value to 1.15 was shown to have a far more beneficial effect than size variation and compositions. But what sort of particles would have a real part of the refractive index as low as this? For solid particles, only hydrogen. But it eventually transpired that the temperatures of the grains could not be low enough, about 2 K, for them to be condensed hydrogen, a position which eventually forced the hypothesis that the grains must be hollow with some 70 percent of their interior a vacuum. This is something that happens inevitably for dried bacteria.

With water evaporated out of bacteria, just about 70 percent of their interiors become a vacuum. So a quite definite calculation, without any adjustable parameters, could be made. Take the laboratory size distribution of a suitable spread of bacteria, say of spore-forming bacteria, and using the known refractive index of biomaterial, calculate the expected interstellar extinction as a function of wavelength. The result obtained by Hoyle and Wickramasinghe, shown in Fig. 2, is far superior to anything which had come out of

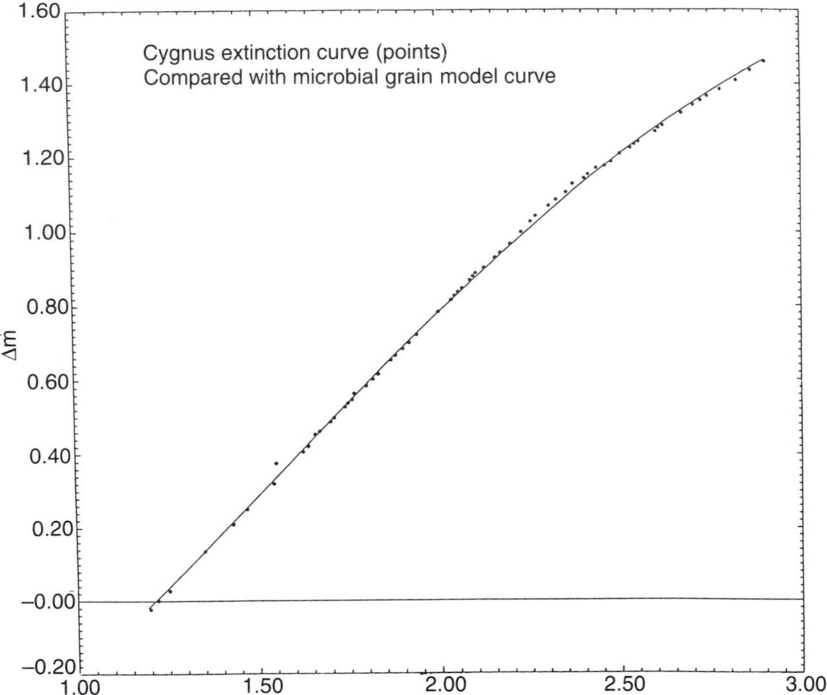


**Fig. 2** Extinction in the visible region of the spectrum (points) compared with a hollow bacterial model.

previous models, despite the many adjustable parameters contained in those models.

There is a continuum source of infrared radiation at an effective temperature of about 1100 K close to the galactic centre.* Passing through the interstellar grains along a path of length about 10 kpc to the Earth, the radiation is subject to absorption by the grains, thereby showing up the absorption in the near infrared of the grain material. Laboratory measurements for bacteria led Al-Mufti, Hoyle and Wickramasinghe to expect the absorption curve shown in Fig. 3, which they advanced as a prediction of what would eventually be observed. Results for the waveband 2.9–4 microns then obtained by D. T. Wickramasinghe and D. A. Allen for the galactic centre GC-IRS7 are shown by the points in Fig. 3.

Laboratory measurements have shown dry bacteria to possess a strong absorption peak slightly short of 2200 *A*, corresponding very closely to the observed absorption peak of extinction in the ultraviolet. And with the addi-

*The data in this paragraph and the next have become available since the lecture was given.

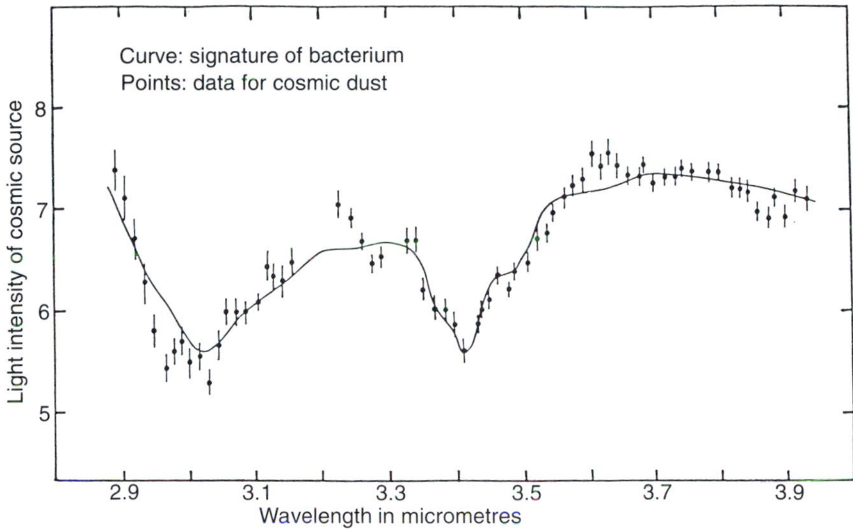


**Fig. 3** Observations of infrared flux from the galactic centre infrared source GC-IRS7 (points) compared with predicted curve for desiccated bacteria.

tion to bacteria of a biological form of silicious material, the absorption curve in the farther infrared shown in Fig. 4 has been obtained. Observational determinations are shown by the points of Fig. 4. Silicious material of biological origin, such as is found in diatoms, sponges, etc. differs from glasses and mineral silicates in that $Si0_4$ tetrahedra are more irregularly orientated, a condition which appears necessary if the expected absorption is to fit the data.

Granted that the general picture is correct, how much more confirmatory evidence than this could be expected from astronomical observations alone? Not much more from the galaxy at large but perhaps a good deal more from the solar system in particular. Following the picture further, since the solar system was formed from interstellar material it contained life from the beginning of its history. Much would be destroyed by the heat developed within the main solar nebula, but some biomaterial would survive on the distant periphery of the system where cometary bodies condensed in the regions and beyond the planets Uranus and Neptune. There is a band of temperature, several hundred degrees centigrade wide, between an upper limit at which life could survive and a lower limit at which water would be liquid, and living cells that happened to be present within this band would be capable of explosive replication. The amount of carbonaceous material involved in this facet of the picture can be estimated at $10^{29-30}$ grams, vastly greater than the $10^{18}$ grams in the Earth's present-day biosphere, sufficient to produce some billions of comets, which according to this picture would contain not merely

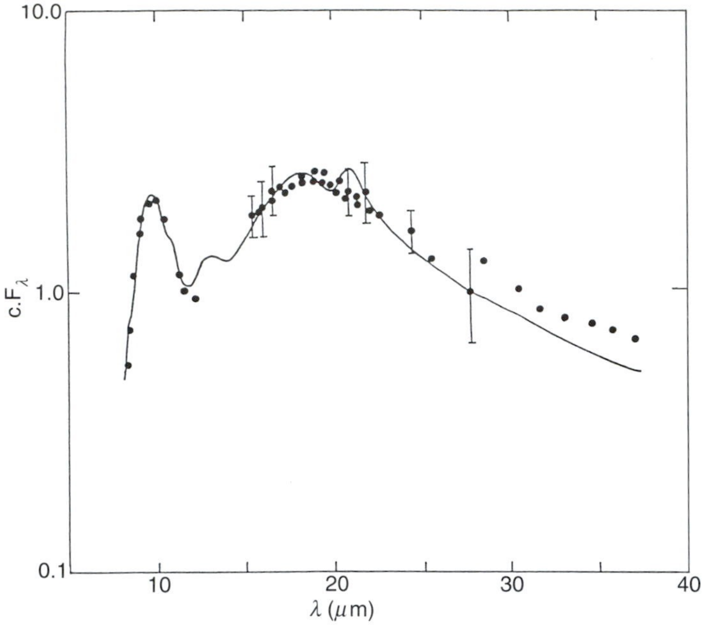


**Fig. 4** Observations of the infrared flux from the Trapezium Nebula (points) compared with a model including a combination of bacteria and silica.

organic material but micro-organisms and other biochemical components of life.

As the protoplanets which eventually became Uranus and Neptune grew in mass, a stage was reached at which they were able to change the orbits of cometary planetesimals (in close encounters) from a generally circular form to highly elliptical orbits. These orbits possessed the radius of the orbit of the protoplanet around the Sun as one of the nodal distances. There were two cases, one with the nodal distance at perihelion and the other at aphelion. The former led to a proportion of comets being sprayed outwards, some far beyond the orbit of Neptune, some even being expelled entirely from the solar system, and so returned back to the interstellar medium from which their material originally came. But this happened after possible biological processing and replication had had ample opportunity to take place, and therefore served to restock the supply of micro-organisms in interstellar space. Cometary bodies in this first group that *were* retained by the solar system became the outer comet cloud often referred to as the Oort cloud. The second group, those with the nodal distance at or close to aphelion, possessed orbits that took them to the inner regions of the solar system. Collisions with the terrestrial planets occurred, which led in particular to the

Earth acquiring its supply of volatile materials, water and carbonaceous material especially. It is likely that it was in this way that the Earth first acquired life, thereby supplying a connection whereby cosmic biology became transferred to our planet.

Do we have any proof of this? Thinking of the entire history of the Earth as a 24-hour clock, the Earth seems to have been subject to intense collisional bombardment from 0.00hr. to about 3.30. On the other hand, there is fossil evidence of the existence of photosynthetic bacteria existing already at 4.00. It follows that the conventional view of life originating here on Earth requires all or most of the astonishingly unlikely biochemical structures described above in Section 2 to have been discovered between 3.30 and 4.00, a thin window of time in which the most remarkable events to have happened anywhere in the universe are supposed, on no supportive evidence at all, to have occurred. This does not seem a speculation in which it would be wise to put much trust. Indeed it does not seem too much to condemn it with some asperity as pre-Copernicanism run rampant. Recent work (subsequent to the lecture) on the $^{12}C/^{13}C$ ratio in rocks with ages going back to 3.30 indicates that photosynthetic life was already present at the earliest moment at which life would have existed on the Earth. It seems to have begun just as soon as the rain of missiles at the Earth's surface permitted it to exist.

## 5 Other proofs

It is a basic tenet of orthodox evolutionary theory that organisms cannot possess any distinctive property that did not aid their survival at some stage in their past history. Yet bacteria do possess such properties, should one seek to argue in pre-Copernican fashion, namely that life originated on and has been confined to the Earth. Provided care is taken to evaporate out water and to exclude high concentrations of free oxygen, bacteria can be taken down to absolute zero and then returned to room temperature without their survival being impaired. They can be taken not only down to zero pressure but up to pressures of at least 10 tons per $cm^2$, again without survival being impaired. They can also be flashed momentarily in the absence of water and oxygen up to about 1000 K. They possess astonishingly complex repair processes against damage by hard radiation, so much so that bacterial cultures given adequate nutriments can grow in the highly radioactive interiors of nuclear reactors.

Likewise they show some remarkable properties when exposed to ultraviolet light. There is a specific dose of ultraviolet light that will serve to inactivate, say a half, of the bacteria in a sample. Then repeat the dose and a further half will be inactivated, and so on, *but* only for a while, a stage eventually being reached at which the remaining cells are substantially unaffected by further doses. It has been shown by numerous investigators that this ultimately remaining fraction is *not* a special radioactive resistant strain which happened to be present among the more numerous members of an otherwise radioactive susceptible culture. It seems rather that those which survive as the radiation dose is repeated, gradually learn to cope. They switch on repair processes, a complex process that would hardly have evolved on a planet whose atmosphere is essentially completely shielded with respect to the radiation in question.

Additionally any dense interstellar cloud of carbonaceous particles possesses an overriding mechanism whereby its interior is automatically shielded against damage by ultraviolet light. This happens because damage leading to the appearance of native carbon occurs first at the outside of the cloud. Then the carbon is highly absorptive of ultraviolet, the outside preventing anything of significance from reaching the central regions.

All these bacterial properties, irrelevant for survival on the Earth, become highly relevant for survival in interstellar space, and for making successful steps from one body or cloud to another, as for instance from comets to the Earth's atmosphere and thence down to the surface of the Earth. This can be seen as indirect proof that bacteria are essentially space travellers, as the above reasoning suggests them to be.

It might be thought feasible to search the Earth's atmosphere for evidence of micro-organisms incident from space, or even to sample interplanetary space directly to obtain samples of material from some comet. But these possibilities are subject to objections, especially as there is a much simpler way to proceed. Micro-organisms from the atmosphere have indeed been obtained, even from altitudes as high as 80 km, the results being bypassed, however, by the claim that the organisms found have come up from the ground. Since this is a claim that is hard to disprove, despite the normal inhibition of vertical air movements in the stratosphere, it becomes unrewarding to attempt to prove the theory in this way. And micro-organisms would be exceedingly hard to recover from interplanetary space, because of high relative speeds between a space vehicle and possible particles, while a mission to an actual comet would be inordinately expensive. Far better then to take a look at the possibility of space-incident viruses and bacteria interacting pathogenically with terrestrial plants and animals. Every plant and animal can be viewed as a sensitive detector for new pathogens, bacteria and viruses that may be incident from space.

This idea has been developed in considerable detail subsequent to the Milne Lecture, notably in the two books *Diseases from space*, 1979, J. M. Dent, London, and *Viruses from space*, 1986, University College Cardiff Press, by Professor N. C. Wickramasinghe and myself. The situation as it was actually described in the Milne Lecture was a somewhat brief synopsis covered by the following verbatim passage:

Reports of the sudden spread of plagues are common in the histories of many countries. The most recent such disaster was the 1918–1919 influenza pandemic in which 30 million people died. Different epidemics, scattered throughout history, bear little or no resemblance one to another. But they all share a common property of afflicting entire cities, countries or even widely separated parts of the Earth in a matter of days or weeks. Thucydides described the plague of Athens in 429 BC thus:

> It is said to have begun in that part of Ethiopia above Egypt . . . On the city of Athens it fell suddenly, and first attacked the men in Piraeus; so that it was even reported by them that the Peloponnesians had thrown poison into the cisterns.

A similar description of sudden onset and rapid global spread is relevant to almost all earlier as well as later epidemics. Such swiftness of transmission is hard to understand if, as is usually supposed, infection can pass only from person to person or be carried by vectors such as lice and ticks. And this explanation is particularly untenable for widespread epidemics which occurred before the advent of air travel, when movement of people across the Earth was a slow and tedious process.

The general belief, which is not well proven, is that major pandemics, such as influenza in recent times, start by random mutation, or genetic recombination, of a virus or bacterium which then spreads by direct person-to-person contact. If this is so, it is somewhat surprising that major pandemics are relatively short-lived, usually lasting about a year, and that they do not eventually affect the entire population, which would not have a specific immunity to the new organism. We contend that primary cometary dust infection is the most lethal, and that secondary person-to-person transmissions have a progressively reduced virulence, so resulting in a declining incidence of the disease over a limited period. Primary infections of a human population could occur directly through contact with infected meteoritic dust, or indirectly through meteoritic infection passing to other creatures such as mosquitoes, rats and lice which act as intermediaries.

The abrupt appearance in the literature of references to particular diseases is also significant in that they probably indicate times of specific invasions. Thus, the first clear description of a disease resembling influenza was in the 17th century AD, while the earliest reference to the common cold was in the 15th centuryAD. Also it is significant that earlier plagues, such as that in Athens, do not have easily recognisable modern counterparts.

Major epidemics of disease could be caused when the Earth crosses the debris of new long-period comets. Relatively minor variants of the 'same' disease—e.g. the common cold—could be due to more frequent, regular passages of the Earth through debris of shorter-period comets.

The factors governing the actual pattern of global incidence for any particular extraterrestrial invasion could be complex. If bacteria or viruses are dispersed in a diffuse cloud of small particles. the incidence of disease may well be global. On the other hand, a smaller disintegrating aggregate of infective grain clumps falling over a limited area of the Earth's surface could provide a geographically more localised invasion. Systematic effects such as air currents over the Earth's surface could also be relevant in controlling the transport and dispersal of particles. In particular, certain latitude belts might well be more favoured than others for either the accumulation and settling of these particles, or their avoidance. Furthermore, spatial variations in settling times, corresponding to variations in atmospheric conditions at different locations, could mimic a situation where an epidemic apparently spreads from a localised focus—the spread having no casual connection whatever with the terrestrial 'focus'.

# Gravitation, ancient eclipses, and mountains

R. A. Lyttleton

When a scientific theory is first put forward, its power to explain and to correlate a range of phenomena, and to make successful predictions, especially this latter, are what constitute its triumphs. But later on when the theory has become well established and is widely relied upon, it is its failures that become its even greater successes. For it is these 'failures' that tell us that we are onto something new, that some feature of the problem has not been recognised, that some new cause may be at work. The paradigm of scientific theories remains Newton's theory of gravitation. It was some 150 years before things began to fail, when the planet Uranus seemed not to conform to the law. There was perhaps more dispute about this than it warranted, but it eventually got cleared up within the theory when the planet Neptune was added to the solar system. Then everything seemed to be all right again and astronomers could relax. But almost immediately there was further difficulty, and this time no such easy solution was to be forthcoming. The trouble arose from the perihelion of Mercury, which by great good fortune moves in a highly eccentric orbit and so has a well-defined perihelion position. This proved to be a failure of the highest importance: the perihelion was found to be out of position by some 40 seconds of arc every century. It is to be emphasized that such a phenomenon could never be recognized without the Newtonian theory: no amount of mere staring at Mercury through a telescope would disclose the difficulty. Numerous suggestions were tried to explain the excess motion within Newtonian theory, but none worked, and as is well known now it was not satisfactorily cleared up until the advent of general-relativity theory, which showed that slight modifications of the Newtonian laws were needed.

But to come to the root of my topic, the story begins with the Moon. The lunar motion is one of the most difficult problems of dynamical astronomy and provides one of the most stringent tests of gravitational theory, because the Moon moves so fast angularly in the sky, and its position is observable with high accuracy. But it began to be suspected about a century ago that the Moon was getting progressively out of position, that is relative to its theoretical place found purely from the inverse-square law. The situation has become more and more acute, and the irregularity abundantly confirmed since then from the records of ancient eclipses. Without going into things too deeply, the nature of the trouble is that on gravitational theory the longitude of the Moon $v$, as seen from the centre of the Earth, ought to be expressible in the form

$$v = nt + \epsilon + [\tfrac{1}{2}\dot{n}t^2] + \text{periodic terms},$$

wherein $2\pi/n$ is a sidereal month, $t$ is the time, and $\epsilon$ is the time, and $\epsilon$ a constant, *without* the term in square brackets. The periodic terms, although highly important in the lunar theory, average out over long periods of time and so have no secular effect. Now the phenomenon that ancient eclipse records reveal is the small *accelerative* term in square brackets depending on $t^2$, and this is quite inexplicable on Newtonian dynamics. For such a term to be present just will not do, since it means that the object of science is not being achieved. There are all sorts of romantic and incorrect definitions of science, but prosaic as it may seem the real definition in the last analysis is that the object of science is to eliminate the time from the equations of motion. It is all right for d$t$ to occur, indeed it is essential that it should, but the presence of $t$ itself is not. So herein lies the puzzle, for a term in $t^2$ could only arise in $v$ if the equations of motion contained $t$, in other words if some *un*appreciated cause were at work that the dynamical equations used have not taken into account.

The general order of the acceleration $\dot{n}$ here concerned is about $\sim 30$ arcsec century$^{-1}$ century$^{-1}$. In one year it would make the longitude wrong by only 0.0015 arcsec but for the most ancient eclipses on record it is as much as 3° or 4°. In 1000 yr it amounts almost to 0.5°, which is just about the diameter of the Moon as seen from the Earth, and of the Sun too by fortunate coincidence. The bodily diameter is just over 2000 miles, 3218 km so the effect is to translate the shadow of the Moon as it makes an eclipse track on the surface of the Earth by huge distances. There are always at least two eclipses of the Sun every year occurring somewhere on Earth. Now the lunar theory can be run backwards in time, and well-recorded eclipses of reliable dates identified. What is found is that instead of having been total along a thin line somewhere in America say, a particular eclipse occurred in Europe. This century, by the way, England is privileged by two total eclipses: one is past, 29 June 1927, and the second will occur on 11 August 1999 and cut across the foot of Cornwall.

A total eclipse means that at some instant the Sun and Moon have exactly the same longitude, and where the shadow of the Moon falls obviously depends on the orientation of the Earth. So ancient eclipses in their location necessarily involve the rotation of the Earth, which has served as the astronomical clock until comparatively recent times. But it has gradually come to be recognized that the rate of rotation of the Earth is subject to all sorts of minor fluctuations: periodic, random, and systematic. It came to be replaced in fairly recent years by so-called ephemeris time, which is defined by means of the orbit of the Earth round the Sun. But even this might not be true dynamical time if the constant of gravitation is changing. In still more recent times, there has been introduced yet a third standard of time called atomic time, which is measured by means of caesium atoms and a gismo called a 'black box', and it is believed to give true dynamical time.

If the Moon is accelerating *and* the rate of rotation of the earth $\omega$ changing as well, then it is the acceleration of the Moon relative to the accelerating Earth that could be measured in effect. The Sun has so much angular momentum in its orbit round the Earth–Moon system, that the puny goings-on in the latter, although of the greatest importance in the problem, have no detectable effect on the Sun, which is not accelerated. If it seems to show acceleration, as it does, this is purely a reflection of the acceleration of the rotation of the Earth. The definitions and numerical values of these so-called *apparent* secular accelerations are

$$\nu\,(\text{Moon}) = \dot{n} - \frac{n}{\omega}\dot{\omega} = 12.10\ \text{arcsec cy}^{-2},\quad \nu'\,(\text{Sun}) = -\frac{n'}{\omega}\dot{\omega} = 2.94\ \text{arcsec cy}^{-2}.$$

Equations of motion can be set up for the effects on the lunar orbit, on the solar orbit, and on the rotation of the Earth, which involve the couple $N$ resulting from tidal interaction of the Earth and Moon, $N'$ that for the Sun, and, extremely important, A, the rate of change of the moment of inertia of the Earth. It is a remarkable property of these equations that the lunar couple $N$ comes out 'clean' from them even if $C$, the moment of inertia of the Earth, is changing (and indeed even if the constant of gravitation $G$ were changing with time).

When the dynamical equations are solved, they show that the lunar couple $N = 4.74 \times 10^{23}$ cgs units, and $N$ alone accelerates the rotation of the Earth by $\dot{\omega}_M = -5.89 \times 10^{-22}\ \text{s}^{-2}$. Now $N$ arises purely tidally, but its value does not reveal where or how the tidal friction is happening: oceans, shallow seas, the body of the Earth: it is rather like weighing a suitcase, the scales give the weight but no indication of the contents. For $N'$, the corresponding purely tidal part can be got from $N$ by scaling, allowing for the different distance and mass of the Sun, but it also has a small thermal part produced by solar heating of the atmosphere, and is in fact accelerative in its effect. Its value can be calculated, so $N'$ can be found, and the total acceleration it produces is $\dot{\omega}_s = -0.68 \times 10^{-22}$ s$^{-2}$. At this point the remarkable result emerges that these two angular accelerations are not sufficient to account for the total $\dot{\omega} = -5.24 \times 10^{-22}$ s$^{-2}$ found from $\nu'$. Thus inescapably arises the question: what else can affect the rotation of the Earth? No other external influences are available, and the simple answer is that the moment of inertia $C$ must be changing. For a rotating body, a changing $C$ gives an intrinsic acceleration of its rotation $\dot{\omega}_i$ to conserve the angular momentum $C\omega$. To find the value of $\dot{\omega}_i$, a balance-sheet for the total acceleration can be constructed, thus

$$\dot{\omega}\,(\text{total}) = \dot{\omega}_i\,(\text{intrinsic}) + \dot{\omega}_M\,(\text{Moon}) + \dot{\omega}_s\,(\text{Sun}),$$

whence follows from the foregoing values $\dot{\omega}_i = +\,1.33 \times 10^{-22}$ s$^{-2}$ for the intrinsic acceleration. Before proceeding, it is of interest that $\dot{\omega}$ (total) implies that the day is lengthening at present at a rate of about 2 s per 100 000 yr. But it is the *positive* term $\dot{\omega}_i$ that has the interesting implication that the Earth *must be contracting*, and raises the question: how can this be?

Modern theories of the origin of the planets show that all the terrestrial group started their existence, not as hot molten bodies, but cool enough to be solid throughout. Mercury and the Moon cannot even in their present state hold any atmosphere, and Mars barely so. So this leads on to the question of how big an all-solid Earth would be. At present it has a liquid core nearly 3500 km in radius containing about 31 per cent of the entire mass of $5.976 \times 10^{27}$ g, and the (spherical) radius is 6371 km. Seismic data, derived from the travel-times of earthquake waves, yield the properties of the material at all depths, but they do not identify it as to chemical composition. These data can be utilized to calculate the radius of an all-solid Earth, and it is 6741 km, *370 km greater than now*.

The high mean density of the Earth was firmly established in the early 1800s when $G$ was first measured in the laboratory, and as it is about twice the density of surface rocks, a high proportion of iron was postulated, which seemed to be supported by the magnetic field as resulting from ferromagnetism. But it is now known that ferromagnetism is destroyed at the temperatures that prevail deep in the earth, so the magnetism cannot result just from plain iron. High iron content seemed to be justified by meteorites, but in reality is not at all. Although most museum specimens *are* irons, this is only because these are much easier to recognize lying around than are stones. But of *meteorites actually seen to fall*, only about 3 per cent are irons. The existence of the liquid core was not discovered until early this century, when it was found that no secondary distortional waves, $S$-waves, transmit through it. Since then it has been widely claimed to be liquid iron plus some nickel. The liquid core also seemed to support the view that the Earth had begun entirely molten, perhaps after passing through an earlier gaseous state, and that the iron had drained to the centre to form the core. However, it is doubtful if an entirely molten Earth would yet have had time to cool sufficiently to become solid down to 2900 km depth.

The question of what the core can be if it is not iron seems first to have been both asked and answered by the late W. H. Ramsey, of Manchester, who proposed that it is a metallic liquid form of mantle-material: a so-called phase change. At 413 km depth in the Earth, there occurs a density discontinuity because of the pressure: the material undergoes a first phase change to a denser crystal form. The two forms are both solid: the required pressure is about 140 000 atmosphere, which can be reached in the laboratory, and this phase change seems to be accepted without any question. Far deeper still at the mantle–core boundary, the pressure has risen to nearly 1.4 million atmosphere, while the temperature, although difficult to settle closely, is several thousand degrees Kelvin. *Static* pressures as high as this are not easy to produce in the laboratory, since the equipment tends to take part in the experiment and give way. However, for hydrogen at low temperatures, it can be proved theoretically that at $0.7 \times 10^6$ atmosphere, the density suddenly jumps from 0.4 g $cm^{-3}$ to 0.8 g $cm^{-3}$, and it takes on a metallic form. Some substances exist in metallic and non-metallic forms at ordinary pressures, as for example arsenic, and also tin.

It is intuitively obvious that matter cannot have infinite strength to resist pressure. Compelling arguments that at sufficient pressure a phase change will occur were put forward by Ramsey in the late 1940s and early 1950s. By discarding the outer shell of electrons in response to high pressure, atomic volume can be much reduced, and the process is aided by high temperature. The work done by the pressure, $pdV$, will more than compensate for the loss of binding energy and, when equilibrium is next reached, the substance will be a metal because of the freed electrons and with higher density. Ramsey showed by general physical arguments that for most substances the critical pressure required for the phase change would be of order $10^6$ atmosphere. But if the Earth started solid throughout, why should anything happen? The answer is *radioactivity*: the temperature in the deep interior gradually rises through radioactive energy release. Hitherto, and to a considerable extent still, radioactivity is regarded just as a complication within the Earth of a non-essential kind, albeit affecting the internal temperature and perhaps causing melting here and there. But in the phase change theory, radioactivity is elevated to the rôle of the driving force causing the Earth to evolve. The elements powering this engine are uranium, potassium and thorium, which have half-lives comparable with the age of the Earth, $4.5 \times 10^9$ yr, and so are still active, continuing to heat up the interior, and as a result the core is still growing. Ramsey proved on certain simplifying assumptions that when in an all-solid Earth some critical temperature, of a few thousand degrees, was reached at the centre where pressure was automatically highest, the onset of the phase change would occur, and a *large* liquid metallic core suddenly form in a matter of minutes (or at most hours if heavy friction were involved), and accompanying this the surface radius would decrease by about 100 km. After this initial collapse, the Earth evolves slowly and steadily but quite stably. The analysis was extended by Lighthill to show quite generally that this sudden collapse occurs if the density of the liquid phase is greater than 1.5 times that of the solid phase. In fact, for the Earth, the seismic data show that the ratio is about 1.8.

A major triumph of the Ramsey theory was in explaining the remarkable sharpness of the mantle–core boundary. If it were a change of composition to an iron core, a gradual transition over a range of depth would be involved, probably as much as 100 km. This would produce diffraction effects on earthquake waves, which are never found; instead the interface reflects waves with mirror-like perfection. The boundary must be a very sharp discontinuity with depth, and the dependence of the phase change on pressure, for a given temperature, accounts for this, as Ramsey showed. Moreover it had been known for some time that the velocities of earthquake waves, both *P* and *S*, behave peculiarly as the interface is approached from the mantle side, showing an unexpected mode of decrease. On the phase change theory, a diminution in the bulk modulus and in the rigidity occurs, and readily accounts for this hitherto unexplained phenomenon. The same thing occurs for the *P* waves at the boundary of the innermost core-within-the-core, so it seems likely that this is a phase change too.

Some 10 years after Ramsey's original studies, computers had become available, and the theory could be developed in accurate numerical detail by introducing the known seismic data for the Earth. Calculations fully confirmed Ramsey's conclusions, and showed that the initial *large* liquid core would have radius just over 2000 km, whereas the present value is almost 3500 km, showing that evolution must have taken place. Also, from the time of initial collapse by about 70 km to the present, the outer surface radius would have contracted by a further 300 km, to the current 6371 km radius. Accompanying this is automatically a decrease in the moment of inertia, and this of course produces an intrinsic acceleration of the rate of rotation. The calculated acceleration $\dot{\omega}$, determined entirely from the seismic data, is found to be closely equal to that already stated as found from ancient-eclipse records. If acting purely on its own, it would have reduced the length of the day by about 20 per cent in the lifetime of the Earth, so it can be seen to be an important contribution to the total rate of change of the rotation. By the way, it is of interest to recall that Ramsey showed that the core material, if the pressure was taken off without change of form, would have density only about 5 g cm$^{-3}$. The later more accurate machine calculations led to a revised somewhat higher value of about 6 g $cm^{-3}$, but his conclusion that it cannot therefore be identified with iron remains unaffected.

We come next to what is probably the most interesting achievement of all, and that is the relevance of the phase change theory to the origin of mountains. For upwards of a century geologists have been crying out for some process that would produce enormous amounts of contraction and compression of the outer layers of the Earth. Here is a plea from a standard work on geology emanating from Princeton: 'One generalization on which all geologists agree is that mountain building involves a reduction in the surface area of the globe, a shortening of the distance between points of the surface, and that all mountain building is the product of a single mechanism—*squeezing by horizontal compression*. But when it comes to what causes the squeezing, there is no general agreement.' In regard to this last statement, naturally enough there could be no agreement if the true cause, the phase change process, had never been suspected. The only mildly serious explanation yet offered is the so-called thermal-contraction theory, the withered-apple story children are told. It requires an initially molten Earth, and this we have seen is an entirely unacceptable postulate. Moreover, if thermal contraction *were* the cause, there should be thrusted and folded mountains on Mars, whereas it is now known for certain from the *Mariner* photography of the whole surface that there are none. (There are volcanic mountains, but these are not formed by compression, and so not concerned.) Similarly none of these types of mountains exists on Mercury or the Moon. In addition, the geologists have maintained all along that thermal contraction would supply a quite inadequate amount of crustal shortening: it might just possibly provide for one era of mountain building, but not for a great many more.

Let it be emphasized at the outset that the real problem of mountain building is not simply to account for the one most recent period, but to find explanations for possibly as many as *twenty* such eras of worldwide orogeny. *The phase change theory can provide this*. What occurs as the phase change mechanism steadily operates is that the interior 99 per cent of the mass of the Earth, all of which material has negligible strength compared with the great pressure it is under, contracts down in unison with the inexorably rising temperature from radioactive energy and the resulting steady growth of the metallic liquid core. To begin with, the outer layers are little affected, for in them the strength dominates compared with the pressure, especially for the outermost layers. But with their support from below steadily being reduced, stress will gradually build up. These outer layers, if unsupported from below like an arched bridge, would need to be able to withstand impossibly high stresses (of order $10^{12}$ dyn $cm^{-2}$), a thousand times the strengths of rocks. After some 100 million years, stresses would have become so great that folding and thrusting of the outer layers would just have to take place. Thrustings of one layer sliding over another by 50–100 km are commonplace geological forms.

Where folding is concerned: granite, for example, can be compressed by 1/800 lengthwise before it fails. Now the terrestrial circumference is 40 000 km, and, if it were a continuous band of rock, a reduction by 50 km would therefore require it to yield and adjust in some way, and folding is a natural and simple means, just as when a slightly starched tablecloth is pushed together endwise, it rucks up into folds and wrinkles. A 300 km reduction in radius means about 2000 km in circumference, and so on this basis, 40 such relief shortenings could be accounted for by contraction. It is still more impressive to consider the surface area that has somehow got to be tucked away. This is $\Delta\ (4\pi r^2)$, and for $\Delta r = 300$ km this means reduction by about *50 million square kilometres*, about a tenth of the present total area. But what is really concerned is the *volume* of rock that has to be redisposed. No solid structure can extend above the surface by more than 5–10 km: there cannot be terrestrial mountains 20 km high, because the floor below has insufficient strength to support them, and they would sink down to such height as could be maintained. The change of volume for a decrease of radius by 300 km (to the present value) is no less than *160 billion* $(10^9)$ *km*$^3$. Shared out for, say, 20 periods of mountain building, it would mean $8 \times 10^9$ km$^3$ for each. As an example, let us consider an idealized mountain range: 5 km high, 200 km broad, 10 000 km long (these numbers are comparable with the dimensions of the Rockies): its volume would be $10^7$ km$^3$. So the figure of $8 \times 10^9$ km$^3$ would allow for 800 such ranges in a single era of mountain building. This is enough to provide for the whole of the present mountains on the surface of the Earth. (To come from the sublime to the ridiculous, on the iron core hypothesis, all radioactivity could do would be to produce slight expansion of the Earth, but would have nothing to offer by way of accounting for mountains at all.)

The reason Mars has no folded and thrust mountains is very simple on the phase change theory. The pressure at the *centre* of Mars, owing to the small mass at only 1/9 of the Earth, is about 1/5 that at the mantle–core boundary in the Earth, and is therefore far too low to bring about any phase change. It is attained at 750 km depth in the Earth, high up in the solid mantle. For the same reason, Mercury cannot have any such mountains, as recent close inspection confirmed. The planet Venus, however, has 0.82 Earth mass, so its central pressure must be higher than at the core–mantle boundary in the Earth, while its internal temperatures at corresponding depths must be several hundred degrees Kelvin higher through its proximity to the Sun plus a strong greenhouse-effect. Venus should therefore possess mountains, but a deep atmosphere impervious to optical light prevents its solid surface being seen. Not so, however, for radar-wavelengths. Spacecraft are now able to examine the surface of Venus by radar-ranging, but the power of resolution is not great and may be inadequate to detect mountains in sufficient detail. Radar-ranging from Earth has already found large-scale irregularities at the surface, but it is early yet to be sure these are mountains. A more ambitious mission, Venus Orbiting Imaging Radar, is hoped to reach Venus in 1985, and will be able to inspect the surface in much greater detail. The acronym for the craft is VOIR, and it is much to be hoped that it lives up to its name. The phase change hypothesis of course predicts that there *will* be found folded and thrust mountains on Venus, but to what extent they may or may not resemble terrestrial mountains, will depend on weathering processes, eroding out of canyons for example, and little if anything is yet known of Venusian weather conditions.

In view of the high pressures concerned, an interesting question is whether it will ever be possible to test experimentally the phase change idea. Bridgman in his high-pressure work did not reach 0.3 million atmospheres, but now in the last few years, at the Geophysics Department, Carnegie Institution of Washington, there has been successfully constructed a Diamond-Anvil-Press that can produce much higher pressure. Already 0.5 million atmospheres has been reached, at which it is found that molecular bondings begin to break down and electrical conductivities greatly increase, a sure sign of free electrons. Naturally enough, pressures are being worked up from below, but soon may reach values sufficient to test out the phase change theory directly. That the experiments will establish its general correctness seems to some almost a foregone conclusion, since this can be inferred with much confidence from the success of the several crucial predictions that it has already made.

Finally let it be mentioned that, quite apart from their awesomely impressive magnificence, mountains are extremely important to us and essential to our continued existence. Through weathering by water and wind and radiation, they are being eroded away comparatively rapidly all the time, and in about 100 million years (if no other process intervened) the resulting detritus would silt up all the seas and ocean basins, for the whole surface to become covered with water to a depth of 2 or 3 km everywhere. Unless mountains were intermittently regenerated, therefore, we ourselves would at best have remained poor fish. But, so long as the solar system remains unendangered from without, there is no cause for alarm. Since the half-lives of U, K, and Th are comparable with the present age of the Earth, they remain active to continue raising the internal temperature, with the consequence that more eras of mountain building and ocean-basin formation are yet to come.

## References

Bell, P. M. and Mao, K. H., 1974, 1975, 1976. High pressure physics, *Carnegie Inst. Wash. Yr Book*, 73, 402 and 507; **74**, 399; and *Science*, **191**, 851.

Lyttleton, R. A., 1965. The phase-change hypothesis for the Earth, *Proc. R. Soc. London A*, **287**, 471.

Lyttleton, R. A., 1972. On the formation of planets from a solar nebula, *Mon. Not. R. astr. Soc.*, **158**, 463.

Lyttleton, R. A. and Fitch, J. P., 1978. On the accelerations of the Moon and Sun, *Moon & Planets*, **18**, 223.

Muller, P. M. and Stephenson, F. R., 1979. Analysis of ancient-eclipse records.

Ramsey, W. H., 1954. Transitions to metallic phases, *Occas. Notes R. astr. Soc.*, **3**, 87.

Ramsey, W. H. and Lighthill, M. J., 1950. On the stability of planetary cores, *Mon. Not. R. astr. Soc.*, **110**, 325.

# E. A. Milne: his part in the development of modern astrophysics

S. Chandrasekhar

This is a precious occasion for me. For, here in Oxford, I used to visit Milne frequently during the early and the mid-thirties. I exchanged letters with him regularly for a period of 20 years; and I have always cherished his friendship and his counselling. In view of my long and sustained friendship with Milne, it has seemed to me proper that I devote this Lecture to an assessment of his rôle in laying the foundations of modern theoretical astrophysics. In making this assessment, I shall try my best to be objective; and, if in the process, I dwell, on occasions, on what appears to me his weaknesses and his failures, it is because I admire Milne sufficiently to be a rationalist about him. Besides, it would not seem to me that I shall be serving his memory by giving a partial account or by adopting anything less than the highest standards.

## I

Milne entered the arena of astrophysics in 1921. At that time, only the barest beginnings had been made in what have since become the two main pillars of modern astrophysics: the theory of stellar atmospheres and the theory of stellar structure.

In 1920, there was only one extant book which may be considered as treating topics in theoretical astrophysics: that was Robert Emden's *Gas Kugeln* —or gas spheres—published in 1907. Emden's book gives a surprisingly complete account of gaseous masses in equilibrium under their own gravitation and in which the pressure is proportional to some power of the density. This is the theory of the polytropic gas spheres—a theory which was to play a key rôle in the subsequent investigations of Eddington and of Milne. Emden's book, besides its more well-known parts dealing with polytropic gas spheres, includes a discussion of the physical conditions in the solar atmosphere with an account, in fact, of Karl Schwarzschild's inferences of 1906 that the outer layers of the Sun cannot be in convective equilibrium but rather be in radiative equilibrium. Schwarzschild had drawn his inferences from the finite brightness of the solar limb—a matter to which I shall return presently. Another landmark of this same period is a paper of Arthur Schuster's in 1905 in which a problem in the theory of radiative transfer, relevant to the formation of absorption lines in the solar and in stellar atmospheres, is treated.

The concept of radiative equilibrium was further analysed by Schwarzschild in 1914. And in 1916, Eddington introduced these same concepts in the larger context of the equilibrium of the stars as a whole and had begun the first of his celebrated series of papers dealing with the internal constitution of the stars. Also in 1918, Eddington had formulated his pulsation theory of stellar variability.

Atomic theory was still very much in its infancy; and Saha's papers dealing with the ionization and excitation of atomic species, at the temperatures and pressures to be expected in stellar atmospheres, were yet to appear.

This was the time when Milne entered the arena of astrophysics. Let me say at once that, more than the particular advances for which he was responsible, his greater contribution was his attitude and his style. I shall say more about them later.

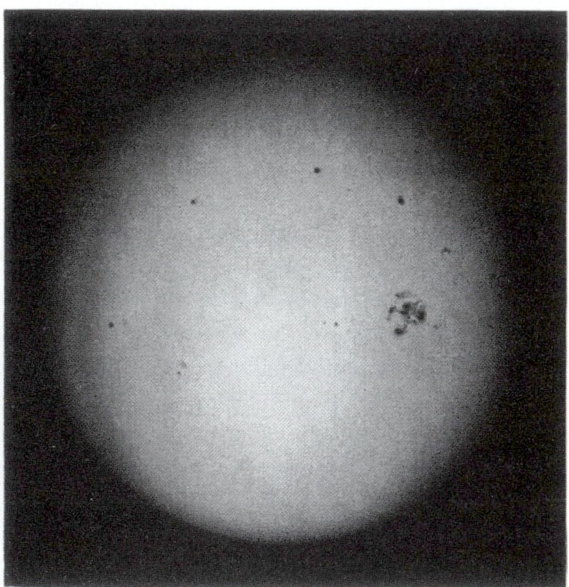

**Fig. 1** Illustration of the darkening of the Sun towards the limb.

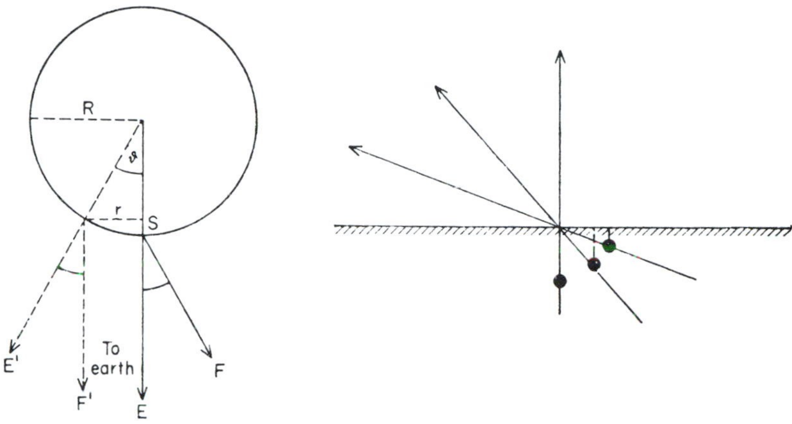


**Fig. 2** The darkening of the Sun results from the angular dependence of the emergent radiation and on the temperature gradient in the atmosphere.

## II

It was fortunate that the problem to which Milne first turned his attention was one that suited his style and his methods admirably. As I shall indicate presently, the results which he derived in these, his first and earliest investigations in astrophysics, have remained essentially unchanged over the years and have provided the basis for certain permanent features of our understanding of the outer layers of the stars. For this reason, I shall consider them in some detail.

The problem which Milne considered was concerned with the interpretation of the variation of the brightness of the Sun across its disc (the phenomenon of the darkening of the Sun towards the limb). This variation of the brightness across the solar disc occurs not only in the total brightness (as exhibited in Fig. 1) but also in the different wavelengths or colours.

It is clear from Fig. 2 that the darkening towards the limb is simply an expression of the angular dependence of the intensity of the emergent radiation. This problem of the darkening of the Sun towards the limb had been considered by Karl Schwarzschild in 1906; and he had related it to the prevalance of radiative equilibrium and to the resulting variation of temperature in the outer layers of the Sun. The interpretation of the darkening, on the basis of these ideas, is very simple.

The basic fact is that the radiation from all depths contributes to the emergent radiation; only the radiation from the deeper layers is increasingly attenuated by the opacity (i.e. the fogginess) of the overlaying material. On this account, we may say that the radiation which emerges from the surface is characteristic of the radiation prevailing at a certain average depth below the surface. We may, in fact, say that there is a depth to which we effectively see. This depth, measured by the extent to which the overlaying layers attenuate the radiation traversing them, must be the same for all wavelengths and for all angles of emergence. In other words, we effectively see down to an optical depth of unity in all cases. (Radiation traversing material of optical depth unity will be attenuated by a factor of approximately 1/4.)

Since radiation emerging at an angle traverses a path which is slanting through the atmosphere, it is clear that such radiation will be representative of the radiation prevailing at a level not as deep as the level which is representative of the radiation which emerges normally from the surface. Since we should expect the deeper layers to be at higher temperatures, it follows that the radiation emerging at an angle will be characteristic of a temperature lower than the temperature characteristic of the radiation emerging normally. Therefore, the intensity of the radiation emerging at an angle must be less than that emerging normally (see Fig. 2). In other words, there must be a darkening towards the limb.

From the foregoing description it is clear that the principal problem, which requires solution before we can account for the phenomenon of darkening, is the distribution of the temperature in the outer layers. Once the temperature distribution has been ascertained, the emergent intensity at any given angle can be directly related to the variation of the opacity (i.e. the absorption coefficient or the absorptive power) of the material for light of different wavelengths.

Suppose $\bar{\kappa}$ is some mean absorption coefficient and let $\tau$ be the optical depths measured in terms of $\bar{\kappa}$. Let $T_\tau$ be the temperature that prevails at depth $\tau$. Then at the depth $\tau$, the spectral distribution of the radiation will be determined by the Planck distribution

$$B_\nu(T_\tau) = \frac{2h\nu^3}{c^2} \frac{1}{\exp(h\nu/kT_\tau) - 1}, \tag{1}$$

where $\nu$, $c$ and $h$ denote the frequency, the velocity of light and Planck's constant, respectively. Accordingly, the intensity of the emergent radiation, at an angle $\theta$ to the normal and with a frequency $\nu$, will be given by

$$I_\nu(\theta) = \int_0^\infty d\tau \, B_\nu(T_\tau) \left(\frac{\kappa_\nu}{\bar{\kappa}}\right) \sec\theta \exp[-(\kappa_\nu/\bar{\kappa})\,\tau \sec\theta], \tag{2}$$

where $\kappa_\nu$ is the absorption coefficient at the particular frequency considered.

It is clear that from a comparison of the observed intensities with those which would follow from equation (2), we can deduce the variation of the absorption coefficient with wavelength (as determined by $\kappa_\nu$/B); and this variation will clearly determine something significant about the constitution of the solar atmosphere.

The problem which I have outlined was solved by Milne with exceptional thoroughness in his early papers; and he deduced from the solar observations the variation of the solar continuous absorption coefficient with the wavelength. His results are shown in Fig. 3.

Milne emphasized two features of the deduced variation: *first*, that the absorption coefficient increases gradually over the entire visual part of the spectrum and attains a very well-defined maximum at about 8000 Å; and *second*, that beyond 8000 Å it decreases to a very deep minimum at about 16 000 Å.

Milne's analysis was repeated by others, in other forms, during the following decades; and they all confirmed his major deductions. I shall consider one such confirmation taken from an investigation of Chalonge and Kourganoff in 1946 (some 25 years after Milne).

Consider a level at some assigned temperature $T$ and ask for the opacity of the overlaying layers in various wavelengths. Clearly, this question can be answered with the aid of Milne's basic theory. The results of the analysis of Chalonge and Kourganoff are exhibited in Fig. 4. The basic deductions of Milne are clearly confirmed.

The simplicity of the analysis leading to the deduced variation of the continuous absorption coefficient with wavelength signifies, unequivocally, that a fundamental constituent of the solar atmosphere is here involved. As to what this constituent may be finally emerged only during the forties with the definite isolation of what could then be truly described as a new fundamental constituent of the solar atmosphere.

Let me briefly trace the history of these developments since it represents the fruition of Milne's basic and earliest researches.

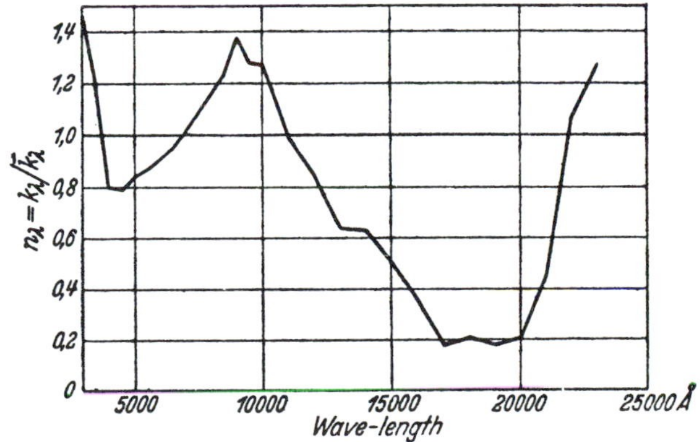


**Fig. 3** Milne's deduced variation of the solar continuous absorption coefficient with wavelength.

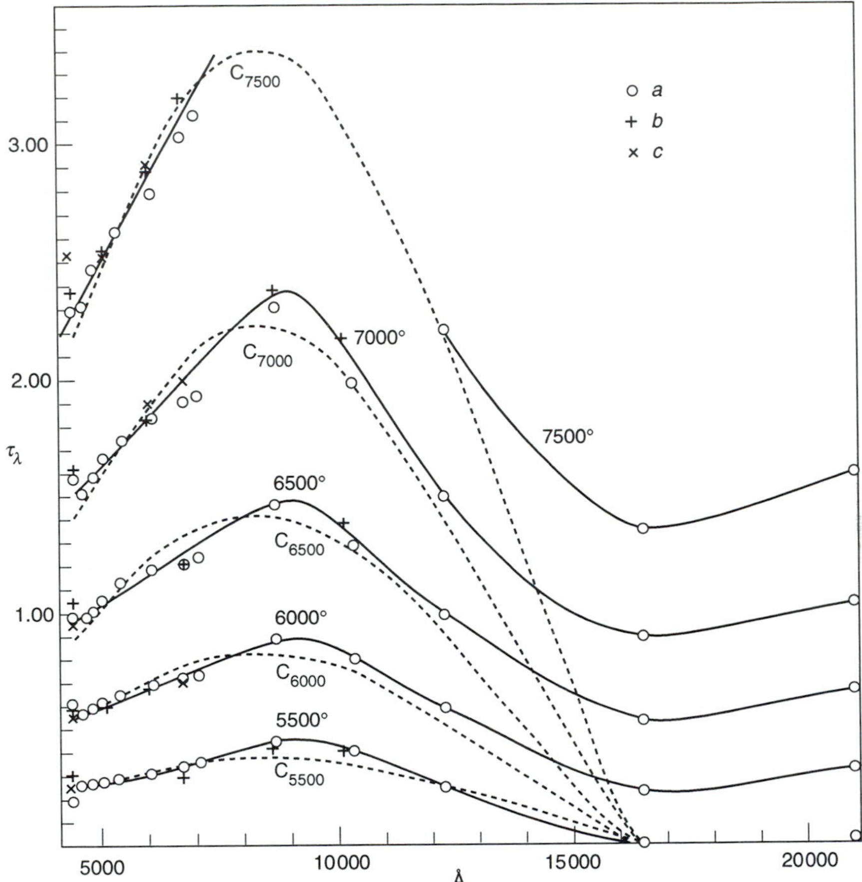


**Fig. 4** Optical depth of photospheric layers at different wavelengths (Chalonge and Kourganoff 1946). Dotted lines are H− absorption; right part of graph explained by free–free transitions.

By an application of the variational method by which one can set upper bounds to the ground-state energies of atomic systems, Hylleras and Bethe had, independently, shown in 1930 that a hydrogen atom can stably bind itself to an electron to form a negative ion with a binding energy exceeding 3.4 eV. But it was only in 1938 that Rupert Wildt, returning to the fundamental problem that had been posed by Milne 16 years earlier and which had been sidestepped during the intervening years, pointed out that the negative ions of hydrogen must be present in substantial concentration in the solar atmosphere if hydrogen is indeed as abundant as other evidences had indicated. This was a most fruitful suggestion: it represented a key discovery which made possible all later developments in the theory of stellar atmospheres. But several difficulties had to be overcome before a definitive identification of the negative ion of hydrogen as the source of the continuous absorption in the solar atmosphere could be made.

The principal difficulty was the theoretical determination of the continuous absorption coefficient of the negative ion of hydrogen. This is neither the place nor the occasion to go into the history of the solution of this problem. For those who might be interested, I may refer to a comprehensive account that has recently been published by Sir David Bates in *Physics Reports*. Figure 5 (taken from Bates's report) suffices to show the years of effort that were required to resolve this problem. It is manifest from this figure that a quantally reliable evaluation of the continuous absorption coefficient of the negative hydrogen-ion reproduces the essential features of the curve deduced by Milne in 1922 (see also Fig. 4). And it should not be overlooked

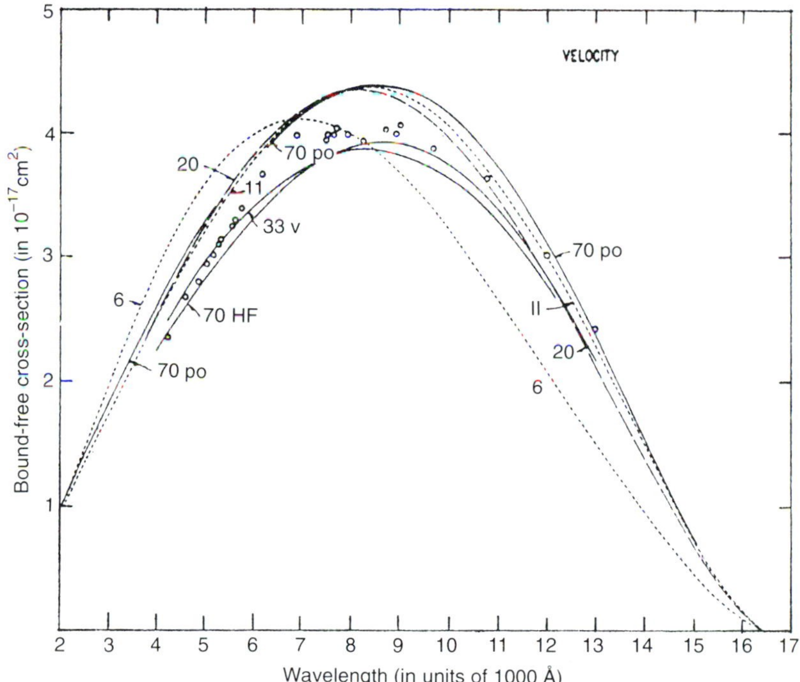


**Fig. 5** Photo-detachment cross-section of H−. Experimental values due to Smith and Burch (1959). Number of variational parameters in bound state wave function is indicated on each curve. Broken lines used if ejected electron is represented by plane wave: 6 (Williamson 1943), 11 (Henrich 1944), 20 (Chandrasekhar 1953). Full lines used if ejected electron is represented by more refined approximation: 70 po, polarized orbital, Bell and Kingston (1967); 70 HF, Hartree–Fock expansion, Doughty *et al.* (1967); 33 v, variational, determined by simplified Kohn–Feshbach method, together with Rotenberg–Stein bound function, Ajmera and Chung (1975). All calculations based on the matrix element of the velocity (Chandrasekhar 1945)—from D. R. Bates, 1978, Other Men's Flowers, *Physics Reports*, **35**, 306.

that at the time this identification was made, the negative ion of hydrogen was a theoretically predicted atomic species and the confirmation of the theoretically predicted continuum by laboratory experiments was still 10 years in the future.

And so we come to the end of one major chapter in the history of modern theoretical astrophysics, which began with Milne's researches.

## III

I now turn to a second chapter. At about the time Milne was working on the problem of the continuous spectrum of the Sun, R. H. Fowler and C. G. Darwin were developing their new approach to statistical mechanics; and Saha's first successful quantitative application of the theory of statistical (or rather, in his case, thermodynamic) equilibrium of stellar reversing layers was appearing in a series of papers in the *Philosophical Magazine* and in the *Proceedings of the Royal Society*.

Saha's theory was based on the following observation. An atmosphere absorbs a different optical spectrum for each stationary stage of ionization, and in fact a different set of lines for each stationary state belonging to each stage; and therefore the relative intensities of the absorption lines of its successive spectra in the spectrum of any star must give some indication of the relative numbers of atoms in the various stages of ionization in the reversing layer, and therefore of the temperature and the pressure. Saha's early application of this idea was based on the points of first and last 'marginal' appearances of particular spectral lines. At such points, Saha had argued that the fraction of the atoms in the reversing layer capable of absorbing the line must be very small. And if the corresponding pressure could be estimated the temperature can be deduced.

The prevision of these early calculations was questionable, owing to the difficulty of formulating conditions for the marginal appearance of particular lines: we do not know how small the 'very small' fraction of atoms must be at marginal appearance; also, the point of marginal appearance will depend on the relative abundance of the element responsible for the line. Fowler and Milne, in a series of papers published during 1923–1925, reformulated the basic problem as follows:

> Other things being equal, the intensity of a given absorption line in a stellar spectrum varies always in the same sense as the concentration of the atoms in the reversing layer capable of absorbing the line.

The difficulties with the concept of marginal appearance are avoided in this formulation; and Fowler and Milne first devoted their attention to the place in the stellar sequence at which a given line attains its *maximum intensity*. On the stated premise the maximum intensity will be attained at the maximum concentration of the atoms capable of absorbing the line; and the conditions for this to happen will involve only the temperature and the pressure. In other words, the temperature at which, for a given pressure, a given line attains its maximum is simply deducible from the properties of the equilibrium state. This was the first satisfactory way of applying Saha's theory quantitatively to fix stellar temperatures and pressures. In this manner, Fowler and Milne established a theoretical temperature-scale for the grand sequence of the Harvard spectral types for the first time: a true landmark in theoretical astrophysics.

In subsequent papers, Milne showed that the concepts of mean pressure and mean temperature, that are at the base of his studies with Fowler, in turn require refinements in at least two directions. First, we must make precise the meaning we are to attach to a phrase such as 'the intensity of an absorption line'. And *second*, we must allow for the variation of the temperature, the pressure, and the various attendant physical parameters through the layers in which the absorption line is formed. We must, in fact, construct model stellar atmospheres. While Milne formulated some of the basic considerations which must be incorporated in such refinements, he did not pursue them in any great depth or detail. They were left for Pannekoek, Unsold, Minnaert and a host of others to analyse and to complete. The construction of model stellar atmospheres has now become a large industry; but it had its beginnings in the heroic efforts of Saha, Fowler and Milne towards a basic physical understanding.

## IV

I now turn to Milne's work bearing on stellar structure.

Already during the years Milne was occupied with problems in the theory of stellar atmospheres, he was turning his attention to problems in the theory of stellar structure. Thus, in a paper published in 1923, he considered the effect of a slow rotation on Eddington's standard model for the stars and on his mass–luminosity relation. This is an altogether exemplary paper in which the mathematics and the physics are scored in counterpoint. (Perhaps, I may be allowed to state here parenthetically that it was this paper of Milne's which stimulated me to develop a complete theory of distorted polytropes some 10 years later.)

However, it was only in 1929 that Milne seriously turned his attention to problems in the theory of stellar structure. But it was begun inauspiciously under the pressure of a bitter controversy with Eddington.

The controversy with Eddington was an unhappy episode which, at least in my judgment, had tragic repercussions on Milne's subsequent work. I shall not say anything more about this episode on this occasion; but it is not possible to avoid overtones of it in any account of Milne's work after 1929.

In 1926, R. H. Fowler had shown in a fundamental paper that the state of matter in the interiors of the white-dwarf stars, such as the companion of Sirius, cannot be a perfect gas governed by the equation of state, $p = \mathscr{R}\rho T$ (where $p$ denotes the pressure, $\rho$ the density, $T$ the temperature and $\mathscr{R}$ the gas constant); and that it should be governed by the equation of state provided by the then new statistical mechanics of Fermi and Dirac and, indeed, in its limiting form when all the energy levels of the free electrons below a certain threshold are occupied and none above it. In other words, matter must be degenerate, as one says.

Fowler's discussion convincingly demonstrated that Eddington's assumption that the stars are wholly gaseous, with the normal equation of state, cannot be universally valid: the white-dwarfs are examples to the contrary. In the white-dwarfs, the matter is degenerate and the relation between the pressure and the density is, in a good approximation, independent of the temperature. It is therefore legitimate and proper to inquire when and under what circumstances degeneracy can develop in the interior of the stars. But Milne's inquiry was not so directed. He started with the premise—at least, he took it as a foregone conclusion—that all stars *must* have domains of degeneracy and that they *must* belong to one or other of two classes which he called centrally-condensed configurations and collapsed configurations, the distinction between them consisting mainly in the extent of the domains of degeneracy.

In his first detailed paper on the subject, published in January 1931, Milne developed some powerful analytical tools for constructing composite stellar configurations in which different relations, between pressure and density, obtain in different parts. Besides, Milne stimulated his long-time friend R. H. Fowler to undertake a systematic study of *all* the solutions of Emden's differential equation governing polytropic distributions.

Parenthetically, may I quote here some remarks of G. H. Hardy's at a meeting of the Royal Astronomical Society in January 1931. Hardy, as some of you may remember, was the Savilian Professor of Geometry here in Oxford during the twenties. He said (with his tongue in his cheek, as he confessed to me later):

> As a mathematician, I don't care two straws what the stars are really like . . . But I am particularly interested in Mr Fowler's paper . . . His paper is probably the only one of the collection which is of lasting value, for he is certainly right, whereas it is extremely likely that everyone else will be shown to be wrong . . .

I am afraid that what Hardy prophesied has mostly come to pass.

To return to Milne's investigations: It was pointed out to him, in fact even before he had communicated his first paper to the Royal Astronomical Society, that the mass of a wholly degenerate star cannot exceed a certain limiting value; and that this fact in turn places an upper limit to the mass that can be contained in the degenerate cores of stars; and finally, that in view of the increasing importance of the radiation pressure in massive stars, sufficiently massive stars cannot possibly develop domains of degeneracy. But Milne would not accept these conclusions. Instead, he wrote:

> If the consequences of quantum mechanics contradict very obvious much more immediate considerations, then something must be wrong either with the principles underlying the equation-of-state-derivation or with the aforementioned general principles. Kelvin's gravitational-age-of-the-Sun calculation was perfectly sound; but it contradicted other considerations which had not then been realized. To me it is clear that matter cannot behave as you predict . . . Your marshalling of authorities such as Bohr, Pauli, Fowler, Wilson, etc., very impressive as it is, leaves me cold.

From the vantage point of today, it is clear that Milne's negative attitude prevented him from realizing that the incorporation, positively, of the consequences of Fermi degeneracy, leads one directly to conclude that massive stars, after they have exhausted their sources of energy must collapse to black holes—a conclusion which Eddington drew but which neither Eddington nor Milne would accept. This failure on their part illustrates the danger of perceiving Nature in the images of one's personal beliefs and faiths.

As I said earlier, in the course of his analysis, Milne developed powerful analytical methods for treating composite stellar models. His methods were ideally suited for exploring stellar models with degenerate cores of the kind that stars can have consistently with their allowed upper limit. Milne could easily have carried out such explorations. That he did not was unfortunate both for Milne and for astrophysics.

## V

Before I turn to Milne's last and largest phase of his work, namely, kinematic relativity and cosmology, I should like to make a reference—if only a brief one—to a beautiful analysis in stellar kinematics which he published in 1935. In this paper, Milne analysed the differential motions that can occur in a stellar system, in the manner Stokes had analysed hydrodynamic fluid-motion into three parts: a rotation, a sheer and an expansion. From the point of view of this analysis, the occurrence of the so-called double-sine wave in the variation, with the galactic longitude, of the radial velocities of stars with an amplitude proportional to the distance of the stars, becomes self-evident. Milne's analysis provided the base for much dynamical discussion carried out subsequently.

## VI

I now come to a phase of Milne's work which he undoubtedly considered as his most important scientific contribution. Thus, referring to his theory of the expanding Universe, he wrote to me in a letter dated 6 July 1943:

> I do not know whether I have ever opened my heart to you on that theory. I only know that the texture of the argumentation in it is something utterly and surprisingly different from usual mathematical physics, and that when it comes to be recognized, it will be regarded as revolutionary. It is not usual to crack up one's own work in this way; but it is all very near my heart . . .

Perhaps it is not fair that I quote what Milne clearly meant only for me. But to the extent that in my assessment, I am unable to give to his theory the same exalted place, it is necessary that I acknowledge it with equal frankness.

In developing his kinematic theory of relativity, Milne took the strong position that a theory of gravitation can do very well without the general theory of relativity. Indeed, 'Gravitation without relativity' is the title of a contribution which he wrote for a collection of essays that was presented to Einstein in 1949 and included in volume 7 of the *Library of living philosophies—Albert Einstein, Philosopher–Scientist* (edited by Paul A. Schilpp). Einstein's reaction of Milne's contribution, in his concluding essay in the volume, was:

> Concerning Milne's ingenious reflections, I can only say that I find their theoretical basis too narrow. From my point or view one cannot arrive, by way of theory, at any at least somewhat reliable results in the field of cosmology, if one makes no use of the principle of general relativity.

In juxtaposition with this view of Einstein's, let me place Milne's view of the general theory of relativity:

> Einstein's law of gravitation is by no means an inevitable consequence of the conceptual basis given by describing phenomena by means of a Riemannian metric. I have never been convinced of its necessity . . . General relativity is like a garden where flowers and weeds grow together. The useless weeds are cut with the desired flowers and separated later!

And Milne goes on to say

> In our garden we grow only flowers.

To be complete, may I be allowed to state my own view. General relativity proceeds on the assumption that a theory of gravitation must reduce to all of the Newtonian laws as they operate in the 'small' as, for example, in determining the motions as they obtain in the solar system; and that only a theory constructed, consistently with the other laws of physics (as incorporated in the principle of equivalence) can, with confidence, be extended to the larger context of the Universe. Milne's procedure is exactly the converse of this. He proceeds on the assumption that gravitation can be understood by first constructing a theory of the Universe in the large and then descending to phenomena manifested in the small. Apart from the fact that Milne did not succeed in completing his programme, it is probable that the programme is an inherently impossible one.

Well! There you have three views of varying authority!

Having stated my overall negative view of this phase of Milne's work, may I say at once that there are some key aspects of his work which are refreshingly original. Thus, Milne's analysis of Lorentz transformations in terms of light-signals exchanged by observers is a model of precision and economy of thought. It deserves much wider knowledge than it enjoys. As Bondi has written:

> I feel that not nearly enough has been said about the deep debt of gratitude that we owe to Milne, who in his work on cosmology, introduced the notion of the radar method of measuring distance.

I now turn to the ideas which Milne contributed to cosmology and which have secured for themselves permanent places in the current literature.

During the late twenties and early thirties, the facts which were perceived as basic for a theory of the Universe as a whole were the following:

1. In a first approximation, the distribution of the extragalactic nebulae is locally homogeneous and isotropic.
2. The galaxies are receding from us and from one another with velocities which are proportional to their mutual distances, as codified in Hubble's law.

The discussion of these facts in the framework of the relativistic cosmological models of Friedmann and Lemaitre, as popularized particularly by Eddington, gave one the impression—intended or not—that general relativity is necessary to incorporate them into a coherent theory. But this perception exaggerated the rôle of general relativity. And Milne was certainly correct in pointing out that the facts considered have a simple interpretation which requires no special appeal to any particular theory.

The observed expansion of the Universe and the Hubble relation imply, only, that all the nebulae we now observe must have been, at one time, close together in a small volume of space.

Suppose, then, that at some initial time, $t_0$, the nebular were all confined to a small volume (see Fig. 6) and that they all had the same speed $V$ but in random directions. Then, after a sufficient length of time $(t-t_0)$, these same nebulae will have moved outward and will be confined to a relatively thin spherical shell of radius $V(t-t_0)$. Now suppose that, in addition to the nebulae with the velocity $V$, the same initial small volume had also contained nebulae with half the velocity $V$. Then, after a lapse of time $(T-t_0)$, these nebulae will be confined to a thin spherical shell of half the radius $1V(T-t_0)$. More generally, it is clear that if the original volume had contained nebulae of all velocities, then after a sufficient length of time, the nebulae, with differing velocities, will become segregated; they will, in fact, arrange themselves at distances from the centre which are proportional to their distances in conformity with Hubble's law. Or, as Milne states: 'the birds of a feather flock together'.

This simple model we have described suggests that the observed expansion of the Universe is simply the result of an early beginning of high mean density —a conclusion no one denies at the present time.

Again, as Milne emphasized, a homogeneous swarm of particles receding from a chosen particle in it, with velocities proportional to their distances from it, has a remarkable property. As a simple application of the parallelogram of velocities shows (see Fig. 7) the description of the motions will be the same with respect to any other particle in the swarm provided one does not go too near the boundary. In such a system, each particle in the swarm can consider itself as at the centre of the swarm with the other particles receding from it radially with velocities proportional to their distances from it, with the same constant of proportionality. In other words, a Universe which is homogeneous and isotropic and in which the motions satisfy a

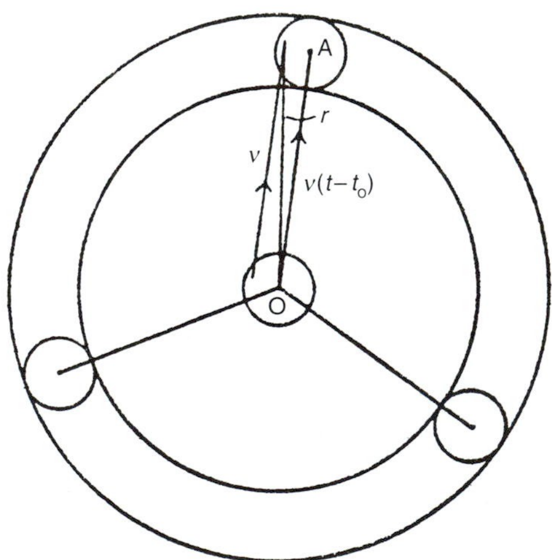


**Fig. 6** The emergence of a Hubble relation for a system initially confined to a small volume.

Hubble relation are related facts which derive their common origin in the requirement that the description of the Universe is the same as viewed from all galaxies. This last requirement was formulated by Milne as a *cosmological principle*. Milne considered this principle as inviolable: it is the centre-piece of his kinematical theory of relativity.

For reasons I have already stated, I shall not go into any detailed assessment of Milne's kinematical theory of relativity. I shall however indicate, following Milne's ideas, how the cosmological principle together with Newtonian laws can be used to derive a description of the Universe locally adequate and which is in agreement with the relativistic models of Friedmann.

It is clear that the cosmological principle requires that the world view of any observer, relative to himself, must have spherical symmetry about himself. Then according to a theorem, which is valid equally in the frameworks of the Newtonian theory and the theory of general relativity, a particle at the boundary of a sphere, in a distribution of matter having spherical symmetry about the origin, will be gravitationally acted upon only by the matter interior to the sphere. Consequently, so long as the velocities of expansion are small compared to the velocity of light and the contribution of the pressure to the inertia can be neglected, we can restrict ourselves to the Newtonian laws of gravitation and to Newtonian concepts in analysing the dynamics of the motions in the system. And we should expect that the results so derived will be valid, also in the wider framework of general relativity, within the stated limitations. Indeed as Milne and, more fully, Milne and McCrea showed, the equations which follow from the Newtonian analysis are in agreement with those which follow from a relativistic analysis, again within the stated limitations. But one must go to the general theory of relativity if the limitations imposed, by ignoring the inertial effects of the pressure and by disallowing velocities comparable to that of light, are to be avoided.

I am doubtful if Milne would have approved of my presentation of his ideas and have allowed the concession to general relativity that I have made.

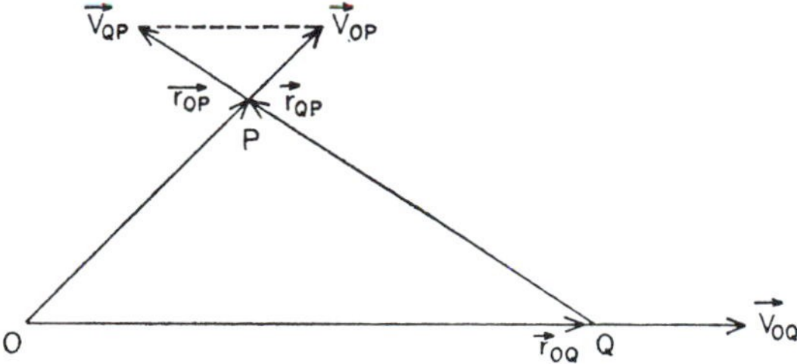


**Fig. 7** A cosmological principle is satisfied in a system in which velocity is proportional to distance. Particles P and Q are receding from point O at velocities proportional to their distance from O, but to an observer moving with particle Q, the only apparent motion of particle P is a recession from Q with a velocity proportional to the distance between P and Q.

Nevertheless, the theory as I have described it, following in the main Milne's ideas, is a part of what every student of cosmology now learns.

## VII

Let me conclude by describing in general terms what manner of a scientist Milne was.

Milne's special strength was in reducing a complex problem to its elements and analysing each element as to its content and as to its meaning. The crisp and vigorous style of his writings are manifestations of his keen analytic intellect. He once told me that often his pen could hardly keep pace with the flow of his ideas. Besides, he took delight in solving his analytical problems with grace and elegance. These admirable traits are discernible in all of his writings, though in much of his later writings they are shrouded, to a larger or smaller extent, by elements of self-defence and controversy. But when the air was free and his thoughts were untrammelled, his obvious enjoyment in the flow of his ideas and in the course and texture of his arguments transports the reader to an equal measure of enjoyment. Nowhere is this transmission of joy more sustained than in his marvellous book on *Vectorial mechanics*. One can also experience the same delight in some of his papers which were not in the main stream of his scientific concerns; and these gems, in many ways, reveal Milne at his best.

If I have to select a paper of Milne's which illustrates his originality, his style, and, above all, his sheer delight in what he is doing, I should select his paper on 'The energetics of non-steady states, with applications to Cepheid variation' published in the *Oxford Quarterly* in 1933. (But even this paper is marred by some unnecessary elements of controversy.) Let me say a few words about this paper.

Milne formulated the ideas contained in this paper during the course of a conversation we were having in my rooms in Cambridge in 1933. Milne was wondering how the phenomenon of Cepheid variability could be grasped in a general theoretical framework without any reference to specific internal parameters such as pressure, temperature, etc. He said that a Cepheid variable is after all a heat engine; and recalling what he had learned from H. F. Newall about Griffith's Heat Engine, he rapidly developed a theoretical framework which led to the functional equation

$$\kappa\phi(t) + \phi(t + b + \phi(t)) = 0 \qquad (3)$$

for the time derivative of the relative amplitude of the light variation. (In equation (3), $\kappa$ and $b$ are certain constants.)

Equation (3) has many remarkable properties. Thus, if $\phi$ takes the value zero at some instant of time, then it must take the value zero at an infinite succession of instants at intervals of time $b$ apart. Further, if $\kappa$ were unity, the solutions are periodic with a period $2b$; and a host of other intriguing properties exist.

At a later time, when Milne gave an account of this work at a meeting of the Royal Astronomical Society he stated, with undisguised delight, that since the functional equation (3)

> gave periodic solutions reproducing some of the features of Cepheid light-curves . . . it should not be beyond the wit of man to devise an analysis [of Cepheid variability] which led to it!

In making an overall assessment of Milne, we have to remember that his early years in Cambridge were interrupted by World War I; that he contracted a fell disease in 1923 which was eventually to prove fatal in 1950; that there were personal tragedies of great magnitude in his life; that his scientific work was interrupted for long years by both World Wars; that during the last several years of his life, he was a sick man; and that, over and above all of these, there was his controversy with Eddington which embittered much of his scientific experience. When we remember all of these and remember also his many solid accomplishments, then we may in truth say, as his long-time friend and colleague Harry Plaskett said,

> . . . he died, as he lived, undefeated.

# Our universe and others

Sir Martin Rees

## Introduction

'The most incomprehensible thing about the Universe is that it is comprehensible' is one of the best known, and indeed most hackneyed, sayings of Einstein. He meant by this that the basic physical laws, which our brains are attuned to understand, have such broad scope that they offer a framework for interpreting not just the everyday world but even the behaviour of the remote cosmos. Einstein was impressed that the large-scale Universe is simpler and more amenable to analysis than there seems any reason to expect. Cosmology—the extent to which we can observe or infer the Universe's overall properties, the extent to which it could have been otherwise—will be the theme of my lecture today. But I shall begin with some comments on 'straight' astronomy, the study of stars and galaxies, to illustrate the different modes of thought involved in our attempts to understand the cosmic scene: one can make deductions or calculations based on known laboratory physics; or one can, contrariwise, use astronomical data to *learn* the relevant basic physics; thirdly, one sometimes has to adopt some general hypoetheses of uniformity or simplicity before getting started at all.

## Understanding stars, etc.

Astrophysicists spend much of their time working in the first of these styles: applying everyday physics, such as Newtonian gravity, atomic physics, and electromagnetism. They use physics as a tool; their mode of thinking is like that of an engineer, trying to construct a working model from known components to meet given specifications. The degree of success is variable. The most substantial and secure progress since Milne's heyday has been in understanding the structure and life cycle of stars, particularly stars like the Sun. Such stars start their lives by condensing gravitationally from interstellar clouds. They then settle down to the so-called main sequence, their energy being derived from fusion of hydrogen into helium in their interiors. When the hydrogen in the core is exhausted, such stars swell up to become red giants, and they eventually settle down to a quiet demise as white dwarfs. Nuclear fusion, a process well understood in theory, provides enough power to keep the Sun shining for 20 billion years.

Study of this life cycle was in fact stimulated by a wish to interpret observations, but it is interesting that the properties of stars could equally well have been deduced by a physicist who had lived on a perpetually cloud-bound planet. He could have posed the question: 'Can one have a gravitationally-confined fusion reactor, and what would it be like?' He would have reasoned like this. Gravity is amazingly weak on an atomic scale. Its weakness is exemplified by the huge number relating the electrical and gravitational forces between electron and proton in a single hydrogen atom:

$$N = \mathrm{e}^2/Gm_\mathrm{p}m_\mathrm{e} \simeq 10^{38}.$$

But if enough atoms are piled together gravity may become important, because

$$\left\{\begin{matrix}\text{Gravitational binding}\\ \text{energy per atom}\end{matrix}\right\} \propto \mathrm{M/R} \propto \mathrm{M}^{2/3} \text{ (for constant density).}$$

Gravity becomes significant when the mass involved reaches

$$\mathrm{M}_* \simeq N^{3/2}\, m_\mathrm{p}.$$

Gravitationally-confined reactors must be massive because gravity is weak; having inferred this, a physicist could calculate the reactor's evolution and lifetime. He could come up with the entire life-cycle of stars, inferring that they are *long lived* because $N$ is so large [1]. Eddington was in fact the first person to express this argument clearly [2]. He then said, 'We draw aside the veil of cloud beneath which our physicist is working and let him look up at the sky. There he will find a thousand million globes of gas, nearly all [with masses in this range]—that is to say, between 1 and 50 times the Sun's mass.'

If our physicist calculated the evolution of heavier stars, he would find a more complex picture. Fusion reactions would proceed all the way up to iron, some stages depending rather delicately on the details of the nuclear physics. The end-point is an explosion where conditions become so extreme that we are not even sure, from our terrestrial basis, of the relevant physics.

The cloud-bound physicist, even if he felt no urge to study or explore the cosmos for its own sake, would be interested in those bits of our environment where conditions are specially extreme. The Universe can then offer a kind of cosmic laboratory, where astronomical data tell us how material behaves under extreme conditions that cannot be simulated terrestrially. Supernova explosions, and the neutron stars they leave as remnants, are the best instances of this. Much of modern high-energy astrophysics (concerning X-ray sources, quasars, galactic nuclei, etc.) is of interest for the same reason. Fifty years ago, a basic understanding of ordinary stars had to await development in atomic and nuclear physics, and there has ever since been a fruitful symbiosis between astrophysicists and nuclear physicists. Likewise, phenomena such as pulsars, quasars and the big bang, sometimes disparagingly assigned to the 'gee-whiz' fringe of astronomy, are likely to be the ones that offer the most interesting input into basic physics in the next decades.

In fact theorists did not get very far by pure deduction, even in fields where the relevant physics was known and understood. They could only echo the Goldwynism: 'with hindsight I could have predicted that'. And when more exotic phenomena are concerned, they certainly need more data to discipline their speculations. In fact the main impetus in astronomy, particularly in the exploratory stages, has generally come not from theory, but from advances in instrumentation. This occurred when the photographic plate and spectroscope were introduced in the nineteenth century. It has been especially so over the last 15 years where the extended scope of astrophysics and the exploitation of new wavebands owes a great deal to space research and exploration. (Optical astronomy, where techniques and detectors have improved dramatically, remains pivotal for studies of stars and aggregates of ordinary stars—it is no accident that stellar surfaces primarily emit photons of energy a few eV, and that this should be the waveband our eyes respond to. But the radiation from more 'exotic' objects spills over into other wavebands. In particular, X-ray emission—a signature of especially high temperatures and particle energies—is detectable only from above the atmosphere.)

## Cosmology

One might expect the prospects for progress in cosmology to be especially bleak. We cannot confidently extrapolate from our laboratory to the whole Universe: on the other hand, relevant data to guide our thoughts are exceedingly sparse. Cosmology, the study of the structure and dynamics of the Universe as a single entity, is in any case a peculiar kind of science. It is by definition the study of a unique object and a unique event. No physicist would happily base a theory on a single unrepeatable experiment. No biologist would formulate general ideas on animal behaviour after observing just one rat, which might have peculiar hang-ups of its own. But we plainly cannot check our cosmological ideas by applying them to other Universes. Nor can we repeat or re-run the past, though the finite speed of light allows us to sample the past by looking at very remote objects.

Despite having all these things stacked against it, scientific cosmology *has* proved possible, but only because the observed Universe, in its large-scale structure, is simpler than we had any right to expect—and than anyone except Milne *would* have expected. Milne was the most celebrated advocate of the view that broad cosmological truths *can* follow from a few unifying principles.

We are talking now not of course on the scale of individual stars and interstellar distances, but on the intergalactic scale. To the astrophysicist, one of the main current challenges is to understand why galaxies exist and have the form they do—why the most conspicuous entities are these self-gravitating aggregates of $10^{11}$ stars. This is still a mystery: we do not know whether it does have a straight physical explanation in terms of Newtonian gravity and gas dynamics (as does the characteristic mass of stars), or whether it is a consequence of initial conditions, so that we have to say 'they are as they are because they were as they were'.

To the cosmologist, entire galaxies are just 'markers' or test particles scattered through space which indicate how the material content of the Universe is distributed. When a cosmologist makes a statement about how homogeneous the Universe is, he is referring to the distribution of galaxies. When he speaks of the expanding Universe he is basically talking about the motions of galaxies.

Galaxies are clustered; some in small groups like our local group, some in big clusters with hundreds of members. But on a really large scale the Universe genuinely does seem smooth. If one imagined a box whose sides were one hundred million light years, dimensions still small compared to the observable Universe, its contents would look about the same wherever we placed it. In other words there is a well-defined sense in which the observable Universe is indeed roughly homogeneous. As one looks at fainter galaxies one probes to greater distances, and the deeper one looks the less evident the clustering is and the smoother the sky appears. Unless we are anti-Copernican and assign ourselves a privileged central position, this implies that the Universe is homogeneous—that all parts evolved in the same way and have the same history (see Fig. 1 and caption).

In 1929 Hubble [3] enunciated his famous law of kinematics: galaxies are receding from us with a speed proportional to their distance. This suggested that we are in a homogeneous expanding Universe where all distances are stretching according to a certain function of time: for instance, the sort of model discussed by Alexander Friedman in 1922 as a solution of Einstein's equations. Hubble's work has now been extended and confirmed by observations of galaxies so far away that they are receding, due to the Universal expansion, at more than half the speed of light.

In a famous textbook on cosmology by Peebles [4], a chapter is entitled

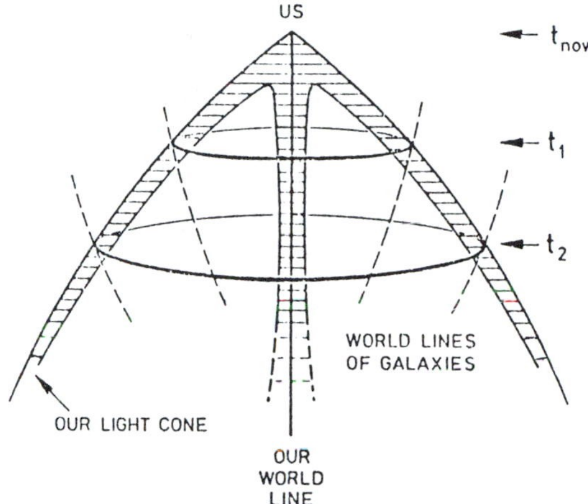


**Fig.1** Schematic space–time diagram showing world line of galaxy and our past light cone. The only regions of space–time concerning which we have direct evidence are those shaded in the diagram, which lie either close to our own world line, (inferences on the chemical and dynamical history of our galaxy, 'geological' evidence, etc.) or along our past light cone (astronomical evidence). It is *only* because of the overall homogeneity that we can assume any resemblance between the distant galaxies whose light is now reaching us and the early history of our own galaxy. In homogeneous universes we can define a natural time coordinate, such that all parts of the universe are similar on hypersurfaces corresponding to a given value of *t*.

'Golden moments in cosmology'. There are really only two of these. The first was Hubble's work. The second, ranking equal in importance, was the serendipitous detection by Penzias and Wilson [5] at the Bell Laboratories of the cosmic microwave background radiation. This radiation is now known to have more or less the spectrum of a thermal black body and a temperature of three degrees above absolute zero. It is almost precisely isotropic, bathing the Earth from all directions, and has no obvious source. It is interpreted [6] as a relic of a so-called fireball phase when the entire Universe was hot, dense and opaque.

According to this concept, about 15 billion years ago all the material in the Universe, all the stuff that now makes up stars and galaxies, constituted an exceedingly compressed and hot gas. The intense radiation in this compressed primordial fireball, though cooled and diluted by the expansion (the wavelengths being stretched and redshifted), would still be around pervading the whole Universe. This is the interpretation of the microwave background that Penzias and Wilson discovered. it is a kind of echo of the explosion which initiated the Universal expansion. The microwave background photons now reaching us have been propagating uninterruptedly since the Universe was a thousand times more compressed and at a temperature of several thousand degrees. The isotropy tells us that since that time the simple isotropic cosmological models are approximately valid. The microwave background introduces a new dimensionless number into cosmology, the ratio $S$ of the number of photons to the number of baryons. Its value, which is essentially constant throughout the expansion, is $\sim 10^8$.

Can we extrapolate back further to higher temperatures and densities? To answer this question I must first digress for a moment to discuss the chemical elements. Complex chemical elements must have been synthesized from hydrogen via the nuclear reactions that provide the power source in the cores of ordinary stars. It is indeed believed that *all* the carbon, nitrogen, oxygen and iron on the Earth, and indeed in our own bodies, could have been manufactured in stars which exhausted their energy supply and exploded before the Sun formed [7]. The Solar System then condensed from gas contaminated by debris ejected from early generations of stars. There is now some quantitative understanding of these processes of cosmic nucleosynthesis, which can account for the relative abundances of different elements—why oxygen is common but gold and uranium are rare—and how they came to be in the Solar System. Pooh-Bah (in *The Mikado*) traced his ancestry back to a 'protoplasmic primordial atomic globule'. We can now do even better. Each atom on Earth can be traced back to stars that died before the Solar System formed. A carbon atom, forged in the core of a massive star and ejected when this explodes as a supernova, may then spend hundreds of millions of years wandering in interstellar space. It then finds itself in a dense cloud which contracts into a new generation of stars. It may then be once again in a stellar interior, where it can be transmuted into a still heavier element. Alternatively it may find itself out on the boundary of a new solar system in a planet, and maybe even eventually in a human cell. (Much of the work on stellar nucleogenesis, incidentally, was pioneered by advocates of the steady-state theory, who believed that all such processes as the production of the elements must necessarily be going on somewhere at the present epoch. Indeed the concepts of stellar nucleogenesis may be that theory's most enduring legacy.)

A long-standing problem was, however, to account in this fashion for helium, which is much more abundant than any other element apart from hydrogen ($\sim$ 25 per cent by mass), and much more uniformly distributed. It was therefore gratifying when the expected composition of material emerging from the big bang, calculated on the simplest assumptions, was found to be about 75 per cent hydrogen and 25 per cent helium [8]. The helium abundance depends on microphysical constants, particularly those involving neutrons and the weak interactions, and also on the expansion rate at the time the reactions occur—when the temperature is about $10^{10}$ degrees, and the Universe has been expanding for only a few seconds. The resulting helium abundance is very sensitive to the expansion rate. if this were boosted by only a factor 2 from the standard model, too much helium would result. There are some cosmological models, involving anisotropic expansion, scalar fields, or a variable gravitational constant, which are consistent with all other data but predict expansion rates at the helium-formation epoch which are up to a million times faster than in the standard model. Thus the measured helium abundance in the Universe strongly supports the simplest version of the hot big bang theory. It also suggests that we are perhaps justified in extrapolating the laws of microphysics back to these epochs.

Helium formation takes us to the era when the Universe was a few seconds old; but can we go back still further? The further one extrapolates, the less confidence one has in the adequacy or applicability of known physics. For instance, the material exceeds nuclear densities for the first $10^{-5}$ seconds. But if one measures time on a logarithmic scale, to ignore these early eras is a severe omission indeed. Within the last 2 years, developments of grand unified theories have emboldened some physicists to go back to the time when $kT$ is $10^{15}$ GeV. This temperature corresponds to the rest mass of the $X$-boson which these theories invoke. When the Universe cools and these annihilate into lighter particles, CP violation leads to favouritism of baryons over anti-baryons. For every $\sim 10^8$ baryons and anti-baryons, there would be one extra baryon. After the Universe has cooled and the pairs have annihilated, this gives $10^8$ photons per baryon, the mysterious ratio $S$ which is found to prevail in our Universe [9]. If this explanation of the baryon content of our Universe is correct, it allows us to test whether a Friedman model was already applicable at epochs as early as $10^{-38}$ seconds—a backward extrapolation from the epoch of primordial nucleogenesis exceeding, in terms of logarithmic time, the extrapolation from the present time back to the nucleogenesis epoch. it would be exciting indeed if we could relate $S$, the hitherto unexplained parameter of big bang cosmology, to experimentally measurable quantities in particle physics, and at the same time vindicate an extrapolation which takes us in a single bound close to the era of quantum cosmology at $10^{-43}$ sec (see Fig. 2 and caption). I must not leave the impression that any of this is too firm. I am myself reluctant to place too much weight on inferences about the first million years, and am still more tentative about the first microsecond. However, there *is* support for the hot big bang concept. It has more than fashion and beauty to commend it and offers a consistent story about the origins of matter and radiation in our Universe. It serves as a framework for interpreting astronomical data, but we must be alert to possible contradictions: our present satisfaction might reflect the paucity of the data rather than the excellence of the theory, and may prove as illusory as that of a Ptolemaic astronomer who has just discovered another epicycle.

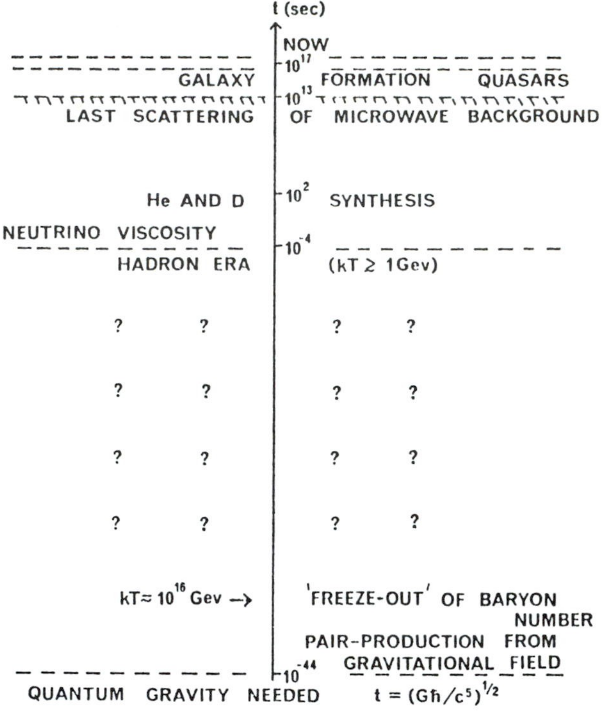


**Fig. 2** This diagram (reproduced from [10]) illustrates, in terms of logarithmic time, various key physical stages in the expansion of a standard 'big bang' model. Sixty 'decades' separate us from the Planck time when quantum gravitational effects would be crucial. Observations of individual sources, even if they extend out to redshifts $z \simeq 5$, permit us to probe only the last decade; the last scattering of the microwave background may have occurred when the Universe had only $\sim 10^{-4}$ of its present age (scale factor compressed by $- 10^3$ from its present value); primordial nucleosynthesis yields evidence on physical conditions when $t \simeq 1$s, Recent ideas that relate entropy production to the consequences of 'grand unification' schemes [9] involve extrapolating back in time by a further 36 orders of magnitude!

## The far future

Having inferred something about how our expanding Universe began, the next question concerns its future and eventual fate. Will the Universe expand for ever and the galaxies fade and disperse? Or, on the other hand, will it re-collapse? The traditional way to try to answer this question has involved extending the work of Hubble to the study of galaxies at greater distances, or (equivalently) to greater look-back times. The redshift of a distant galaxy tells us its speed as the light set out on its journey towards us. So in principle one can compare the expansion of the Universe at early times with the present rate, and thereby infer how much it is decelerating. In practice this type of work is bedevilled by various observational difficulties and uncertain corrections, which have as yet prevented it from yielding a reliable answer.

But there is another more indirect way of trying to determine whether the Universal expansion is destined to stop and go into reverse. Imagine that a big sphere, or asteroid, is shattered by an explosion, the debris flying off in all directions. Each fragment feels the gravitational pull of all the others, and this causes the expansion to decelerate. If the expansion were sufficiently violent, then the debris would fly apart for ever. But if the fragments were not moving quite so fast, gravity might bind them together strongly enough to bring the expansion to a halt. The material would then fall together again. According to general relativity, essentially the same argument holds for the Universe.

In the case of the galaxies, which for the purposes of this argument are regarded as 'fragments' of the expanding Universe, we know the expansion velocity. What we do not know so well is the amount of gravitating matter that tends to brake the expansion. It is easy to calculate how much will be needed to bring it to a halt: it works out at about three atoms per cubic metre. If the average concentration were below this critical density, we would expect the Universe to continue expanding for ever; but if the mean density exceeded the critical density, the Universe would seem destined eventually to recontract.

It is straightforward to estimate how densely the galaxies are packed in the Universe. There is, on average, about one in every $10^{21}$ cubic light-years. The amount of material in the stars one sees in a typical galaxy, spread out over this volume, falls short by a factor 30 of the critical density. But there may be a larger amount of *unseen* mass. Studies of individual galaxies yield dynamical evidence for a distribution of gravitating mass far less concentrated than the light which we see, and extending way beyond the visible image. Dynamical studies of clusters or groups of galaxies which appear gravitationally bound, imply that there is 10 times as much dark material as is actually seen [11]. So more than 90 per cent of the mass in the Universe may be in some unknown form, not necessarily stars. Possibilities include 'dead' stars or black holes of pregalactic origin. This mass may not necessarily be baryons. A much ballyhooed idea currently is that neutrinos have a rest mass of a few electron volts. Even for such a small mass they would be dynamically important because in the primordial fireball they would be as numerous as photons, outnumbering baryons by a factor $S \simeq 10^8$.

There is no reason to assume that everything in the Universe necessarily shines conspicuously; what we have already observed or inferred may be a small and atypical fraction of what actually exists. Our present inventory of what is in the Universe may still be biased and incomplete. It should quench optimism of quick progress that more than 90 per cent of the Universe is in entities whose individual masses could be anywhere from 10 eV ($10^{-32}$ g) up to $\sim 10^6$ solar masses ($\sim 2 \times 10^{39}$ g)—an uncertainty of 70 orders of magnitude! Some forms of matter are so elusive that they could provide the critical density without there being any hope that we would ever detect them directly. Absence of evidence is not evidence of absence. So it may be that there *is* enough material to make the Universe collapse.

Some of what I have just said has implicitly assumed general relativity. In this theory, there is a relation between the mean density of gravitating stuff in the Universe, and its deceleration and eventual fate. In the early days, Friedman's relativistic cosmology was just one among several model Universes. Milne himself did not accept conventional general relativity. But general relativity has enhanced its standing since his time: theoretical developments over the last 20 years have highlighted its internal integrity; experimental work, admittedly relating only to the weak-field limit, has refuted all rival theories except those specially constructed *a posteriori*. Consequently, most of us are now devotees of general relativity, and regard it as a supreme example of what Eugene Wigner called the 'unreasonable effectiveness of mathematics in the physical sciences'. However, it would be a crucial test of relativity in cosmology to measure *both* the density *and* the deceleration of the Universe and see if they are indeed related in the way the theory predicts.

What will happen if the Universe recollapses? The redshifts of distant galaxies would be replaced by blueshifts, and galaxies would crowd together again. Space is already becoming more and more 'punctured' as isolated regions—dead stars and galactic nuclei—collapse to black holes; but this would then be just a precursor of a Universal collapse, a 'big crunch' that engulfs everything (Fig. 3). Some key stages in the 'countdown' are shown in the diagram. The final state would be a fireball like that which initiated the Universal expansion, though it would be somewhat more lumpy and unsynchronized. According to Penrose [12], this asymmetry is connected with the origin of 'time's arrow'.

The most recent discussion of the future of the Universe, eschatology, is a lengthy article in *Reviews of Modern Physics* by Freeman Dyson [13]. He berates earlier writers on this subject [14, 15], myself included, for not taking themselves seriously enough; and then embarks on boisterous and entertaining speculations of his own. He says rather little about the recollapsing Universe (the mere idea gives him claustrophobia!), but addresses in detail the far future of an ever-expanding Universe.

A one-sentence answer to the question 'what is happening in the Universe?' might go like this: gravitational binding energy is being released as stars, galaxies and clusters progressively contract, this inexorable trend being delayed by rotation, nuclear energy, and the sheer scale of astronomical systems, which make things happen slowly and stave off gravity's final victory. Dyson discusses what will happen if the Universe expands indefinitely and

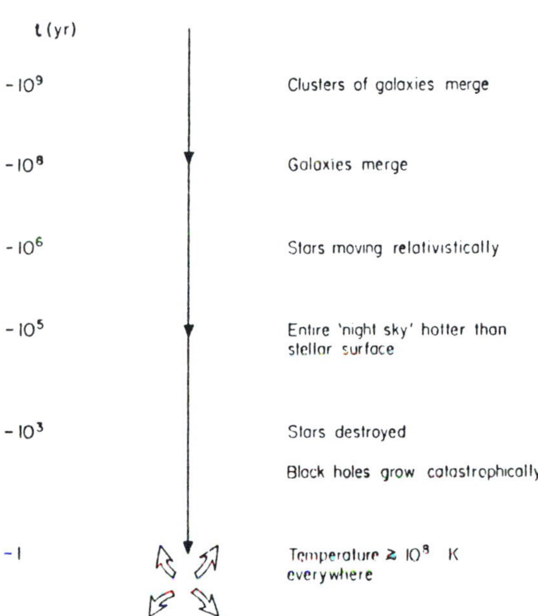


**Fig. 3** Stages in the contraction of a recollapsing Universe (*cf.* [14]). Times are measured backwards from the 'crunch'.

there *is* enough time to attain a terminal state. Figure 4 illustrates various key stages. Everything eventually decays back to radiation, either via proton decay (which some grand unified theories [9] permit) or else by black hole formation and subsequent evaporation. Dyson contemplates the outlook for intelligent life. Can it survive and develop intellectually for ever, thinking infinite thoughts (including a unified field theory!), and storing or communicating an ever-increasing body of information, on finite energy reserves? He shows that in principle this can be done: as the background temperature falls one must keep cooler, think progressively more slowly and 'hibernate' for long intervals.

## Is our universe 'special'? Coincidences and consequences

Which Universe is ours? Will our descendants need to follow Dyson's conservationist maxims to survive an infinite future? Or will they fry in the big crunch a few times $10^{10}$ years hence? These two outcomes are very different. But the initial conditions which could have led to anything like our present

THE FAR FUTURE OF AN EVER-EXPANDING UNIVERSE

| | |
|---|---|
| $10^{14}$ yr | Ordinary stellar activity completed |
| $10^{17}$ yr | Significant dynamical relaxation in galaxies |
| $10^{20}$ yr | Gravitational radiation effects in galaxies |
| [$10^{31}$–$10^{36}$ yr | Proton decay] |
| $10^{64}(m/m_{\odot})^3$ yr | Quantum evaporation of black holes |
| $10^{1600}$ yr | White dwarfs → neutron stars* |
| $10^{10^{26}}$–$10^{10^{76}}$ | Neutron stars undergo quantum* tunnelling to black holes, which then 'quickly' evaporate |

*If proton decay does not occur

**Fig. 4** Timescales for an ever-expanding Universe, as elaborated by Dyson [13].

Universe are actually *very restrictive*, compared to the range of possibilities that could have been set up: the fact that the Universe has gone on for $10^{10}$ years without recollapsing, but is expanding sufficiently slowly to permit galaxies to form, in itself requires 'fine tuning' (see Fig. 5 and caption). Moreover, our Universe looks even more special if we recall that the whole family illustrated in Fig. 5 are just the Friedman models which are isotropic, their dynamics being described by the time-dependence of a single scale-factor $R$. Further free parameters can be introduced by admitting anisotropy, or large amplitude inhomogeneity: but would generally make the Universe less conducive to galaxy formation. The mystery is why the Universe is so homogeneous—why it has not availed itself of the other macroscopic degrees of freedom—and why the initial kinetic energy and gravitational energy were so closely balanced as its present properties indicate. The homogeneity was not *too* perfect, however. To get galaxies at all, to avoid having a Universe which is still after $10^{10}$ years composed of amorphous gas and radiation, there must have been some initial fluctuations—regions of enhanced density which lagged behind and condensed out from the expanding background. The early

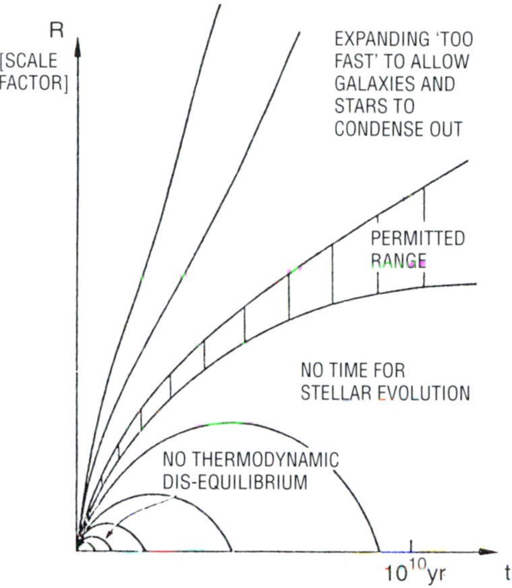


**Fig. 5** In the isotropic Friedman models, the scale factor $R(t)$ evolves essentially in the same manner as in 'Newtonian cosmology': the Universe either expands for ever (positive total energy) or recollapses. We know that our Universe is still expanding after $10^{10}$ yr; if it had recollapsed sooner, there would have been no time for stars to evolve; if it collapsed after less than $\sim 10^6$ yr, it would have remained opaque and close to thermal equilibrium throughout its life. The expansion rate cannot, however, be *too* much faster than 'parabolic': otherwise gravitational instability would have been ineffective and bound systems would not have condensed out. (This is equivalent to the statement that the present density is not orders-of-magnitude below the critical density.) There is therefore a sense in which the dynamics of the early Universe must have been 'finely tuned': in Newtonian terms, the fractional difference between the initial potential and kinetic energies of any spherical region must have been very small: in terms of the Robertson–Walker metric, the curvature of the hypersurfaces of homogeneity must have been very small compared to the particle horizon at early times. As mentioned further in the text, our Universe seems even more 'special' when we realize that the introduction of anisotropy, large-scale inhomogeneity, etc., offers many more degrees of freedom than the Friedman models.

Universe must have been smooth and uniform only in the same sense as a golf ball or an orange: an overall smooth curvature, with small amplitude wrinkles superposed on it. The amplitude of these curvature fluctuations is no more than one part in $10^5$. It could be that $10^{-5}$ amplitude curvature fluctuations needed for galaxies, and still quite unexplained, have a spectrum extending over all mass scales; these would be undulations extending out to the Hubble radius or even beyond. *Why* the initial conditions were like this is something that known physics does not allow us to explain.

Any analysis of the early Universe, or of the physics of gravitational collapse or the end of the Universe, leads us to believe that conditions must develop so extreme that they transcend our present understanding. We know physics is incomplete in that there is still no adequate theory of quantum gravity. The two great foundations of twentieth-century physics are the quantum principle and Einstein's formulation of general relativity. The theoretical superstructures erected on these foundations are still disjoint. There is no real overlap between their respective domains of relevance. Quantum effects are crucial on the microscopic level of the single elementary particle; but gravitational forces between individual particles are negligible, weaker by a colossal factor $N \simeq 10^{40}$ than electromagnetic forces. Gravitational effects are manifested only on the scale of planets, stars and galaxies. (The gravitational forces between large numbers of particles reinforce each other, whereas electrical forces cancel out because electrons are neutralized by an almost identical number of positive charges.) On the astronomical scale where gravity is important, a classical approach to dynamics, ignoring quantum processes, is entirely adequate.

But our present understanding, however adequate it may appear for interpreting observed phenomena, will remain conceptually unsatisfactory until gravitation has been quantized. Answers to such questions as: Why is the Universe so large? Why does it contain fluctuations? What happens at the singularity predicted by Einstein's theory? Can a collapsing Universe rebound Phoenix-like into another cycle? must all await the quantization of gravity. Conversely, maybe the *only tests* of such a theory will be offered by cosmological observations. Quantum gravity is important at the Planck time of $10^{-43}$ s (*cf.* Fig. 2). Naive ideas of space and time, and even 'before' and 'after', break down when the Universe is younger than this. (The usual demarcation in physics between the differential equations that determine how things evolve, and the initial data which depend on how they were set up, is particularly constricting in cosmology, and this distinction would be blurred in the complex space–time structure at the Planck time.)

I began this lecture by recalling some phenomena, such as ordinary stars, where we feel fairly confident that we know the relevant physics. When physical conditions are more extreme, we are less confident, though it is amazing how far we can go without running up against a contradiction. But in considering the early Universe we are led towards a set of issues where we know for sure that we do *not* know enough physics. Professional scientists do not spend all their time pondering the really deep 'fundamental' questions: it is sensible methodology to work on bite-size topics where there is hope of progress, rather than to worry excessively about problems which are not timely or tractable. but these latter are generally the broader and more significant ones, and our piecemeal efforts are just steps towards the goal of elucidating them. The occupational risk of astrophysicists, indeed of all scientists, is to become so preoccupied with the bit of the puzzle they are focusing attention on, that they forget they are wearing blinkers and that there *are* broader questions at issue. For instance: What came before the big bang? How were the fluctuations that gave rise to galaxies imprinted in the early

Universe, and why is the Universe rough on small scales but smooth in the large? Why are we in a space–time where there are three spatial dimensions and one of time? And why are the constants of Nature so universal? Above all, why does the Universe have the overall symmetry and simplicity that is a prerequisite for any worthwhile progress in cosmology?

These questions trigger further speculations. For instance, all the features of the everyday world and the astronomical scene are essentially determined by a few basic physical laws and constants, such as the masses of the elementary particles and the relative strengths of the basic forces that operate between them. In many cases a rather delicate balance seems to prevail. For instance, if the nuclear forces were slightly stronger than they actually are relative to electromagnetism, the diproton would be stable, ordinary hydrogen would not exist, and stars would evolve very differently. If nuclear forces were slightly weaker, no chemical elements other than hydrogen would be stable and chemistry would be very dull indeed. The details of stellar nucleosynthesis are sensitive [16] to apparent 'accidents', for instance the fact that there is a particular resonance in the carbon nucleus, which allows carbon to form from $^4$He plus $^8$Be despite the instability of $^8$Be. There are two contexts where the cross-sections for weak (neutrino) interactions are crucial. The first is in supernova explosions, where a neutrino-driven shock blows off the outer layers of the star and recycles chemically enriched debris back into the interstellar medium. The second is primordial helium production. Decreasing these cross-sections would yield a universe composed almost entirely of helium, where no supernovae could explode [17].

One can in imagination tinker with cosmological models in other ways. For instance if $G$ were stronger, and $N$ were (say) $10^{30}$ rather than nearly $10^{40}$, we could have a small-scale speeded-up universe, in which stars (gravitationally-bound fusion reactors) had $10^{-15}$ the Sun's mass, and lived for about one year. This might not allow enough time for complex systems to evolve: there would be fewer orders of magnitude between astrophysical timescales and the basic microscopic timescale for physical or chemical reactions. There would also be a more stringent maximum size imposed on complex structures that do not themselves get crushed by gravity. Although gravity dominates on a large scale in our Universe, it is because it is actually so weak compared to other forces that very large and long-lived systems can exist. Our Universe is large and diffuse *because* gravity is so weak [1]. Maybe its extravagant scale is necessary to provide enough time for interesting evolutionary processes to occur.

Perhaps the contingency that we are here to ponder such matters itself poses some constraints on what the physical laws can be like. In other words, given that we know that our cosmic environment permits observers to exist, maybe we should not take the Copernican principle too far. We would not feel justified in assigning ourselves a central position in the cosmos, but it may be equally unrealistic to deny that our situation can be privileged in any sense. Carter [18] has called this line of argument the 'anthropic principle', and has coined for it the neo-Cartesian slogan 'cogito ergo mundus talis est' ('I think, therefore the world is as it is').

Some people take this concept very seriously. Wheeler [19] imagines an ensemble of universes with different physical laws and different fundamental ratios, all 'laid down' at the initial singularity, on which a kind of evolution by natural selection operates. Most of the universes are 'stillborn', in the sense that the prevailing laws do not permit anything interesting to happen in them. But maybe in some of them complex structures *can* evolve, and one would have achieved something if one could show that any such universe had to possess features which our actual Universe *does* possess. One must of course not be too anthropomorphic, nor too restrictive in envisaging the requirements for the emergence of a conscious observer: maybe neither stars nor heavy elements are absolutely necessary. However, some degree of thermodynamic disequilibrium would, for instance, seem essential.

The anthropic principle cannot obviously provide a scientific explanation in the proper sense. At best it can offer a stop-gap satisfaction of our curiosity regarding phenomena for which we cannot yet obtain a genuine physical explanation. For example, the world would be very different, and perhaps even 'uncognizable', if the relative strengths of the strong (nuclear) and electromagnetic interactions were somewhat altered, but one still hopes for a unified physical theory that predicts or relates the actual coupling constants. For instance, a century ago one might have imagined varying the electrical and magnetic forces and the speed of light, but the work of Maxwell showed us how these were interconnected. Likewise, Weinberg and Salam have taken a further step in incorporating weak (neutrino) forces in an integrated scheme. By extension, we may hope eventually for a unified theory, maybe some version of supergravity, which predicts the ratios of all fundamental forces [20]. In the ideal theory there is no arbitrariness: everything not forbidden is compulsory. All the physical constants are determined by basic mathematical formulae. But, were a theory ever devised which accounted for the constants of Nature as we know them, it would still seem coincidental or providential that, at energy levels infinitesimal compared to the unification energy of $10^{15}$ GeV, the theory should prescribe a physical world propitious for life.

Let me terminate these speculations and briefly conclude. The proverbial rational man who loses his keys at night searches only under the street lamps, not because that is necessarily where he dropped them, but because his quest is otherwise quite certain to fail. Cosmologists approach their subject in a similar way. They start by using the physics that is validated locally and making simplifying assumptions about symmetry, homogeneity, etc. There seems no reason why the Universe should be so ordered that this permits any real progress, but unless there is a firm link with local physics, cosmology risks degeneration into *ad hoc* explanations on the level of 'Just so' stories. What does seem amazing is that this has led to some progress—that the Universe is comprehensible. Questions such as how the Universe began and how it will end can now be addressed scientifically, and not just in our unprofessional moments. Speculation is at least somewhat bridled, and I hope I have conveyed some of the flavour of the subject in this lecture.

## References

[1] Dicke, R. H., 1961. *Nature, Lond.*, **192**, 40.

[2] Eddington, A. S., 1926. *Internal constitution of the stars.* Cambridge University Press.

[3] Hubble, E., 1929. *Proc. Nat. Acad. Sci.*, **15**, 168.

[4] Peebles, P. J. E., 1971. *Physical cosmology*. Princeton University Press.

[5] Penzias, A. A. and Wilson, R. W., 1965. *Astrophys. J.*, **142**, 419.

[6] Gamow, G., 1948. *Nature, Lond.*, **162**, 680; Alpher, R. A. and Herman R. C., 1949. *Phys. Rev.*, **75**, 1089; Dicke, R. H., Peebles, P. J. E., Roll, P. G. and Wilkinson, D. T., 1965. *Astrophys. J.*, ***142***, 414.

[7] Burbidge, E. M., Burbidge G. R., Fowler, W. A. and Hoyle, F., 1957. *Rev. Mod. Phys.*, **29**, 547; Trimble, V. L., 1975. *Rev. Mod. Phys.*, **49**, 877 and references cited therein.

[8] Hoyle, F. and Tayler, R. J., 1964. *Nature, Lond.*, **203**, 1108; Peebles, P. J. E., 1965. *Astrophys. J.*, **146**, 542; Wagoner, R. V., Fowler, W. A. and Hoyle, F., 1967. *Astrophys. J.*, **148**, 3.

[9] Yoshimura, M., 1978. *Phys. Rev. Lett.*, **41**, 381; Dinopoulos, S. and Susskind, L., 1978. *Phys. Rev. D.*, **18**, 4500; Toussant, D., Treiman, S. B., Willizek, F. and Zee, A., 1979. *Phys. Rev. D.*, **19**, 1036; Sakharov, A. D., 1979; *Sov. Phys. J.E.T.P.*, **76**, 1172; Ellis, J., Gaillard, M. K. and Nanopoulis, D. V., 1979. *Phys. Lett.*, **80B**, 360; Weinberg, S., 1979. *Phys. Rev. Lett.*, **42**, 850; Kolb, E. W. and Wolfram, S., 1980. *Nucl. Phys. B.*, **172**, 224.

[10] Rees, M. J., 1980. *Phys. Scripta*, **21**, 614.

[11] Einasto, J., Kaasik, A. and Saar, E., 1974. *Nature, Lond.*, **250**, 309; Ostriker, J. P., Peebles, P. J. E. and Yahil, A., 1974. *Astrophys. J.* (lett.), **193**, L1.

[12] Penrose, R., 1979. In: *General relativity: an Einstein centenary survey* (eds S. W. Hawking and W. Israel). Cambridge University Press.

[13] Dyson, F. J., 1979. *Rev. Mod. Phys.*, **51**, 447.

[14] Rees, M. J., 1969. *Observatory*, **89**, 193.

[15] Davies, P. C. W., 1973. *Mon. Not. R. Astr. Soc.*, **161**, 1; Islam, J. N., 1977. *Q. J. R. Astr. Soc.*, **18**, 3.

[16] Hoyle, F., 1954. *Astrophys. J.* (Suppl.), **1**, 121.

[17] Carr, B. J. and Rees, M. J., 1979. *Nature, Lond.*, **278**, 605.

[18] Carter, B., 1974. In: *Confrontation of cosmological theories with observation* (ed. M. S. Longair) p. 291. Reidel, Dordrecht.

[19] Wheeler, J. A., 1974. Chapter 19 of *Black holes, gravitational waves and cosmology* by M. J. Rees, R. Ruffini and J. A. Wheeler. Gordon and Breach, New York and London; and in: *Proc. Princeton Einstein Centennial Symposium* (1979) (ed. H. Woolf), p. 339. Addison-Wesley, New York.

[20] Hawking, S. W., 1980. *Is the end in sight for theoretical physics?* Cambridge University Press.

# Astrology, religion and science

T. G. Cowling

I first met Professor E. A. Milne in January 1929, when he took up duties as the first Rouse Ball professor. I had the honour of being his first research student at Oxford. Perhaps, therefore, I may be permitted a few minutes to describe the man as I came to know him.

He was small of stature, intensely active, and usually in a hurry. I have been told that, when he lectured at Cambridge, by the end of a lecture his audience was out of breath as well as he. From my Oxford experience I can well believe this. He once told me that when he really got going at research he could have kept two pens busy. His temperament was mercurial: he was a delightful conversationalist and normally cheerful, but a natural sensitivity made him feel keenly the attacks he experienced when advancing unorthodox ideas. For example, he fell out with Eddington when trying to provide an alternative to Eddington's theory of stellar structure. He told me, in the course of advice that I should not treat scientists with wrong ideas as morally sinful, that he maintained the friendliest personal relations with Eddington despite their scientific differences: I suspect, however, that this may have represented the ideal that he strove after rather than actually attained. I found him inspiring and generous, and he did a lot for me, though I did not always appreciate it at the time (I tended to side with Eddington).

He of course had no truck with astrology, and you may ask why I chose this as subject for a Milne lecture. Astrology is of interest to the historian of science: the terms of foundation of the Rouse Ball chair asked the holder to pay attention to the wider aspects of mathematics, historical and philosophical. Astrology in its birth relied on a primitive cosmology: Milne spent some years trying to construct a rival cosmology to that of Einstein. The first astrologers were priests, trying to ascertain the will of the gods from the stars: Milne had a simple religious faith, and a friend who became rector of a North Oxford parish told me that Milne was nearly as much help to him as a curate would have been.

But why, you may ask, should astrology be regarded a suitable subject for a lecture to mathematicians? The answer is that for many centuries mathematician meant astrologer. Often when the Roman authorities felt that interest in astrology was getting out of hand, they would issue a decree expelling all mathematicians from Rome. It was therefore possible for the Roman emperor Diocletian (about AD 300) to say

> It is to the public interest that one learns and practises the art of geometry. But mathematical art is damnable: it is absolutely forbidden.

After these preliminaries, I turn to the history of astrology. Our astrology had its origin in Mesopotamia, among the Babylonians and Assyrians. Much of our knowledge of this 'Chaldaean' astrology comes from a library of clay tablets collected for the Assyrian king Ashurbanipal. The date of the library is about 650 BC, but many of the tablets are copies of much older ones. The actual birth of astrology may date back to before 2000 BC. This does not mean that (as is often claimed) the rules of modern astrology are based on centuries of observations in 'Chaldaea'. Modern astrology differs from that of Chaldaea in essential respects.

Chaldaean astrology was inextricably bound up with religion. The heavenly bodies were regarded as gods, or at least the abodes of gods, able and ready to influence men on Earth. At first the stars were supposed to move on a domed vault over the Earth, and to travel from east to west each night (though what happened to them between their setting in the west and their next rising in the east was left obscure). Later it was recognized as more reasonable to regard the stars as fixed to a great sphere which surrounded a stationary Earth and rotated about a diameter once in roughly a day; this was the firmament. The five planets then known—Mercury, Venus, Mars, Jupiter, Saturn—were observed not to share the motion of the stars exactly, appearing to crawl over the surface of the firmament; these must be more powerful gods, able to take an independent line. The same was true of Sun and Moon; the Sun described each year a great circle on the firmament (the Zodiacal circle, inclined to the celestial equator): the Moon described each month an irregular path, roughly along the Zodiacal circle, but wobbling on either side of it. For brevity I shall call these seven bodies (luminaries plus planets) the Wanderers. To each star-god were ascribed powers consistent with its name. Modern astrology retains the powers, but discards the gods whose names determined the powers.

Chaldaean astrologers were not concerned with the fortunes of ordinary mortals, but only with the fate of kings and kingdoms. They based their predictions mainly on untoward events seen in the heavens, like eclipses of the Sun or Moon, shooting stars, changes of colour of the heavenly bodies, the near approach of two Wanderers, etc. They at first attached only minor importance to the Signs of the Zodiac (twelve constellations evenly spaced out along the Zodiacal circle); they used stars from all over the sky and at first they knew only eleven Signs. However, when one of the Wanderers was seen as crawling through a particular constellation on the firmament, it was natural to assert that its influence was affected by the company it was keeping. The Signs ultimately acquired importance because the Moon and the planets, though moving irregularly, were never far from the Zodiacal circle. Present-day newspaper astrologers, with their emphasis on the Sign under which a person is born, are relying only on the effect on the Sun of its background Sign.

As stated above, each of the seven Wanderers was endowed with a nature consonant with its name and appearance. The Sun, though favourable as the bringer of light, could be the reverse as the bringer of scorching heat; the Moon, associated with humidity, was likewise usually but not always favourable. Venus (Ishtar in Chaldaea) was favourable as the mother and life goddess; Jupiter (Marduk) similarly as representing the creator of the world. Both of these shone with a clear white light: Mars (Nergal) was red, the colour of blood and fire, and so was baneful. Saturn with a greyish colour was the same, save when it took over as deputy for the absent Sun. Mercury, rapidly alternating between leading and following the Sun, was inconstant, even in sex. (The idea that one heavenly body could deputize for another was not limited to Saturn; each planet had a star or stars which could deputize for it, apparently because of a similar colour. The Greeks later gave more 'logical' reasons for the properties of the planets, with much the same conclusions—the disadvantage of knowing the answer beforehand?)

As examples of Chaldaean astrology I quote the following:

'When Mars is dim, it is lucky; when bright unlucky. When Mars follows Jupiter, that year is lucky.' (Jupiter here is said to capture Mars, and add his strength to his own.)

'When a halo surrounds the Moon and Jupiter stands within it, the king will be besieged. The halo was interrupted; it does not point to evil.' (The breach in the halo represented an escape route for the king.)

'An eclipse has occurred, but it was not visible in the capital. As the eclipse approached, the great gods who dwell in the king's city overcast the sky and did not let the eclipse be seen. So let the king know that the eclipse is not directed against him or his land.' (What is not seen does not count, a principle which if strictly adhered to would make nonsense of taking horoscopes. In any case, there were specific rules to decide to which local kingdom a given portent applied.)

'When on the first day of Nisan the rising Sun is red like a torch, while clouds rise from it and the wind blows from the east, then the Sun will be eclipsed on the 28th or 29th of the month, the king will die in that month and his son will ascend the throne.' (A prophecy after the event?)

Chaldaean astrology began as an attempt to interpret the will of arbitrary gods. It began to be scientific when the priestly observers noted that certain celestial events tended to repeat themselves. There were periodicities in the motion of planets, and even eclipses had regular features. After a few centuries of observation it was possible to use the periodicities to construct tables giving future positions of the planets with an accuracy comparable with that of the observations. Lunar tables also were constructed, but there is no evidence whether these were actually used to predict eclipses. The cause of eclipses was not known, and the priests were concerned with direct deductions from the observations, not with deep theory. But at least a first suspicion should have arisen that the Universe is not ruled by blind chance or arbitrary acts of the gods, but by eternal unchanging laws.

The conquests of Alexander the Great opened up the Greek world to eastern astrology. The way had been prepared by Pythagoras and his school (sixth century BC). These were natural philosophers who taught the unity of Heaven and Earth, so that it would be natural to suppose that events in the heavens were portents of events to come on Earth. Also, though it is not certain if Pythagoras was the actual originator of the theorem that bears his name, the Pythagoreans had more than a touch of mathematical mysticism. They are credited with the first assertion that the Earth should be a sphere—a 'perfect' shape; also they found evidence that the laws of Nature are mathematical in the fact that a stretched string of variable length gives an unchanged note only if its tension is proportional to its length.

Similar ideas are met in Plato, though he regarded an intellectual search for heavenly harmonies as far more important than mere observations. His cosmology, and that of Aristotle after him, was based on a mystical belief in the perfection of the heavens, which meant that the motions of the Wanderers must be explicable in terms of a machinery of perfect spheres. Aristotle wanted a physical explanation, and so his spheres had to be real and composed of perfectly transparent crystal. Astronomers, concerned only with the kinematics of the motions, used the spheres only as a calculating device—the poor man's Fourier analysis.

Belief in a mystic unity of the Universe is not confined to religion; it is found also in scientists like Kepler, Newton and Einstein. Coupled with the general belief that the Earth is at rest at the centre of the Universe, it prepared the Greeks after Alexander to accept the central tenet of astrology, that influences from the heavenly bodies are continually raining down on the Earth. Astrology invaded the Greek world along two main routes, through a school set up on the Greek island of Chios by Berossos, a priest from Babylon, in the third century BC, and through an Egyptian school whose teaching culminated about 160 BC in a book attributed to two probably mythical characters, Nechepso and Petosiris. Despite opposition from a few sceptical philosophers, astrology spread like wildfire among the Greeks.

However, the astrology which they took over differed in material aspects from that of the Assyrian library. It now concerned itself with the fates of ordinary mortals, not only with kings and kingdoms. (A cynic might suggest that the reason for this was that astrologers, made redundant when lesser kingdoms were merged in the one Persian empire, had to find an alternative market for their wares.) Whatever the reason, in order to satisfy the demands of a mass market some simplification of method was needed: thus, since the positions of all the stars in the sky could be found if those of stars in the zodiacal belt were known, it was reasonable to use only the Signs of the Zodiac. Again, the planets were given the names of Greek gods with attributes as far as possible identical with those of their Chaldaean originals. However, whereas the Chaldaean gods were the embodiment of forces of Nature, stories about the Greek gods depicted them as all too human in their failings and inconsistent in their behaviour, so that different suppositions about planetary influences were possible. Finally, the Chaldaeans had an essentially two-dimensional picture of the heavens, with the Wanderers effectively crawling on the firmament sphere: the Greeks had already begun to think of the Wanderers as attached to celestial spheres well inside the firmament. Thus instead of saying that the influence of a Wanderer was affected by the company it was keeping, the influence had to be supposed to be affected by the influences of background stars on their way to the Earth; what mattered was the background *as seen from the Earth*. The Greeks swallowed this idea without too much difficulty: it had too much prestige to be jettisoned.

Being mathematicians in the modern sense, Greek astrologers sought to replace earlier disjointed rules by a more systematic set. The first step was to replace older irregular Signs by twelve regular ones dividing the 360° of the zodiacal circle into twelve equal parts each subtending exactly 30° at the centre. These agreed with the earlier Signs as far as possible, but to get twelve of them it was necessary to replace the claws of the Scorpion by a new constellation, Libra (the Scales). The final delimitation of the Signs is dated 430 BC, after which the position of a Wanderer at any time was specified as so many degrees in the Sign. The earliest Greek horoscope on record is dated 410 BC.

The names given to the Signs were mainly traditional, based on a fancied resemblance between the star pattern and the shape of an animal, or the weather of the month when the Sun was in the Sign, or some other 'logical' argument. The Sun was in the Scales at the autumnal equinox, when day and night are evenly balanced; because of the equinoctial gales, it was a windy constellation. Nature's awakening in spring was associated with the skittish Ram, and so on. By a sort of sympathetic magic, the name given to a Sign usually determined its characteristics—but not always.

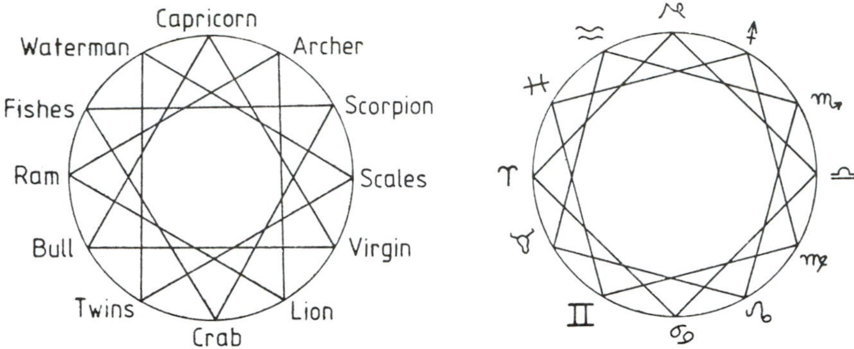


**Fig. 1** Aspects of the Signs; triangles, squares. The left-hand diagram shows the names of the Signs, the right-hand one their astrological symbols.

For example, the Waterman (mid-January to mid-February) might be expected to be a watery constellation: it actually is windy, because of a geometric association with the Scales. The Greeks' delight in geometry led them to connect up the twelve Signs, set out along the zodiacal circle (Fig. 1), to form four equilateral triangles, each associated with one of the four elements, fire, air, water and earth. Signs in the same triangle were supposed to reinforce each other. The Signs could also be connected into three squares: Signs in the same square were hostile to each other. Ptolemy, the great Alexandrian astronomer and geographer of the second century AD, also wrote a treatise on astrology, which he tried to purge of its more irrational elements. He explained the rules for triangles and squares in terms of 'aspects', i.e. of the way that the Signs regarded each other. It was as if they could see Signs of the same triangle without difficulty, but had to look askance to see their neighbours in a square, and glowered at the Signs directly opposite. Such ideas had some appeal to geometers, but little else.

Greek astrologers divided the celestial sphere into twelve equal 'houses' like the sections of an orange, meeting in the horizontal north–south line of the observer. The houses were non-rotating in the frame of the observer: the first house began at the point where stars of the zodiacal belt were just rising above the horizon. The houses provided a convenient framework for displaying in summary form the positions of both the Wanderers and the Signs at the instant considered (Fig. 2)—for horoscopes, the instant of birth. (From now on, I shall concentrate on horoscopes, ignoring applications of astrology to medicine, meteorology, the fixing of lucky days, etc. Even so, I cannot more than hint at the way that the simple horoscopic diagram was used to predict fortunes.)

A Sign had a strength depending on the house where it was found; by the rules of sympathetic magic, the greatest influence at birth was that of the Sign

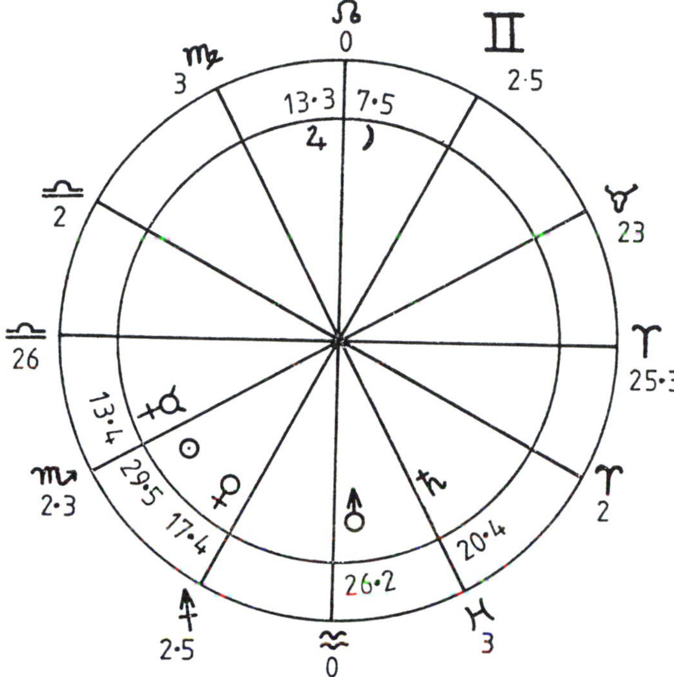


**Fig. 2** A horoscopic diagram.

seen rising above the horizon, and so being born. Because of the variations of power with position in the sky, the time of birth had to be known fairly accurately. The astrological rules made the influence of a Wanderer depend not only on the Sign in which it was seen, but also on the presence of another Wanderer in its vicinity or in a Sign connected with its own by a triangle or a square. The occupants of the different houses at birth determined parts of a baby's future, as shown in Fig. 3. These were not limited to a simple statement of good or bad luck, but also referred to a man's character, family, business, friends, honours, etc. The rules again were based on sympathetic magic depending on names, etc. but could be trimmed to improve logical consistency. Since many of the Greeks asserted that the laws of Nature could be uncovered by pure thought, they saw nothing wrong in such trimming. However, because of it, predictions could vary somewhat with the astrologer.

The development of the rules took place mainly in the 300 years taken by astrology to invade the Greek world. The invasion was resisted by sceptical philosophers like Carneades. He raised three objections. First, he said, twins with the same horoscope often had very different fates: the same was true of the son of a king and the son of a slave born at the same instant. Next, he asked, should not all who fall in the same great battle have the same horoscope?

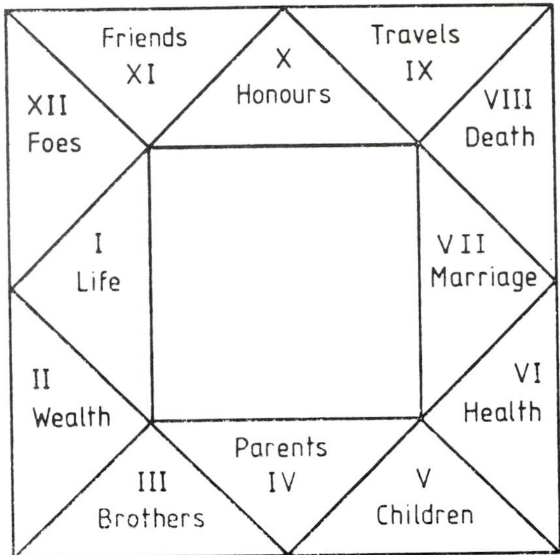


**Fig. 3** Houses in square representation, and what they govern,

Again, why do (say) Jews born in different countries have the same national characteristics though they could not have had identical horoscopes? Of these objections, the first was clearly the most important. Those perishing in the same battle had to have converging but not identical horoscopes, and the third objection could be met by arguing, with Ptolemy, that certain determining factors were beyond the power of the stars to alter.

The first objection could be met by positing a fine structure of the houses, whereby a few minutes difference in the time of birth could lead to a totally different fate. Many later astrologers experimented with such a fine structure, but the remedy was worse than the disease. The time of birth was usually known only approximately, and such a fine structure would destroy the possibility of any accurate predictions. So the objection remained, to be raised regularly by successive generations of sceptics, and as regularly smoothed over by the generations of believers.

Another not altogether academic question was this: If a horoscope depends only on the state of the heavens at the instant when life began, should not that instant be that of conception, not that of birth? After much discussion, believers came to the comforting conclusion that since both instants embody the will of the gods, either instant must yield the same result (and the birth instant was easier to use).

Again, for believers there was the question of determinism versus free will; was a man's fate immutably fixed by the stars at birth or not? The Stoics were determinists, and readily wove astrology into their framework. Others objected 'What is the use of a warning from the stars if it is impossible to avert ills by repentance?' A common attitude was summed up in the tag 'Inclinant astra, non necessitant' (the stars tilt one's actions, but do not compel).

The coming of Christianity introduced a further complication. Augustine pointed out that astrology left no room for the operation of God's grace, so that if predictions could be made through a stellar god, this god must be different from the Christian's God of grace. Being unwilling to accept more than one god, he argued that if successful predictions could be made through astrology (which he doubted) they must succeed through the intervention of the Devil. Other fathers of the early church adopted a similar point of view.

Others thought differently. They took astrology to be a genuine science, because it had led to discoveries in astronomy and had developed a logically consistent system of rules, like any other science. Astrology demanded the operation of spiritual forces: Christianity spoke of a Holy Spirit. Harmonization of the two therefore seemed no impossible task.

Both astronomy and astrology reached a peak with Ptolemy. After him, both declined *as sciences*: both virtually disappeared in Western Europe after the fall of Rome (AD 410). They began to return about AD 1000, chiefly through the Arabs in Spain, and Arabic translations of Ptolemy, etc. Knowledge of the two grew as more direct translations became available, but it was not until about AD 1400 that general knowledge of them in Western Europe attained the level attained in ancient Greece. Because of the mode of transmission, the astronomy and astrology re-acquired were essentially those of ancient Greece. There were sceptics about astrology, but even these were content to repeat Carneades' objections.

During the Renaissance all seemed to be going well for astrology. It had returned to Western Europe hand-in-hand with astronomy, and profited from the reputation of astronomy as a science. On the religious side, though the official attitude still was to discountenance astrology, many churchmen were believers. Pope Leo X founded a chair in astrology (which probably included astronomy) in the papal university, and other universities followed suit. One could hardly move along city streets without encountering astrological symbols. The Reformation made little difference. Luther described astrology as a 'comic fantasy' when not a 'wicked cult', but Melanchthon lectured on astrology at Wittenberg.

Astrology reached Britain somewhat later than the continent, and at first chiefly infected court circles. Chaucer was careful to give his Canterbury pilgrims characteristics consistent with the Signs under which they were born. Shakespeare, e.g. in *Julius Caesar* and *King Lear*, often refers to the power of the stars. As you probably already know, a horoscope was cast for the young King Edward VI by the mathematician Cardano: it prophesied long life for him, but he actually died within six months.

As Ptolemy had said much earlier, such failures could be blamed on the ignorance or venality of self-styled astrologers, the inadequate length of astrological records, and errors of calculation. Thus, despite the failures, in AD 1500 there seemed to be no reason why astrology should not persist indefinitely as an acceptable branch of science. In actual fact, belief in it had largely ceased in cultured circles by AD 1700. I have seen the collapse attributed by churchmen to scientific discoveries, and by scientists to the increasing hostility of the church. I think that there is something in both of these. Churchmen realised that the spiritual powers invoked by astrology had little in common with the spirit of compassion that drives people like Mother Teresa of Calcutta to sacrifice themselves in the service of others. They may also have been put off by the resemblance of astrology to village magic, and by the way that, in times of civil strife, each party produced astrologers claiming that the stars predicted the success of their own side. This change of attitude by the church was reinforced by new objections based on science.

The scientists who contributed to the downfall of astrology were mostly themselves believers, with no intention of being iconoclasts. The first of them was, of course, Copernicus (AD 1541). He moved the centre of the Universe from the Earth to the Sun: he made the firmament of fixed stars to stand still, the rotation being that of the Earth; and he made the Earth to orbit round the Sun, *like the other planets*. This left no reason for regarding the Earth as a special target of influences raining on it from all sides. Also, since no perceptible changes in direction of the fixed stars could be observed as the Earth moved from one side of its orbit to the other, these stars must be enormously distant, with a correspondingly increased difficulty for them to exercise much influence at the Earth.

Tycho Brahe was the greatest observing astronomer before the coming of the telescope. In 1574 he produced a defence of astrology in terms that Ptolemy would not have disowned. He argued that God would never have created the countless heavenly bodies without a purpose. The Sun was responsible for the four seasons, the Moon for tides, and the fixed stars were useful in navigation. The near approach of two planets had on occasion been attended by storms or pestilence; clearly, then, the use of the planets was for predictive purposes.

Copernicus' world picture was based on no new observations, and when he got down to mathematical details his anxiety to preserve Aristotle's crystal spheres in all their perfection led to his theory being little less complicated than that of Ptolemy, and his ideas had no ready acceptance. Tycho was so appalled by the enormous size of the firmament demanded by an orbiting Earth that he proposed a compromise system, in which the Sun orbited round a stationary Earth, and the other planets orbited round the Sun. His system explained the appearances just as well as that of Copernicus, but was an awful headache to anyone seeking a physical explanation. He is better remembered for having shown from the unimpeded motion of comets that there is no room in the Solar System for Aristotle's crystal spheres. Immaterial spheres could still be used, but only as computing aids.

Then came Galileo and the invention of the telescope (about AD 1610). Galileo found that the Milky Way was populated by innumerable faint stars: ought not these to affect astrological predictions? Again, how were arguments involving the mystic number seven affected when to the official seven stars of the Pleiades were added 40 more? More importantly, Galileo found in the motion of the satellites of Jupiter a perfect representation of the Copernican motion of the planets, suggesting again that the Earth has no especially favoured status. Galileo also began to construct a theory of parabolic motion. This came closer to embodying the Pythagorean idea of a universe governed by mathematical laws than did the rules of astrology, even of a mathematical type.

The case of Kepler is especially interesting, both because he was a practising astrologer and because he saw his scientific work as trying to uncover the harmonies that God had hidden in His universe. As regards the latter, he first tried to attach the planets to a nest of the five regular polyhedra, but the distances refused to come out right. Later he sought parallels between musical notes and planetary periods, in Pythagorean fashion: the only survivor from his results was his important third law of planetary motion, that the square of a planet's orbital period is proportional to the cube of the orbital radius. His other two laws (law of areas, elliptic orbits) he found by a painstaking fitting of observational numbers to a succession of models, during the course of which he discarded the last relics of Ptolemy's spheres. He was prepared to introduce any number of 'spiritual' forces to explain his results: he differed from others who did the same by insisting on numerical agreement with the observations.

As an astrologer he made one or two strikingly successful and important predictions, but most of his work was humdrum, coping with everyday chores. Two features of his horoscopes are noteworthy. First, he made scant use of the Signs of the Zodiac, insisting that human fates could not depend on geometrical lines drawn in the heavens by human hands, or on properties ascribed to the Signs in virtue of names given them by humans. There was no future for astrology as normally practised if these arguments were accepted, but he did not perceive this. Secondly, he acknowledged that in telling a fortune one did not only use the state of the heavens, but also any relevant information about one's client that came one's way. This was an honest confession of what astrologers regularly did, but often concealed.

Kepler's astronomical results were wholly kinematic: he speculated about possible 'magnetic' driving forces, but no more. However, 'occult' forces were increasingly abandoned by natural philosophers during the seventeenth century. This perhaps was part of the reason why Newton paid little regard to Kepler's first two laws until he was able to provide a dynamical basis for them in terms of inverse-square gravitation; he then in turn met criticism because such a force was 'occult'. However, when his ideas were better comprehended, it was realised that they left no room for the supposed arbitrary actions of the gods which provided the *raison d'être* of Chadlaean astrology.

Modern astrology has perforce to accept a universe moving according to exact mechanical laws, but still asserts that radiations from the heavenly bodies influence men on Earth. It even claims that the existence of such radiations is supported by modern physics, which lives on radiations; but whereas the physicist's radiations can be detected and measured by appropriate instruments, those of the astrologer cannot. The rules of astrology make sense only with the cosmos of the ancients. When distances to the fixed stars began at last to be measured (about AD 1840) they showed beyond question that the Signs of the Zodiac (and the other constellations) are appearances without physical reality: they consist of stars some of which are many times as distant as others, and do not form a close group capable of exerting a common influence. Aspects of the Signs (triangles, etc.) are likewise only creatures of the imagination.

All the arguments against astrology would, of course, have come to naught had there been a steady and reliable stream of events conforming to the rules of astrology. Such a stream is not found; astrology has many failures. What one hears in its support is in remarks like 'There must be something in astrology: I am an Aquarian, and the prediction for Aquarians in today's newspaper exactly matches what has happened to me.' But the rules of astrology are not based on masses of observations, but on imagination helped out with logical arguments.

Recently some students of astrology have sought to provide a statistical justification of its rules. For example, a French investigator, Gauquelin, has claimed to find a correlation between a person's occupation and the Sign under which he was born, in agreement with the old astrological rules. One would appreciate an independent check, but even if his results are confirmed this does not mean that the stars have anything to do with it. The Signs are introduced only to represent ranges of dates, and if a correlation is found it need not be more than a correlation with terrestrial dates. Moreover, if a correlation with the stars is admitted, this does not confirm the astrologers' rules. Because of the precession of the equinoxes, the Signs have moved one place along the zodiacal belt in the last 2000 years and, since the astrologers retain the old dating, any present-day correlation refers not to the actual Signs, but to Signs one place out.

What justification, then, can one offer for the original founding of astrology? Here Milne must be given the last word. In his 1929 inaugural lecture here he describes the task of a theoretical physicist not as to provide immediate explanations but to analyse in detail the consequences of assuming a given model of Nature, and comparing these with what is actually observed. If the comparison proves unsatisfactory one has at least found what route not to follow. If it proves satisfactory, this does not immediately show that one's model is correct: there may be many models yielding the same final result. It is the duty of the natural philosopher to consider as many rational models as reasonably possible and to test their basic assumptions by the degree of conformity with Nature of the results derived from them.* The result is to delimit the range in which acceptable models are to be found (though sometimes the discovery of fresh facts may lead to a total realignment of our ideas). Among the Chaldaeans, in the state of their knowledge and traditions, astrology provided a reasonable model. Unfortunately the need to carry out a searching comparison with Nature at each fresh step was too little appreciated. In consequence, astrology now 'is the stale, superstitious relic of what was once a great pantheistic religion and a glorious philosophical attempt to understand and rationally explain the universe'.

## Appendix: the order of days in the week

The later Greeks supposed the Wanderers, in order of increasing distance from the Earth, to be Moon, Mercury, Venus, Sun, Mars, Jupiter, and Saturn. This order was derived on the reasonable assumption that bodies seen to move fastest across the firmament were the closest.

The days of the week are named after the Wanderers. This is not clear in English, which makes use of Norse deities as well as Roman. It is clear in French, in which the names are, in order, Lundi, (Moon) Mardi (Mars), Mercredi (Mercury), Jeudi (Jupiter), Vendredi (Venus), Samedi (Saturn) Dimanche (Sun). To get this order, let the names of the Wanderers be written round the circumference of a circle, in the order of the supposed distances from the Earth. Then to get the order of days of the week one simply connects these names with an heptagram (seven-pointed star) and follows its sides round (Fig. 4).

This trick has little connection with astrology save for its dependence on the mystic number seven. This number had some importance astrologically, but this was reduced by the discovery of extra planets.

---

* *Cf.* Karl Popper. 'The essence of a good mathematical model is that it should embody the bold ideas, unjustified assumptions and speculations which are our only means of interpreting Nature. The good scientist then puts his model to the hazard of refutation.'

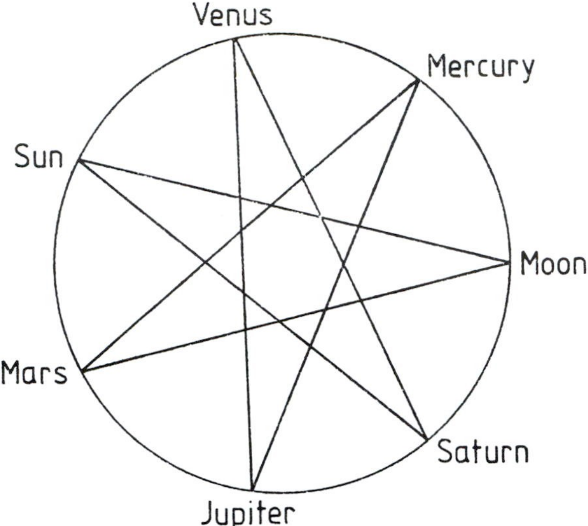


**Fig. 4** Heptagram giving order of days in the week.

## Further reading

It is difficult to suggest references for further reading; recent discussions of astrology from the standpoint of a scientist are few. The standard authority on Greek astrology still is *L'astrologie Grecque*, by A. Bouché-Leclercq (1899). There is no equivalent on Chaldaean astrology, but A. Pannekoek's *History of astronomy*, Chapter 4, is helpful, especially in that it quotes all the significant parts of a translation of cuneiform texts by R. C. Thompson (1900). A helpful account of astrology as a whole is given in *Sternglaube und Sterndeutung*, by F. Boll, C. Bezold and W. Gundel (1966, but the reprint of a 1928 edition).

*Corrigendum* (printed in *Quarterly Journal*, **24** (1983), 368)

In my paper entitled 'Astrology, religion and science' (*Q, Jl R. Astr. Soc.*, **23**, 515–526, 1982), I gave a rather inadequate list of suggestions for further reading. Since the paper went to the printers I have found an addition to that list in a book by R. B. Culver and P. A. Ianna entitled *The Gemini syndrome: Star Wars of the oldest kind* (Pachart Publishing House, PO Box 35549, Tucson, Arizona 85740). In a review Owen Gingerich says of it 'the present volume is one of the best anti-astrology books available'. Much of it is devoted to the systematic search for correlations between events on Earth and those in the Heavens which might provide a justification for the laws of astrology. It finds that no correlation exists that is more than marginal, and such correlations are found only in a small minority of cases, and are not those required to justify astrology. In particular, Gauquelin's results (as pointed out to me by Owen Gingerich and others) are massively contrary to astrology. I must apologize for my error in what I said in my paper.

# The origin of cosmic rays

Sir Arnold Wolfendale

E. A. Milne (1896–1950) was a Mathematician and Natural Philosopher of brilliance and it is an honour for me to give this Lecture named after him.

Milne's contributions to Astronomical Science were many and varied but three strands [1] can be discerned: the thermodynamics of the stars (to use the title of his *Handbuch der Astrophysik* article of 1930), the theory of stellar structure (an interesting by product being his theory of the ejection of high-speed particles from the Sun) and, most important, the introduction and development of the theory of kinematic relativity. It is apparent that there is some entanglement of these strands with the topic of this lecture; in fact, the subject of 'The origin of cosmic rays' is one that would have appealed to Milne. It is indeed a pity that the experimental features of cosmic rays—and their astronomical significance—were not realised in his time to such an extent as to have aroused his interest.

## 1 Introduction

Cosmic rays were discovered by Victor Hess [2] in 1912 and it is true to say that today, 70 years later, the origin of the bulk of the radiation, or, to be more specific, the manner and location of the acceleration of the particles, is still largely a matter of speculation. The purpose of this lecture is to endeavour to set the stage by briefly describing the astrophysical aspects of cosmic rays—particle type, energy, etc., and, after reviewing possible origins, to give the author's own predilictions.

It should be stated immediately that the cosmic radiation does not represent some tiny phenomenon on the cosmic scale; the energy density in the primary radiation above the atmosphere is about the same as that in starlight (and in other astronomical entities too, as we shall see). At the rather parochial level of interaction with human beings one can remark that there are about five secondary comic rays (mainly muons) passing through our heads every second. We really ought to know where they are coming from.

## 2 Properties of the primary cosmic rays

### 2.1 Energy spectrum and mass composition

The energy spectra of the major components of the cosmic radiation are shown in Fig. 1. Starting with nuclei, not surprisingly, protons and helium predominate and many other nuclei have also been identified, with lower intensities (see later). The shape of the energy spectrum carries with it information about particle production and propagation. Starting at the lowest energies, there is curvature below about $10^{10}$ eV nucleon$^{-1}$ due to the modulating effect of the solar wind and the intensity in this region is sensitive to solar activity. Continuing to higher energies there is evidence for a power law with constant exponent ($N(E)dE \propto E^{-\gamma}dE$, with $\gamma \sim 2.65$) up to about $10^{15}$ eV nucleon$^{-1}$, above which $\gamma$ increases to about 3.15. Finally, there is good evidence for a remarkable flattening of the spectrum above about $10^{19}$ eV.

The mass composition is known quite well below about several times $10^{10}$ eV nucleon$^{-1}$ and Fig. 2 shows a comparison of the elemental abundances of the cosmic rays with the 'Solar System' or 'universal' abundances. It will be noticed that there are strong similarities, especially when it is realized that the large excesses in cosmic rays (CR) in the region of Li, Be, B and Sc, V and Mn (and some others) can be easily explained in terms of the fragmentation of nearby, heavier nuclei in their passage through the interstellar medium (see Section 2.3).

Turning to electrons, both particles and anti-particles have been seen; indeed anti-protons have also been observed.

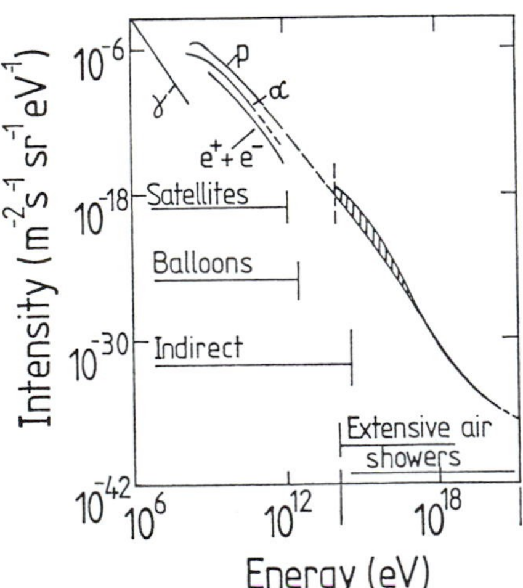


**Fig. 1** Energy spectra of the major components of the cosmic radiation. Indication is given of the techniques used. The mass composition is very uncertain in the shaded region but there is general agreement that the particles are mainly protons at the highest energies.

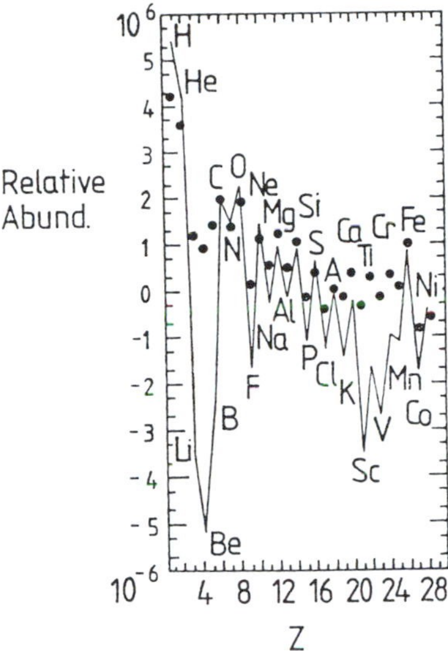


**Fig. 2** Elemental abundances of cosmic rays (filled circles) compared with the Solar System (universal) abundances (joined by full lines). The two have been normalized at carbon. The data refer to the range 70–280 MeV nucleon$^{-1}$ and are from the summary by Meyer [3].

Gamma rays have been detected, both a general continuum and fluxes from specific discrete sources. It is interesting to recall that initially it was considered by many that the 'radiation' comprised some form of ultra-$\gamma$-radiation (hence the use of the term cosmic *radiation*) but in fact, in the region of a few times $10^9$ eV (where most of the energy of CR resides) the ratio of $\gamma$-intensity to particle intensity is only $\simeq 10^{-6}$. Despite its low intensity, I believe that the $\gamma$-radiation has important things to say about the 'origin problem'.

## 2.2 Energy densities

Much of the later discussion will be taken up with the question of particle energy and energy densities and it is useful at this stage to summarize the energy densities of the more important components. These are given in Table 1 and Fig. 3.

## 2.3 Spatial extent of cosmic rays: interactions in the interstellar medium (ISM)

The question of the interactions of CR with the ISM and their effect on the characteristics of the detected particles can be considered by moving directly to a basic problem: do cosmic rays represent a local phenomenon in the Galaxy?

**Table 1** Energy densities of the cosmic ray components near the Earth

| Component | (eV) | Energy density (eV/cm$^{-3}$) |
|---|---|---|
| Protons and heavier nuclei | Above $10^9$ | $\sim 5 \times 10^{-1}$ |
| | $10^{12}$ | $2 \times 10^{-2}$ |
| | $10^{15}$ | $10^{-6}$ |
| | $10^{18}$ | $10^{-8}$ |
| Electrons and positrons | Above $10^3$ | $\sim 6 \times 10^{-3}$ |
| | $10^{10}$ | $1 \times 10^{-3}$ |
| | $10^{11}$ | $2 \times 10^{-4}$ |
| γ-rays: diffuse background | Above $10^7$ | $\sim 1\times 10^{-5}$ |
| | $10^8$ | $2\times 10^{-6}$ |

For the electron component the answer is immediately *no* because the distribution of a radio continuum, due almost certainly to synchrotron radiation from CR electrons spiralling in the magnetic field in the Galaxy, is widespread. In fact, data indicate the generation of electrons in specific sources in various parts of the Galaxy and of course emission has been detected from other galaxies, too.

The situation with the predominant nuclear component (p, α . . . ) is more difficult and the answer to the question is clearly bound up with the general origin problem. Here we can make some preliminary comments. The scale of distance in the Galaxy from the CR point of view is set by the Larmor radius ($\rho(\text{cm}) = pc(\text{eV})/(300\ H(\text{gauss})Z)$) and in so far as a typical field in the ISM is $\simeq 3$ μgauss then $\rho \simeq p(10^{15}\ \text{eV}/c)/3Z$ parsec. The observation of particles with momentum as high as $10^{20}$ eV/$c$ thus implies that these particles at least are likely to be widespread) ($\rho \simeq$ kpc even if $Z \gg 1$), and to cover the

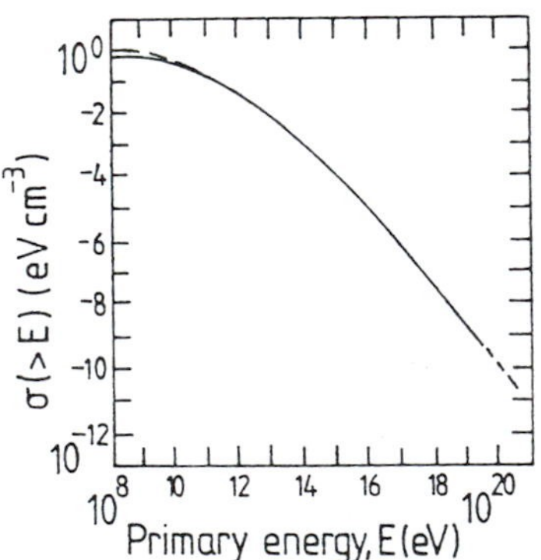


**Fig. 3** Energy densities for the nuclear component of cosmic rays.

Galaxy. This argument can clearly be extended downwards to perhaps $10^{16}$ eV or so but in the important region $10^9$–$10^{10}$ eV where $\rho \simeq 10^{-6}$ pc (for protons) it has no validity of course, and it could be that these particles are generated locally. However, a number of arguments can be made which give the contrary view and these can be itemised:

(i) The long-term near-constancy of the radiation (a variation of less then $\simeq 2$ over $10^9$ yr [4]) during which time the Solar System has made about one-quarter of an orbit round the Galaxy, with respect to the general star population.

(ii) The likelihood of sources of nuclei being distributed in a similar fashion to those of electrons, even if the sources are not identical.

(iii) The possibility of explaining a number of features of the radiation in terms of the interaction of CR nuclei with the nuclei of the ISM during long passages ($> 10^7$ yr) through the tenuous gas.

It is this last argument which gives really the strongest support for an extended distribution of the bulk of the CR. For example, the flux of positrons can be explained in terms of secondaries, as can much of the apparently diffuse $\gamma$-radiation (this topic will be taken up in more detail later). Concerning the mass composition of the nuclei, as has already been pointed out, it is easy to account for the presence of such apparent anomalies in the primary beam as the grossly excessive (by Universal Abundance standards) flux of Li, Be and B nuclei in terms of fragmentation products of heavier nuclei (e.g. oxygen) in the ISM. In this connection it is important to mention an interesting feature related to the relative numbers of secondary nuclei (e.g. Li) and true primary nuclei. The ratio is found to fall steadily with increasing energy, roughly as $E^{-0.5}$ [3] over the range 1–100 GeV. The extent to which the above information tells us just where the cosmic rays originate is another question, however, because of uncertainties in the manner of propagation in the Galaxy and in the whereabouts of the gas traversed by the particles (is it in the general ISM in the disc, the halo, the high density region surrounding specific sources, the intergalactic region, etc.?).

There can be no question of 'overkill' in the discussion on the Galactic distribution of nuclei because there has been no demonstration (to my knowledge) of a characteristic 'proton signature' elsewhere in the Galaxy. The problem is that electrons are so efficient at producing quanta, by virtue of their low mass, in comparison with protons and heavier nuclei, that the electron component usually dominates. It had been hoped that $\gamma$-ray astronomy would provide such a signature by virtue of the detection of a maximum in the $\gamma$-ray flux at 67 MeV ($m_{\pi^0}c^2/2$) arising from $\pi^0$-decay, the $\pi^0$-mesons having been produced in CR–ISM nucleus interactions. Such a maximum has not been seen and, although one can (and does, see Section 5.1) interpret some of the γ-ray flux as being due to CR-nucleus interactions, there is residual worry.

## 2.4 The anisotropy of arrival directions

This section on the properties of the primaries can be concluded by a brief examination of the lack of isotropy of the incident particles. After many years' effort by a substantial number of researchers a concensus is emerging concerning the magnitude of the anisotropy and phase (of the maximum) and its dependence on energy. Fig. 4 comes from a recent summary and would probably be acceptable to most workers in the field. It will be noticed that measurements do not start until ~ $10^{12}$ eV, the reason being confusion caused by solar modulation at low energies. Obviously the form of δ and ϕ is related to origin and propagation and an interpretation will be given later; equally obvious is the fact that fate is again conspiring to prevent a view of the important region below $10^{11}$ eV where most of the particles and energy lie.

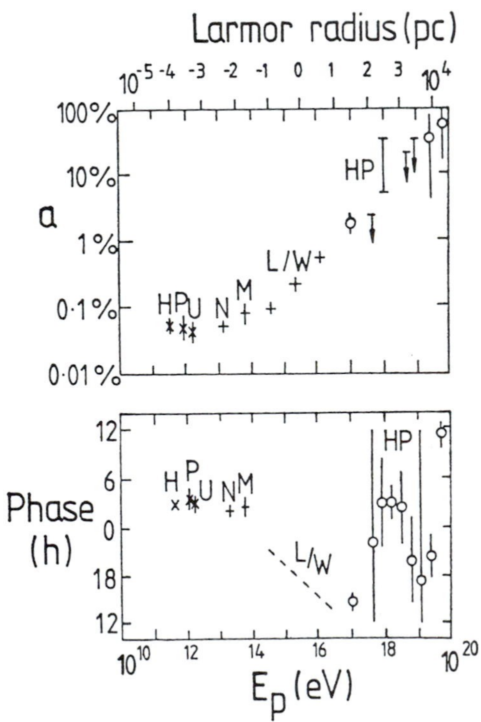


**Fig. 4** Amplitude and phase of the first harmonic anisotropy of cosmic rays. The near constancy of amplitude and phase below $10^{14}$ eV is marked. The Larmor radius in a typical interstellar magnetic field of 3 μG is indicated.

# 3 Energy density considerations

## 3.1 Galactic versus extragalactic origin

We have already seen that there is comparatively direct evidence for CR electrons generated in sources in the Galaxy and evidence against a significant flux of extragalactic (EG) electrons is strengthened greatly by the fact that the relict radiation (2.7 K), presumed universal, shields the Galaxy (Inverse Compton Interactions attenuate electrons considerably above some tens of MeV). It has already been mentioned that positrons are quite well explained as secondaries generated from collisions of nuclei with the ISM, and the subsequent chain: $\pi^+ \rightarrow \mu^+ \rightarrow e^+$, so that in what follows we can concentrate largely on the nuclear component.

Nuclei are not attenuated by the 2.7 K radiation (until $10^{18}$ eV or so) so that in principle at least they could be derived from EG sources. In fact, the observation of a gross imbalance in the number of protons and electrons ($p : e^- \simeq 30:1$, see Fig. 1) makes it rather natural to assume that the nuclei are coming largely from EG sources. The lack of clear identification of galactic sources of nuclei, the comparative quietness of our Galaxy and the presence of other very active galaxies add to the attractiveness of an EG model. We can regard the problem of distinguishing between galactic (G) and EG models as the first and perhaps the most important on our list in the search for the origin of cosmic rays.

We have already estimated the 'local' energy density in cosmic ray nuclei ($\simeq 1$ eV cm$^{-3}$) and this can be used as a datum with which to compare energy densities that might reasonably be expected to result from various processes. At this stage we ignore the fact that we know the composition of the cosmic rays in question and clearly a postulated origin model must give the correct answer for this parameter, too. However, in some cases at least the energy requirement is more serious.

The likely (?) energy densities can be considered for the alternative basic models—extragalactic and galactic—by taking values for the energy density present in a phenomenon and estimating what might be expected to be a reasonable efficiency ($\eta$) for converting this energy to cosmic ray energy.

We start with galactic energy densities.

## 3.2 Galactic energy densities

(1) *Rest energy of total mass*, i.e. $\Sigma mc^2$. $\sim 10^9$ eV cm$^{-3}$
High efficiency of conversion to CR (say 10 per cent) for black holes but mass in black holes (BH) (at GC) probably $<10^7 M_\odot$ so effective $\eta \gtrsim 10^{-5}$ leading to: $\epsilon_1 \sim 10^4$ eV cm$^{-3}$

(2) *Gratitational potential energy of Galaxy as a whole*, i.e. $GM_G^2/R_G$ where $M_G$ is galactic mass and $R_G$ is the effective galactic radius. A few per cent of this energy may have gone into CR when the Galaxy formed: $\sim 3 \times 10^3$ eV cm$^{-3}$ $\epsilon_2 \approx 30$ eV cm$^{-3}$

(3) *Gravitational energy of stars in the Galaxy*, i.e. $\Sigma GM_s^2/R_s$ where $M_s$ and $R_s$ refer to star. Situation as for (2): $\sim 2 \times 10^4$ eV cm$^{-3}$ $\epsilon_3 \approx 2 \times 10^2$ eV cm$^{-3}$

(Note, for (1), (2) and (3), it is assumed that most of the CR are still in the Galaxy.)

(4) *Magnetic field in Galaxy* ($B^2/8\pi$). $\sim 1$ eV cm$^{-3}$
Equipartition could result and so $\eta \simeq 1$: $\epsilon_4 \sim 1$ eV cm$^{-3}$

(5) *Kinetic energy of gas motion* ($\frac{1}{2}\rho v^2$). $\sim 1$ eV cm$^{-3}$
Equipartition could result and $\eta \simeq 1$: $\epsilon_5 \sim 1$ eV cm$^{-3}$

(6) *Starlight* ($\int I_\nu d\nu$). $\sim 1$ eV cm$^{-3}$
Difficult to see any direct connection with CR—the Sun has $\eta \sim 10^{-9}$ only, although CR trapping by a factor probably $\sim 10^3$ increases this to $\sim 10^{-6}$. However, some other stars are certainly much more efficient (see later).

It is apparent from the above that there are several possibilities for galactic origin, at least from the standpoint of energetics.

## 3.3 Extragalactic energy densities

(1) *Rest energy of total mass* (i.e. $mc^2$) (assuming $\Omega = 0.1$). $\simeq 10^3$ eV cm$^{-3}$

*If* black holes also carry a mass of this order and if their efficiency is high, $\sim 0.1$, say, then $\eta$ could be $\approx 0.1$, leading to: $\epsilon'_1 \simeq 10^2$ eV cm$^{-3}$

(2) *Energy of 2.7 K radiation* ($h\nu$). $\simeq 0.24$ eV cm$^{-3}$
Although it is hard to think of a specific model which would give equipartition of CR with the relict radiation such a model cannot be ruled out.

(3) *Energy of intergalactic starlight* ($h\nu$). $\simeq 10^{-2}$ eV cm$^{-3}$
Hard to think of a specific model which would give equipartition of CR with intergalactic starlight (situation as for galactic starlight).

(4) *Gravitational PE of galaxies (i.e.* $\Sigma G M_G^2/R_G$, where $M_G$ is galactic mass and $R_G$ is the effective galactic radius). A few per cent of this energy may have gone into CR when galaxies formed, the CR then escaping. $\simeq 10^{-4}$ eV cm$^{-3}$ $\epsilon'_4 \simeq 10^{-5}$ eV cm$^{-3}$

If CR are confined to clusters of galaxies this value can be raised considerably.

Although it is true that the energy densities from extragalatic sources are less than those from the Galaxy it can be seen that there are possible sources, in particular truly cosmological sources, in which a significant fraction of the total available energy in the Universe finds its way into cosmic rays (cases 1 and 2). Furthermore, there is the possibility referred to under [4] that, although extragalactic, the cosmic rays at the Earth are not universal in their extent but rather concentrated in the local cluster of galaxies. Such a possibility is attractive if the cluster contains a small number of galaxies with very high CR efficiencies (M87?). One is then deriving a contribution from the energy referred to under (1).

The possibility of an extragalactic origin is seen to be sufficiently strong that the idea must be taken seriously and experimental checks are necessary. We continue by looking at the ramifications of an extragalactic origin.

## 4 Extragalactic origin of cosmic rays

### 4.1 Interactions of CR with radiation and matter

The interactions of CR in extragalactic space have importance as do those in our own Galaxy discussed earlier. The interactions are relevant for two reasons; first, they may under suitable situations cause the Universe to be opaque and thus forbid contemporary CR from originating in the postulated manner, and secondly their interaction products may be directly detectable at present—one thinks immediately here of X-rays and $\gamma$-rays.

### 4.2 Interactions with radiation

Two interactions are important: photonuclear reactions, leading to fragmentation of nuclei and pion production for protons (and for heavier nuclei at higher energies), and electron pair production for all nuclei. Starting with $e^+e^-$ production, the threshold energy is $2\,m_ec^2$ so that for the contemporary

2.7 K radiation ($<h\nu> \sim 7 \times 10^{-4}$ eV) the threshold Lorentz factor for a CR particle to suffer energy loss is $\gamma \simeq 1 \times 10^6/7 \times 10^{-4} \simeq 10^9$, i.e. an energy of $\simeq 10^{18}$ eV for a proton. For a nucleus of charge $Z$ the Lorentz factor for fragmentation (say an interaction energy of 15 MeV) is $\gamma \simeq 1.5 \times 10^7/7 \times 10^{-4} \simeq 2 \times 10^{10}$, i.e. an energy of $\simeq 2Z \times 10^{19}$ eV. It is very clear that although these energies have relevance to the behaviour at the top end of the cosmic ray spectrum they have none of the energies of immediate concern here, $10^9$–$10^{10}$ eV.

With cosmological theories of origin, however, they rapidly assume importance because of the increase in temperature of the relict radiation with redshift, $z : T = 2.7\ (1 + z)$ K. As is well known, matter and radiation are assumed to have decoupled at $z \simeq 10^3$ where $T \sim 5 \times 10^3$ and hydrogen starts to become ionized, thus ending the transparency of the Universe to radiation (the free electrons scatter very readily). Similarly, there is a value of $z$ ($z_0$) behind which the interactions of protons with the relict radiation cause the Universe to be opaque to energetic cosmic rays, i.e. CR produced before $z_0$ will not survive to the present. Choosing $10^{10}$ eV (contemporary energy) and allowing for the loss of energy by redshift ($E \propto (1 + z)^{-1}$) it can be seen that by $z \simeq 10^4$ the threshold for catastrophic energy losses has been reached. An accurate estimate can be derived from related calculations by Strong, Wdowczyk and Wolfendale [6] (to be considered later); these show that for $z \lesssim 2 \times 10^3$ the energy losses are so great by way of $\gamma$p reactions that CR protons produced before that time would not have survived. Clearly, a similar value will be applicable to heavier nuclei.

Interest now moves to the possibility of CR having been accelerated at smaller values of $z$ and, immediately, production during galaxy formation springs to mind. In fact, if such a model were valid then the mode of production would have had to be such that a significant fraction of the rest energy ($10^3$ eV cm$^{-3}$: see Section 3.3) would have been utilized.

Such a model has been considered already by Hillas [7]. The aim of the model was to explain the change of slope of the primary CR spectrum (Fig. 1) at $\sim 10^{15}$ eV in terms of relict radiation—CR interactions in the past. Use of the arguments given earlier in this section gives a value of $z \simeq 15$ when coupled with a source efficiency varying as $(1 + z)^{4.3}$ to give agreement with the contemporary spectrum.

An interesting development of our own [6] was the derivation of the expected energy spectrum of isotropic $\gamma$-rays expected at the present time generated by the electron-pairs (and electrons from $\pi$-mesons) resulting from the CR-relict radiation photons. The ensuing cascade is one of some complexity involving as it does $\gamma_{bb}$–p, $\gamma_{bb}$–e, $\gamma$–$\gamma_s$, and e–$\gamma_{bb}$ interactions (where $\gamma_s$ is a starlight photon) but the resulting spectrum is conditioned largely by the energy made available in the initial $\gamma_{bb}$–p interactions.

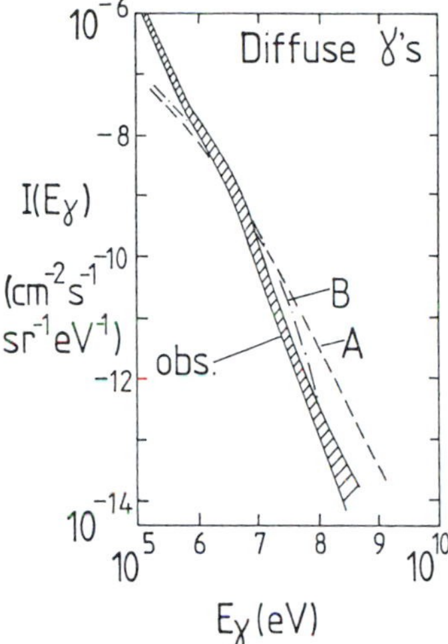


**Fig. 5** Energy spectrum of diffuse, extragalactic, γ-rays from the work of Fichtel *et al.* [8] and others—shaded area. Comparison is made with our predictions [6] for a model in which CR nuclei are extragalactic and generated in galaxies at $z \simeq 15$. A and B represent extreme predictions.

Figure 5 gives the result of our calculations and a comparison with the experimentally measured hard X-ray and γ-ray spectrum [8]. It is most remarkable that the two are so close—it must be stressed that there is no question of normalization here, the normalization was applied to the proton spectrum and there is no reason to expect a fit in Fig. 5 unless the model is correct (or chance has intervened).

Thus, an extragalactic (EG) model, as prescribed, is still in with a fighting chance.

## 4.3 Interactions with matter

Gas in the Universe is also efficient as manifesting the presence of CR by way of photon production (γ,X . . . ). As was the case for interactions with the relict radiation at high $z$-values, there will be a limit where CR–gas nucleus collisions cause a catastrophic reduction in the CR intensity. The calculations depend on the value of the contemporary gas density, but taking $\Omega = 0.1$, as before, we find that opacity again occurs for $z \lesssim 10^3$.

This argument can be pressed further by estimating the flux of γ-rays to be expected from EG cosmic rays by adopting contemporary estimates of gas density and different models for the spatial distribution of the EG cosmic ray flux. Ideally, if we wished to put paid to all EG models, with production occurring at any value of $z$, we would find that the intensity of γ-rays seen now from interactions of EG CRs with the known gas would be higher than observed. We have made such a study [9] and this can be described briefly.

The key question is, of course, the amount of 'free' gas in the Universe available for CR to interact with. The gas content of individual galaxies, probably averaging a few per cent of the total mass, is too small to be of significance (this is clear from the fact that the diffuse $\gamma$-ray intensity is some 20 times what would have been expected from galaxies like our own, the Galaxy producing most of its $\gamma$-rays from CR–gas interactions [10]). In principle, the intergalactic medium could be essentially devoid of gas, the process of galaxy formation having been so efficient as to mop up all available gas so that it is not obvious that there is extra gas over and above galactic gas. However, there is much evidence now for the existence of large amounts of gas in galaxy clusters, this evidence coming from a variety of phenomena; X-ray emission from extended regions [11], the diminution of the cosmic microwave background in the direction of clusters [12] and the 'tails' left by radiogalaxies as they pass through the cluster medium [13]. Studies have been made of the X-ray flux and the amount of incandescent gas needed for its explanation and it is these which we have used; it will become apparent that this work is still in its infancy and there are many gaps in our knowledge. Nevertheless it is possible to draw some conclusions.

Figure 6 summarizes the situation for what can be regarded as the limiting (and probably unphysical) situation of a constant cosmic ray intensity over the Universe at the present time. Results for an increasing scale of cluster gas are presented, the point being that evidence for gas is strongest for rich clusters only and the weight of evidence diminishes as one proceeds onwards to 'all clusters' and 'haloes'. The conclusion to be drawn at this stage is that although the majority of the situations rule out EG origin, such an origin can just be preserved if gas is confined to rich clusters alone and the CR intensity is uniform across the Universe.

Here we have yet another example of Nature guarding its secrets by preventing us from being certain about the answer to such a simple question.

It remains to examine whether a uniform CR distribution is likely. There can be no question of being absolutely sure but it appears to be most unlikely. We have seen that models involving CR production at $z > 10^3$ are untenable and following the arguments in Section 4.2 even production at $z > 15$ would probably give too many $\gamma$-rays. Inevitably, then, an acceptable model would involve production of CR associated with galaxy formation and immediately we are into the problem of how galaxies form and, particularly, the stage at which clustering takes place. A crucial question now is the extent to which significant magnetic fields are associated with the cluster region at early epochs. The situation regarding contemporary fields is that we would expect them to exist, if only as leakage fields from constituent galaxies, and there is in fact some slight direct evidence for their existence [14]. Furthermore, if there is any semblance of equipartition between magnetic

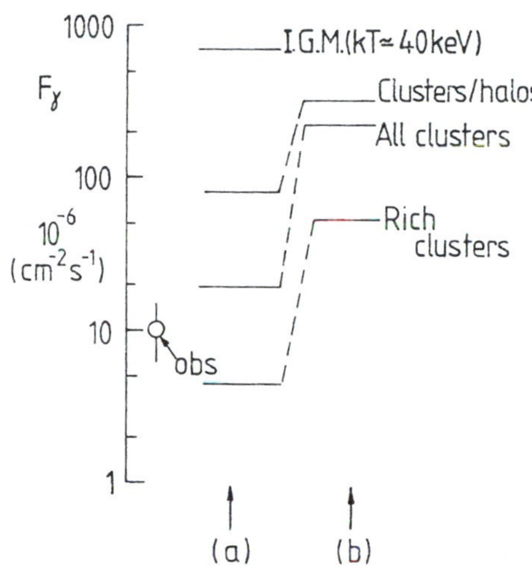


**Fig. 6** Comparison of the flux of extragalactic γ-rays above 100 MeV with expectation from interactions with gas in the IGM. An ascending scale of mass of gas in the Universe is indicated on the right-hand side, from gas being confined to rich galaxy clusters alone to gas being present universally to such an extent as to explain the whole of the X-ray background. The horizontal lines above (a) represent the situation where the CR intensity is constant throughout the Universe; those above (b) correspond to clumping of CR in galaxy clusters.

energy density in cluster space and the energy density associated with gas motion (*cf.* the situation locally in our Galaxy, Section 3.2) then these fields will be remarkably high ($\simeq 1\ \mu$G). If the fields are significant now then presumably they were even higher at the time of particle acceleration and there is an immediate difficulty for the cosmic rays of getting away from the clusters. Bearing in mind the Larmor radii referred to in Section 2.3, $\rho \simeq 10^{-6}$ pc for a CR of energy $3 \times 10^9$ eV in a 1 μG field, we see that the magnetic field can be many orders of magnitude less than 1 μG and the outward drift of particles will still be very slow in what is almost certainly a very tangled field.

Said *et al.* [9] consider that a reasonable distribution of CR in contemporary clusters is one with $I_{CR}(r) \propto r^{-1}$, where $r$ is the distance from the cluster centre. The effect of concentrating the CR as well as the gas is immediately obvious: the yield of γ-rays increases markedly and the case against EG origin strengthens considerably (Fig. 6(b)). In fact, the γ-yield is so high that the flux from the nearby Virgo cluster would be 5–9 times bigger than the present upper limit.

Unless the gas densities have been grievously overestimated, then, it does look as though the majority of extragalactic origin models can be ruled out; the only one remaining being essentially that in which CR were produced when galaxies were formed (with remarkable high efficiency) and were able to diffuse rapidly away from these galaxies—the magnetic fields in the IGM being quite negligible ($\ll 10^{-15}$ G: a Larmor radius of 10 kpc for a proton of $10^{10}$ eV $c^{-1}$). Paradoxically, the model described in Section 4.2—which gives a tolerable explanation of the flux of extragalactic $\gamma$-rays—satisfies the conditions. Thus, although EG origin is almost ruled out there can be no claim yet for proof positive.

# 5 Galactic origin of cosmic rays

## 5.1 The evidence from $\gamma$-rays

The use of $\gamma$-rays, with their rectilinear propagation, in probing the distribution of their parent cosmic ray particles in the Galaxy is an obvious technique and one that is being actively pursued (see [15, 16] for recent summaries). In so far as the current measurements give best results for $E_y \lesssim 100$ Mev, the parent particles (protons and other nuclei in the range 1–10 GeV, together with electrons of several hundred MeV) are just in the region of interest to our search for an answer to the basic origin question. Not surprisingly, a number of problems arise and these can be listed:

(1) The contribution for discrete sources, as distinct from interactions in the interstellar medium (ISM) (*thought* to be only 10–20 per cent).

(2) Electrons are important contributors, probably 50 per cent of the $\gamma$s above 100 MeV are derived from electrons and even more at lower energies—there is thus a problem in deriving the contribution of the more important proton component.

(3) Surprisingly, perhaps, the distribution of the target gas in the ISM is not known well, the density of $H_2$ inside the solar circle being particularly uncertain.

Despite the problems we succeeded in showing in 1975 [17] that there was evidence for a 'cosmic ray gradient' in the Galaxy, at least in its outer parts, that is that the CR intensity falls slowly with increasing distance, $R$, from the Galactic Centre. The virtue of looking in the anticentre direction is that the discrete source contribution is surely very small there and the contribution to the gas density from the uncertain $H_2$ component is also small. Our original analysis used the results of the *SAS II* satellite experiment and there has since been another rather similar detector put in orbit (*COS-B* [18]). This instrument has given an order of magnitude more data and enabled the analysis to be honed up (although because of a higher instrumental background there are difficulties). The general consensus at present is that there is probably a gradient (again, most certain in the anticentre) but it may be

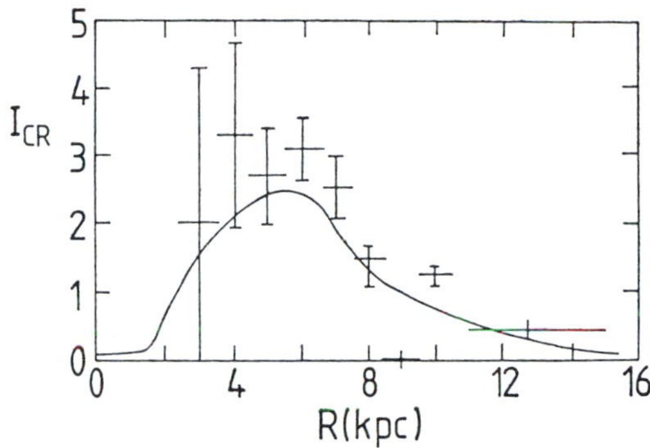


**Fig. 7** Relative cosmic ray intensity as a function of galactocentric distance, *R*, for the mixture of cosmic rays ($\simeq$ 50 per cent electrons and 50 per cent nucleons) responsible for producing cosmic $\gamma$-rays with energy above 100 MeV. The results are from the analysis by *SAS II* and *COS-B* data by Issa *et al.* [19]. The full line represents an attempt to smooth out fluctuations and to allow for unresolved discrete sources.

somewhat smaller than derived previously. A further important point is that the derived gradient does not depend much on $\gamma$-ray energy [19] from the lowest energies (35–100 MeV), where electrons predominate as parents, to energies above several hundred MeV where protons and nuclei are doing most of the work.

Figure 7 gives what we [19] regard as the best estimate at the present time. The assumption of radial symmetry for the CR distribution, such that we can simply write $I_{CR}(R)$, is clearly a gross oversimplification and there is indeed strong evidence for a dependence on longitude; for example, the third quadrant gives particularly low cosmic ray intensities [20].

The significance of the 'CR-gradient' for CR origin will be clear. If CR were extragalactic and received no acceleration in the Galaxy one would expect a uniform distribution throughout the galaxy but if the sources were in the Galaxy *and* if the particles did not diffuse too far from their sources before escape (say $\lesssim$ few kpc) then a non-uniform distribution would result. As some sort of test of this hypothesis one can consider the electron component and its manifestation—synchrotron radiation in the 100s of MHz region. Under the simplifying and very reasonable assumption that the magnetic field falls off with increasing $R$ in a somewhat similar fashion then a form for $I_e(R)$ can be derived which is very similar to that derived by us for CR [21]. Without wanting to get into too convoluted a discussion it can be pointed out that *if* instead the electron distribution were uniform (with the attendant problem of the necessary very long mean free path for scattering) the magnetic field would need to fall rapidly with increasing $R$; this in itself would destroy any semblance of equipartition of energy density in the field with that of the CR particles unless $I_{CR}(R)$ fell rapidly too. Finally, if $I_{CR}(R)$

did fall rapidly yet $I_e(R)$ were constant there would be the unlikely situation of very different sources for the electrons and the CR.

The author regards the above as fairly strong evidence for a galactic origin of most of the cosmic ray particles (say those below $10^{10}$–$10^{11}$ eV).

## 5.2 Possible galactic sources

### 5.2.1 Energetics

Earlier, in Section 3.2, rather general comments were made about galactic energy densities and it was shown that there was no shortage of galactic processes which could, in principle, provide the energy content of the cosmic radiation. Now we must examine the question in more detail.

A general division can be made into what might be termed spatially continuous and discontinuous processes. In the first category we have general acceleration by shocks in the ISM, and the agreement of energy densities of CR, magnetic field and gas motion (Section 3.2) gives support to such a mechanism. The problem is that the secondary nuclei generated by fragmentation of true primaries in the ISM would be expected to be accelerated after production and it is very hard to see how the fall-off in secondary/primary ratio with increasing energy can be generated (Section 2.3). The evidence all seems to point to most, at least, of the acceleration occurring in a 'source' covering a restricted region, with the energetic particles then being propagated through the ISM, where most of the 'grammage' is accumulated. The 'spatially discontinuous' process is, therefore, to be preferred.

Restricting attention to this category, division can again be made into two types, depending on whether the phenomena are essentially comparatively rapid violent events or whether they occur continuously (but in restricted volumes). Explosive events such as novae and supernovae are in the first category and steady stellar winds are in the second.

### 5.2.2 Violent events

Figure 8 summarizes the situation for violent events and these will be considered first. Starting at the lowest energy per event (solar-type stars) there is the well-known fact, e.g. [22] based on observations of solar flares, that there is insufficient energy available. In fact, using our knowledge of solar flare CR there is a shortage of energy by about 5 orders of magnitude; even if the shock energy were eventually converted, very efficiently, to CR there would still be a gap of at least 3 orders.

Flare stars are more attractive and it has been suggested that the bulk of the low energy particles ($E \lesssim 1$ GeV) might be accelerated this way [23]. Unfortunately, however, several authors, e.g. [24], have argued that there is

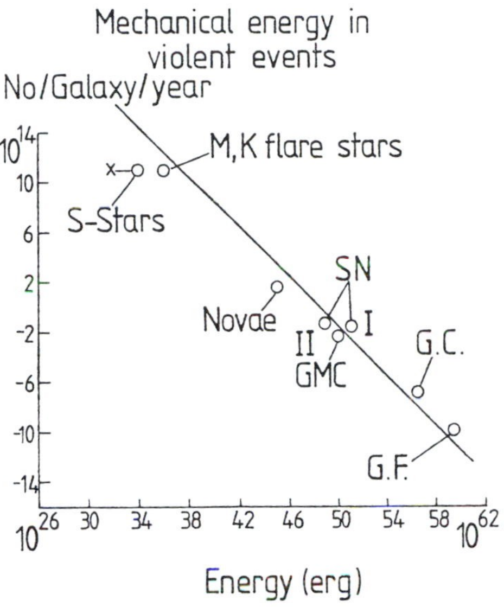


**Fig. 8** Mechanical energy in violent events. The line represents the CR energy input needed. Most of the symbols are self-explanatory; for the others: S-stars represents solar-type stars (and the cross indicates the actual energy going into CR in a solar flare); GMC—giant molecular clouds; GC—Galactic Centre explosions; GF—Galactic Formation.

insufficient energy here, too, and they seem to be right. Continuing upwards in energy, novae are 'getting warm' but the desired line is not reached until supernovae (SN) are encountered. Such violent and impressive objects have long been a favourite source, ever since the pioneering suggestion of Baade and Zwicky [25] and it is true that they have attractive features. There are worries, though, as follows:

(1) The efficiency of conversion of mechanical energy to CR must be very high.

(2) Adiabatic energy losses are thought to be large for particles in the expanding SN shell (a resulting efficiency of only 1 per cent is often quoted, e.g. [26]).

(3) The $\gamma$-ray data shows no evidence for a 'cloud' of CR round supernovae remnants.

(4) For the CRAB nebula, although electrons are certainly being accelerated, an upper limit to the flux of protons is only a factor 3 higher (there is *no* evidence in fact for protons); where then does the p:e ratio of 30:1 come from?

Thus, in my view, we cannot yet put a tick against supernovae origin.

An interesting possibility, also shown in Fig. 8, relates to giant molecular clouds, objects which have become increasingly studied in the last few years. Our interest here stems from the fact that $\gamma$-rays have been detected coming from such clouds and although in many cases [27] my view is that the CR intensity in the clouds is the same as that in the general ISM, there does seem to be evidence [28] that *some* need an enhanced CR intensity. Of course, it is possible that discrete sources inside the clouds are generating the CR ([28, 29] and see later) but it is also possible that a significant fraction of the gravitational collapse energy of the cloud finds its way into CR. The difficulties are surprisingly similar to those for SN production, such as the difficulty of CR extraction, but this type of model needs more study.

The possibility of pulsar acceleration can also be considered in this section, although it is arguable as to whether a pulsar should be regarded as a violent event. There is no doubt that pulsars accelerate electrons—the radio emission shows that—and doubtless much of the observable energy of SNRs comes from the central pulsars, but their contribution to cosmic rays is very uncertain.

Even the pulsar contribution to the CR electron flux is probably small because pulsars would be expected to generate energetic $e^+e^-$ pairs, whereas experimentally $e^-$ predominate and such $e^+$ as are seen can be explained simply in terms of interaction products (e.g. the chain $\pi^+ \rightarrow \mu^+ \rightarrow e^+$). Considering now the nuclear component, the problem is the observed mass composition. True pulsar models (as distinct from those which use the shock working on the ISM at remote positions) involve ion emission from the neutron star surface and the ions will surely be very different in composition from those in the cosmic radiation. The surface is presumably iron and its products formed from bombardment, and I concur with Arons's view that 'I see no possibility for the composition that results from grinding up iron being a decent model for cosmic rays' [30].

We march on. 'GC' in Fig. 8 relates to the most violent event that the mature Galaxy has suffered, or might have suffered—namely an explosion at the Galactic Centre, the word 'mature' is used because a larger energy was presumably available when the Galaxy first formed (see Section 3.2). The evidence with respect to the GC explosion is equivocal, with some astronomers for and some against. The observation of an expanding molecular ring round the GC can be explained in terms of an explosion some $10^6$ yr ago [31] with perhaps a repetition rate of once every $10^7$ yr, the corresponding energy release being as shown in the figure. The Moscow group [32] have favoured this model for some years and my own group [33] has contributed by making more elaborate calculations. Rather surprisingly, perhaps, quite a good case can be made, the model's biggest triumph being that the change of slope of the energy spectrum at $\simeq 10^{15}$ eV appears in a rather natural way as a result of diffusive motion from an intermittent source. As usual, though, there are shortcomings—the energetics are tight when allowance is made for inevitable early particle losses, and although the mass composition is perhaps understandable in terms of a violent shock accelerating the ISM near the GC a worrying deficit of small grammages is predicted. It is possible to circumvent this [34] but only by postulating a rather *ad hoc* propagation model.

Finally, in this section, there is the 'Galaxy Formation' (GF) model. It is hard to eliminate it completely but its difficulties are similar to those for 'GC'. A time approaching $10^{10}$ yr is a very long time to store cosmic rays in the Galaxy and it is hard to see why a CR gradient should appear, to mention only two problems.

### 5.2.3 Stellar winds

As mentioned earlier, quite large amounts of energy are being put into the ISM by steady stellar winds. Figure 9 shows what are probably the two most important stellar classes—Wolf–Rayet and OB associations; again we see that the energy available is about right if the efficiency of conversion to cosmic rays is high. Also shown for completeness is a very approximate estimate for the stellar wind of solar-type stars.

Having now (probably) exhausted the evidence on energetics we can turn our attention to other features of CR which give clues as to their origin.

# 6 Evidence on origin from mass composition and gamma ray sources

## 6.1 The evidence of the mass composition

It has been mentioned already that the CR composition is not too difficult

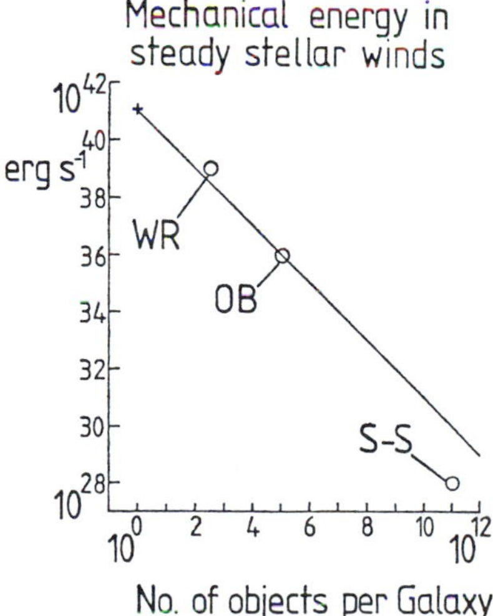


**Fig. 9** Mechanical energy in steady stellar winds. The line represents the CR energy needed. WR—Wolf–Rayet stars; S-S—solar-type stars.

from that in the Solar System, which in turn approximates to the 'local galactic abundance' (LG) when corrections are made for fragmentation in the ISM. This fact immediately eliminates certain models, as we have seen. Thus, the apparently attractive supernova origin model fails because of the very different composition of the material ejected, at least in model calculations, and similarly the pulsar model is unsatisfactory.

Closer inspection of the composition does, however, reveal some differences, of which the most interesting is probably the well-documented relation between the CR abundance (with respect to LG or SS) and the first ionization potential (see Fig. 10). This observation suggests a selective injection mechanism based upon the *atomic* properties of the elements and in turn that the temperature of the primary 'source' is not excessive, say $T \simeq 10^4$ K.

Further, interesting differences are encountered when the isotope ratios are examined. The most dramatic difference between CR and LG abundance ratios concerns $^{22}Ne/^{20}Ne$ which is higher by a factor of 4 and there appears to be an excess, by about 60 per cent, in $^{25,26}Mg$ relative to $^{24}Mg$ and $^{29,30}Si$ relative to $^{28}Si$ (see the summary by Cassé [35]). Various suggestions have been made to account for the isotope differences and they are worthy of study. Woosley and Weaver [36] speculate that most of the local CR come from metal-rich regions of the Galaxy, several kpc nearer to the GC. They point out that SN produce $^{22}Ne$ and other light neutron-rich isotopes in amounts proportional to the initial metallicity of the progenitor and thus, in the inner Galaxy, where the 'metallicity gradient' enhances the metallicity by a factor of 2 or 3 in comparison with the local value, an excess of $^{22}Ne$ will result. The Weaver–Woosley model is unlikely to be correct in its entirety because of the differences in mass composition from SN already referred to,

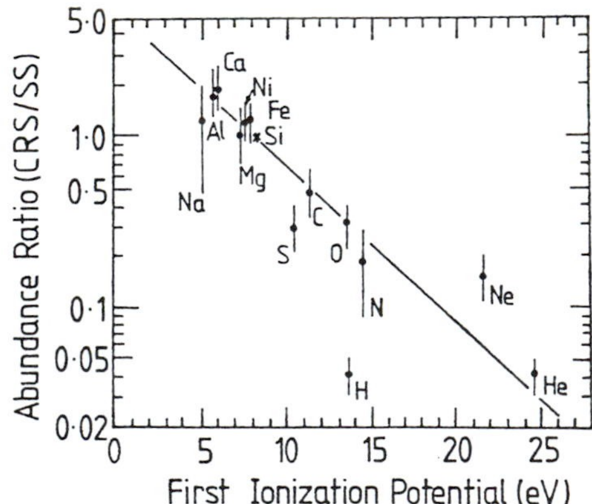


**Fig. 10** Ratio of CR source to Solar System abundance (CRS/SS) versus first ionization potential (after Wefel [26]).

but it does draw attention to the importance of considering sites for CR acceleration removed from the Sun where the conditions are different.

Another suggestion of importance has come from Cassé [35] who again is at pains to explain the $^{22}$Ne excess. Wolf–Rayet stars are expected to expose surfaces with $^{22}$Ne/$^{20}$Ne ratios of order 120 during their carbon-rich phase (a likely model of evolution of these stars is: a massive O-star (10–50 $M_\odot$) transforms successively into a red supergiant, a nitrogen-rich W–R star, a carbon-rich W–R star and finally a SN). Coupled with the strong stellar wind which both exposes the 'fresh' surface and provides considerable energy (see Fig. 9) there are the immediate ingredients for a satisfactory explanation of the $^{22}$Ne excess. Cassé argues that because of the very high ratio (120) expected for $^{22}$Ne/$^{20}$Ne in comparison with the factor 3–4 in cosmic rays only some 2 per cent of CR are generated in this way, but it appears to the present author that various loss processes could allow a bigger fraction of CR to arise in this fashion.

## 6.2 The evidence of $\gamma$-ray sources

The rôle of cosmic $\gamma$-rays in providing evidence for a gradient of cosmic ray intensity in the Galaxy, and thereby a galactic origin for the particles, has been mentioned already (Section 5.1). It now remains to inspect the data from the standpoint of identifying sources of particles, specifically protons and other nuclei. Ideally, we should see the proton signature—a $\gamma$-ray peak at 67 MeV coming from $\pi^0$-decay–and an intensity of $\gamma$-rays well above what would have been expected from the ambient CR particles interacting in the known mass of gas surrounding the potential source.

The present situation is that the $\pi^0$-peak has not been seen, presumably because of the presence of CR electrons, interacting by way of bremsstrahlung, which fill in the low energy side of the peak, although there is curvature in the $\gamma$-ray spectrum which is *probably* due to protons. However, there does seem to be some evidence for CR excesses. The situation with $\gamma$-ray 'sources' is as follows. In the most comprehensive set of measurements to date [37] using the *COS-B* satellite, some 25 'sources' have been detected (the 2CG catalogue), the definition of a source being that its width is such as to be not inconsistent with the region of emission being point-like and its significance above the local background being at the level of there being only $2 \times 10^{-2}$ spurious sources expected in the whole sky (the significance being calculated in a particular way). The minimum flux level comes out at $1.0 \times 10^{-6}$ cm$^{-2}$ s$^{-1}$ for $E_\gamma$ above 100 MeV. As might be expected, in view of the poor angular resolution of the contemporary $\gamma$-ray detectors (typically, 32 per cent of particle trajectories lie beyond 3° of the true direction) there is much argument about the nature of many of the sources.

Starting with those for which there is agreement, the Crab and Vela are certain sources, in that pulsed γ-rays are observed at the respective pulsar frequencies. The nearby quaser 3C 273 is also a source and 2CG 195 + 4 is a strong source which also has a good chance of being discrete (an unresolved pulsar?). The problem arises with the others and the extent to which the lumpy ISM irradiated by the ambient CR intensity throws up γ-ray peaks which satisfy the source requirements.

Our own view is that about half of the 2CG catalogue sources are simply CR-irradiated molecular clouds, the CR intensity being roughly the local value. It is a number of others which are of such interest in the present situation because the γ-ray peaks are towards known molecular clouds and their masses seem inadequate to explain the fluxes unless the CR intensity is raised appreciably. The problem of determining cloud masses is well known to be a difficult one but there is some confidence in the mass estimates from the fact that the nearby molecular clouds Orion A and B, which have been detected in γ-rays by both the *COS-B* and *SAS II* satellites, give roughly the expected γ-ray fluxes (assuming that they are inert from the CR point of view, [38, 15]). It is interesting to note that the ratio of the fluxes from A and B is equal to the ratio of the γ-fluxes [39] a result that is independent of the mass calibration and one which indicates that T-Tauri stars, which are much more abundant in Orion A than B, are unlikely to accelerate many cosmic rays.

Returning to the clouds which appear to show an excess, the situation is summarized in Fig. 11. In fact, it has been known for some time (e.g. [29]) that there are coincidences between γ-ray sources and OB associations, and in the nature of things, OB associations and giant molecular clouds are often coincident; what is new is the attempt to quantify the excess CR intensity needed. Before continuing, it is necessary to sound a warning—the distribution of molecular gas in the Galaxy is still not known very well and the argument that we are dealing with proton-initiated γ-rays (in part at least) rather than merely electrons, is not very strong. However, we continue. The evidence favouring an increase in γ-ray luminosity with increasing optical luminosity is interesting, as is the fact that we need an increase in CR luminosity over what we observe in the Sun of magnitude $\sim 10^4$; referring back to Fig. 8 it is seen that this is just the order of magnitude needed from the point of view of energetics in order to explain cosmic ray origin.

Reference should be made to one source (2CG 288–00) which we have identified with the Carina Nebula. It has been pointed out [39] that this region is one of the most active in the whole Galaxy, housing Wolf–Rayet stars and OB-associations. Particularly notable is the presence of the enigmatic Eta Carina object, thought by many to be a very massive and luminous star which gave several outbursts in the last century and may currently be in

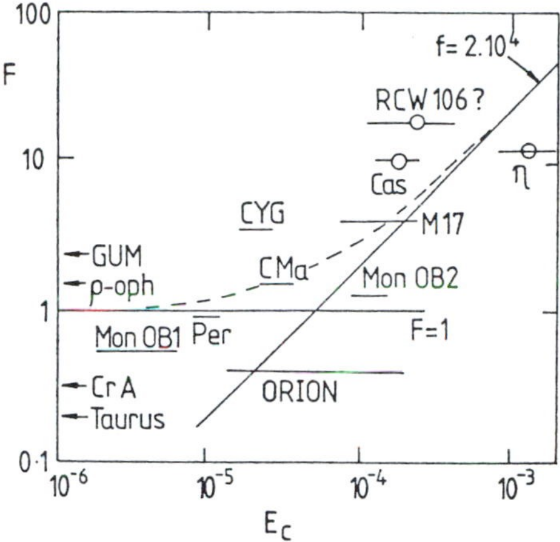


**Fig. 11** Relationship between CR enhancement factor, $F$, and the quantity $E_c = L\tau V^{-1}$ (with $L$ the luminosity, $\tau$ the lifetime of the cloud and $V$ its volume) from the $\gamma$-ray analysis of Issa *et al.* [28]. $F = 1$ represents inert clouds. The inclined line is for the situation where the high luminosity stars emit cosmic rays at a rate $f = 2 \times 10^4$ times higher than the Sun (which emits $\approx 10^{24}$ erg/s$^{-1}$ in CR—see Fig. 8) per unit of luminosity. The circles represent *COS-B* $\gamma$-ray sources [37].

the pre-supernova phase. Montmerle *et al.* [29] estimate that the current mechanical power output is $\simeq 2 \times 10^{39}$ erg s$^{-1}$ and if *all* the energy were to go into $\gamma$-rays above 100 MeV the flux would be $\simeq 10^4$ times the measured flux of $1.6 \times 10^{-76}$ cm$^{-2}$ s$^{-1}$. The efficiency for conversion of proton energy to $\gamma$-rays is about 10 per cent so that we only require the efficiency for conversion of mechanical energy to protons above 1 GeV to be greater than 0.1 per cent (or 1 per cent if 90 per cent of the protons escape before interacting with the gas in the vicinity of the star). Such an efficiency is not unreasonable and here then at least is one likely cosmic ray source.

# 7 A model for cosmic ray origin

## 7.1 Particles below $10^{11}$ eV

In the previous sections we have presented evidence in favour of a galactic origin for the bulk of the cosmic ray particles detected at the Earth; it is very difficult indeed to escape this conclusion. Of the many galactic possibilities we prefer composite sources in the sense that pre-acceleration of a somewhat perturbed solar system abundance is followed by acceleration by shock waves local to the sources. The bulk of the acceleration occurs near the source, with little later in the career of the particle—a requirement forced by the observation of a gradual reduction in the ratio of secondary to primary particle numbers with increasing energy (Section 2.3).

The sources themselves appear to be a mixture of many objects—WR stars, OB stars, novae, supernovae with probably none dominating (Figs 8 and 9). The final acceleration then occurs in the associated stellar winds and shocks.

The non-uniform distribution in the Galaxy of stars of all types and the CR sources in particular means that the bulk of CR probably originate in the annular ring of galactocentric radius at $R \simeq 5$ kpc (Section 5.1). A necessary corollary is that the diffusion coefficient is quite large (mean free path $\simeq$ kpc) and the ensuing CR radial gradient will be small—as observed (Section 5.1).

It remains to estimate the expected energy density of CR in extragalactic space as a result of particles which have escaped from galaxies. If the lifetime of CR in our galaxy is $\sim 10^7$ yr then, for equality of production by all galaxies, the average energy density in EG space will be $\simeq 10^{-4}$ eV cm$^{-3}$, using the date of Table 1. In fact, somewhat higher yields are expected for the 'average' galaxy and this figure could well be $\simeq 3 \times 10^{-3}$ eV cm$^{-3}$. If, as is more than likely, following earlier arguments, the CR are concentrated in galaxy clusters by tangled magnetic fields then Fig. 6 shows that the situation for the gas in clusters is that the predicted flux does not exceed the measures extra-galactic $\gamma$-ray flux. Here, then, is some further (slight) support for galactic origin. It should be added in parentheses that the presence of sufficient ionized gas in the Universe to give the diffuse extragalactic X-ray flux would not result in too high a $\gamma$-ray flux.

## 7.2 Particles above $10^{11}$ eV

The absence of abrupt spectral changes, at least until $10^{15}$ eV is reached, supports the idea that galactic sources continue to provide the bulk of the cosmic rays. Evidence comes from the measured anisotropy and phase (Fig. 4). Indeed, it is tempting to attribute the increase in slope of the primary energy spectrum at about $3 \times 10^{15}$ eV as simply due to a more rapid decrease of galactic lifetime with energy, there being astronomical evidence [40] for a change in trapping at just such an energy.

The situation above about $10^{16}$ eV is less clear. The anisotropy continues to increase, as expected for a galactic origin, but above about $10^{18}$ eV difficulties occur. Here, the radii of curvature of particle trajectories are becoming comparable to galactic dimensions: initially, the disc thickness, and eventually, at the highest energies ($10^{20}$ eV) to the galactic radius, if, as seems very likely [41] the particles are mainly protons. Above $10^{19}$ eV on the basis of a galactic origin, and near rectilinear propagation, we would expect most of the arrival directions to be collimated along the galactic plane, whereas in fact there is a preference for high galactic latitudes [42]. A possible explanation put forward by the author and his colleagues [43] (following the earlier suggestion of Brecher and Burbidge [44]) is that these particles are extragalactic, having come from sources in the central region of the Virgo supercluster. The energy requirement is not excessive ($\simeq 7 \times 10^{44}$ erg $s^{-1}$) and the observed shape of the energy spectrum can be explained in a rather natural way.

## References


[1] McCrea, W. H., 1951. *Obit. Not. Fell. R. Soc. Lond.*, **7**, 421.

[2] Hess, V. F., 1912. *Phys. Z*, **13**, 1084.

[3] Meyer, P., 1981. *IUPAP/IAU Symp. 94, Origin of cosmic rays*, p. 7, Reidel, Dordrecht.

[4] Schaeffer, O. A., 1975. *Proc. 14th int. Cosmic Ray Conf.*, **11**, 3508.

[5] Wolfendale, A. W., 1982. *Q. Jl R. astr. Soc.*, **23**, 325.

[6] Strong, A. W., Wdowczyk, J. and Wolfendale, A. W., 1974. *J. Phys. A*, **7**, 120.

[7] Hillas, A. M., 1968. *Can. J. Phys.*, **46**, S623.

[8] Fichtel, C. E. *et al.*, 1978. *Astrophys. J.*, **222**, 833.

[9] Said, S. S., Wolfendale, A. W., Giler, M. and Wdowczyk, J., 1982. *J. Phys. G*, **8**, 383.

[10] Issa, M. R. *et al.*, 1981. *J. Phys. G*, L187.

[11] Mushotzky, R. F., Serlemitsos, P. J., Smith, B. W., Boldt, E. A. and Holt S. S., 1978. *Astrophys. J.*, **225**, 21.

[12] Birkinshaw, M., Gull, S. F. and Northover, K. J. E., 1978. *Mon. Not. R. astr. Soc.*, **185**, 245.

[13] Miley, G. K., Perola, G. C., van der Kruit, P. C. and van der Laan, H., 1972. *Nature*, **237**, 269.

[14] Thomson, R. C. and Nelson, A. H., 1982. *Mon. Not. R. astr. Soc.*, **201**, 365.

[15] Wolfendale, A. W., 1981. *IUPAP/IAU Symp. 94, Origin of Cosmic Rays*, p. 309, Reidel, Dordrecht.

[16] Strong, A. W. and Lebrun, F., 1982. *Astr. Astrophys.*, **105**, 159.

[17] Dodds, D., Strong, A. W. and Wolfendale, A. W., 1975. *Mon. Not. R. astr. Soc.*, **171**, 569.

[18] Mayer-Hasselwander, H. A. *et al.*, 1982. *Astr. Astrophys.*, **105**, 164.

[19] Issa, M. R., Riley, P. A., Strong, A. W. and Wolfendale, A. W., 1981. *J. Phys G*, **7**, 973.

[20] Li, T. P., Riley, P. A. and Wolfendale, A. W., 1983. *Mon. Not. R. astr. Soc.*, **203**, 87.

[21] Phillipps, S., Kearsey, S., Osborne, J. L., Haslam, C. G. T. and Stoffel, H., 1981. *Astr. Astrophys.*, **98**, 286; **103**, 405.

[22] Ginzburg, V. L. and Syrovatsky S. J., 1964. *The origin of cosmic rays*, Pergamon Press, Oxford.

[23] Lovell, A. C. B., 1974. *Phil. Trans. R. Soc. A*, **277**, 489.

[24] Mullan, D. J., 1979. *Proc. 16th int. Cosmic Ray Conf.*, **2**, 92.

[25] Baade, W. and Zwicky, F., 1934. *Proc. Natn. Acad. Sci. U.S.A.*, **20**, 259.

[26] Wefel, J. P., 1981. *IUPAP/IAU Symp. 94, Origin of cosmic rays*, p. 39, Reidel, Dordrecht.

[27] Issa, M. R. and Wolfendale, A. W., 1981. *Nature*, **292**, 430.

[28] Issa, M. R., Riley, P. A., Li, T. P. and Wolfendale, A. W., 1981. *Proc. 17th int. Cosmic Ray Conf.*, **1**, 150.

[29] Montmerle, Th., 1979. *Astrophys. J.*, **231**, 95.

[30] Arons, J., 1981. *IUPAP/IAU Symp. 94, Origin of Cosmic Rays*, p. 175, Reidel, Dordrecht.

[31] Sanders, G. C., 1981. *Nature*, **294**, 427.

[32] Khazan, Y. M. and Ptuskin, V. S., 1977. *Proc. 15th int. Cosmic Ray Conf.*, **2**, 4.

[33] Said, S., Wolfendale, A. W., Giler, M. and Wdowczyk, J., 1981. *17th int. Cosmic Ray Conf.*, **2**, 344.

[34] Giler, M., 1983. *J. Phys. G*, **9**, 1139.

[35] Cassé, M., 1982. In *Composition and origin of cosmic rays*, ed. M. M. Shapiro. D. Reidel Publ. Co., Dordrecht, Holland, p. 193.

[36] Woosley, S. E. and Weaver, T. A., 1982. *Essays in Nuclear Astrophys*, eds Barnes, C. A., Schramm, D. N. and Clayton, D. D.

[37] Swanenburg, B. N. *et al.*, 1978. *Nature*, **275**, 298.

[38] Caraveo, P., 1981. *Phil. Trans. R. Soc. A*, **301**, 569.

[39] Issa, M. R. and Wolfendale, A. W., 1981. *J. Phys. G*, **7**, L187.

[40] Bell, M. C., Kota, J. and Wolfendale, A. W., 1974. *J. Phys. A*, **7**, 420.

[41] Walker, R. and Watson, A. A., 1981. *J. Phys. G*, **7**, 1297.

[42] Lloyd-Evans, J., Pollock, A. M. T. and Watson, A. A., 1979. *Proc. 16th int. Cosmic Ray Conf.*, **13**, 130.

[43] Giler, M., Wdowczyk, J. and Wolfendale, A. W., 1980. *J. Phys. G*, **6**, 1561.

[44] Brecher, B. and Burbidge, G. R., 1972. *Astrophys. J.*, **174**, 253.

# Statistics, geometry and the cosmos

David G. Kendall

E. A. Milne was born in Hull in the East Riding of Yorkshire in 1896, and died in Dublin in 1950 at the relatively early age of 54. I first met him within a few days of my arrival in Oxford in 1936. Through the kindness of Frank Twyman of Adam Hilger Ltd I had been given a letter of introduction to the Director of the University Observatory, H. H. Plaskett, who promptly invited me to the Observatory Colloquium (4.30 on Wednesdays, with a cup of tea, and cake also if fresh offprints by an Observatory author had arrived that week). This I attended without fail throughout my three years as an undergraduate. On that first day Milne immediately came over and talked to me, and continued to show me great kindness until his death, as indeed did Plaskett also. I need not say that it was somewhat alarming to find oneself thrust into such company, but Milne's warm personality and infectious excitement about the latest discoveries quickly melted away most of my shyness. After that I also attended his own seminar regularly, and was a frequent visitor at 19 Northmoor Road. I remember helping to decorate the house and garden with bunting for the Coronation in 1937, with the assistance of two delightful small girls to whom I send warm greetings today.

Milne was a vivid lecturer and always communicated at least his enthusiasm to the audience. His undergraduate course on Vectorial Mechanics was a *tour de force*, but of his seminar lectures I remember now only his insistence that 'mass is just a constant of integration'. This seminar went on from year to year without recapitulation, having started I suppose in 1929 or thereabouts. If anything notable had happened in the preceding week (like the death of Rutherford, or Dirac's letter to *Nature* on 'big numbers') Milne would abandon the course and give his immediate reactions. I recall his saying that there were good days and bad days with Rutherford. The good days were immediately recognizable by his coming into the Lab whistling *Onward, Christian Soldiers*.

My own research interest was then in such matters as radiative equilibrium and the Milne integral equations (of the Wiener–Hopf type subsequently to prove of importance in statistics), and I was encouraged in this by Milne, Plaskett and E. C. Titchmarsh. Milne was very proud of his training as an analyst by G. H. Hardy. It was Hardy who in 1916 introduced Milne to A. V. Hill, and so became responsible for Milne's membership of the famous Anti-Aircraft Research group organized and directed by Hill. Later (1939–1944) he became a Member of the Ordnance Board, an extraordinary body of medieval origin, and brought me also within its orbit in a very minor capacity. One piece of experimental work which I carried out rather against the wishes of some of my superiors ultimately found its way on to Milne's desk as an unsigned report, and was then used by him in the context of very remarkable but still unpublished work which involved some quite delicate analysis. Milne himself considered this to be the best thing he had ever done. When later Milne realized that it was my data he had used, he was delighted, and so was I.

Throughout the war Milne continued writing his constant stream of papers on kinematic relativity. When I asked him how he found the time, the answer was: 'I write them before breakfast'.

After the war I returned to Oxford as a novice Fellow of Magdalen, and Milne helped me to step out of the egg-shell into academic life. By then my interests had swung away from astronomy towards statistics, but I was very pleased to be able to tell Milne that Giordano Bruno too had espoused the Cosmological Principle. 'If so,' Milne later wrote, 'I am in good company. Bruno was burnt at the stake.'

In 1950 Milne was President of the Royal Astronomical Society, and it may have been by his wish that the Society met that year in Dublin; what should have been a very happy occasion was saddened by his collapse and death. These events impressed themselves on my mind, and when, many years later, I found myself President of the London Mathematical Society, I took the opportunity of encouraging it too to have an Irish meeting, which was much enjoyed and appreciated by all. Reflecting on this again, it occurs to me that it might be thought very appropriate for the Milne Society to meet in Dublin from time to time.

It is a privilege and a pleasure to be invited to give this lecture in memory of a great man who was also a good friend.

Statistics and geometry are obviously necessary tools for the study of the cosmos, but it has not been usual to apply them together. Classical statistics was primarily tailored to fit problems in Euclidean space, and despite two early pioneer papers by R. von Mises [8] and R. A. Fisher [3] it is only recently that statistical problems on manifolds have received systematic attention. When one passes from statistics on manifolds to stochastic motions on manifolds, then the global as well as the local geometry becomes very important. Thus it will not suffice to employ analysis posing as geometry in a dress of upper and lower suffixes. The global geometrical structure becomes itself of prime significance.

My own interest in such problems has an archaeological origin. When one maps the positions of a given class of archaeological sites, there will always be some who claim that they are approximately positioned on a framework of unseen straight lines (or perhaps circles, etc.). Here of course it is understood that each such straight line must, to avoid trivialities, pass through (or nearly through) at least three points, so the basic question becomes one of deciding how many triplet collinearities one can expect to find by chance in a given set of $n$ points in the plane. To make sense of the question one must be precise about the 'nearly'. As a first shot we could say that three points are $\epsilon$-collinear when they form a triangle with largest angle greater than $\pi - \epsilon$. Here $\epsilon$ is to be a small angle, given in advance. But if we take this view then what we are doing is to consider the whole set of $\binom{n}{3}$ triangles and ask whether an unduly large number of them have a *shape* which meets the test of $\epsilon$-collinearity. From this point onwards one can proceed in various ways [4], but an attractively general procedure is to carry out the investigation within a general theory of shape [5] for sets of $k$ points in $m$ dimensions, where in the example just mentioned $k$ is 3 and $m$ is 2.

Of course it is essential to make clear what we mean by 'shape'. So let us say that shape is what is left when the effects of location, size, and rotation have been filtered out. Even this is unsatisfactory, however, because 'size' could be assessed in various ways. What seems to me natural is to assess size (or rather 'size squared') by calculating the sum of the squares of the distances of the points from their centroid. This has the advantage of being meaningful for all $k$ and all $m$. A quite different theory, developed independently by my friend Ruben Ambartzumian (of an astronomically famous family), assesses size when $k = 3$ and $m = 2$ by the length of the perimeter of the triangle formed by the three points. It is not obvious how that should be generalized to larger values of $k$.

If we proceed as I have suggested here, we find that in two dimensions each shape (that is, each equivalence class of $k$-ads in $\mathbb{R}^2$ equivalent under displacements, rotations, and magnifications) can be identified as a point on a $2k-4$ dimensional manifold called $\mathbb{C}P^{k-2}$ (complex projective space). It can be shown that the representation is topologically correct, and indeed more is true. To understand what this means we need to introduce the further idea of the distance between two such shapes. Here the inspiration comes from classical mythology. We follow Procrustes (who so treated those unwise enough to accept his hospitality when travelling from the Peloponnese to Athens) by forcibly matching one $k$-ad against another by suitable rotations, translations, and changes of scale. (Procrustes went further, and allowed himself the additional operation of truncation, not appropriate here.) A coarse and natural measure of the discrepancy between two $k$-ads is the sum of the squares of the distances from each point of one $k$-ad to the correspondingly labelled point of the other. Obviously we ought to standardize that discrepancy for a common change of scale, so we divide it by the sum of the squares of the distances of all the $2k$ points from their pooled centroid. The ratio so formed is now dimensionless, and will have a minimum if we consider all allowable procrustean transformations of the two $k$-ads. If this minimum value is $D$, we define $\varrho = \arccos(1 - D)$, which is a distance-function, to be the distance between the two shapes. When this programme is carried out, our shape-manifold becomes a Riemannian one, and it turns out that as a Riemannian manifold it is actually isometric with the algebraic geometers' complex projective space up to the choice of a unit of length. Thus we have reduced the study of the shapes of $k$-ads in $\mathbb{R}^2$ to a classical geometric situation.

The particular case $k = 3$ is of special importance, and then the shape-manifold is isometric with a 2-D sphere of radius ½ (this being the appropriate version of $\mathbb{C}P^1$). When $k$ exceeds 3 the situation is just as attractive, save only that the shape-space will then have at least four dimensions and cannot be visualized so easily.

Now the shapes of *exactly* collinear $k$-ads can be regarded as forming a special set of points lying in the shape-manifold. This set can be identified as a version of real projective space $\mathbb{R}P^{k-2}$. In the important case when $k = 3$, it is a specific great circle on our sphere $S^2$ (½). We can see it in Fig. 1 as the 'equator' running across the centre of the figure. The remaining curves in the diagram show other contours for a function-of-shape which is max $(A, B, C)$, where $A$, $B$ and $C$ are the angles of the triangle. Thus the set of *nearly* collinear shapes is represented by a union of narrow lenticular regions close to the 'equator' (there are three of these, of which one can be seen, and two partly seen).

So far we have avoided introducing statistical ideas. But suppose we remedy this by agreeing to think of our three points as a sample from a Gaussian (normal) distribution of circular symmetry. This extra degree of specification will provide us with a probability-measure sitting on the shape-manifold, and so will enable us to make probability statements about shapes. In particular we can compute the probability

$$\text{pr(the triangle } ABC \text{ is } \epsilon\text{-collinear)}$$

by integrating the above probability-measure over the three lenticular regions corresponding to $\epsilon$-collinearity. Now the probability-measure in the Gaussian situation turns out to be the uniform distribution over the spherical surface, and so the original problem has been reduced to spherical trigonometry.

In numerical terms a main result (obtained earlier for a similar problem by

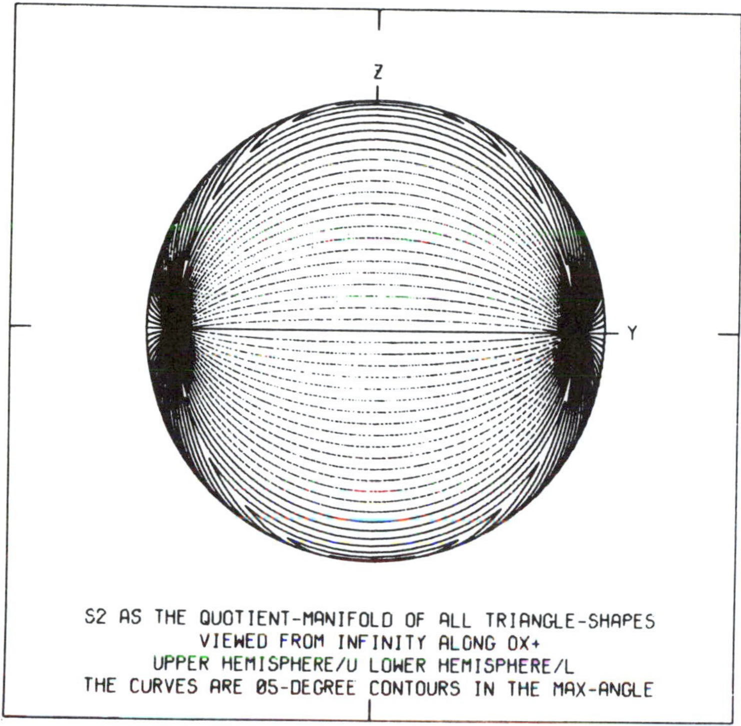


**Fig. 1** The shape-manifold $S^2(\frac{1}{2})$ for the shapes of labelled triads of points in the plane, carrying contours for the maximum angle of the triangle formed by the points (reproduced, by permission, from [5]).

S. R. Broadbent using direct computations) tells us that each such set of 50 points in the plane will turn out to have, among its $\binom{50}{3} = 19\,600$ triplets, about 130 triplets that are collinear for a value of $\epsilon$ corresponding to 1°. That this result is beyond our intuition is easily tested, by asking any audience, professional or not, to guess the expected number of such nearly collinear triplets. Wherever in the world I put this question, I get answers ranging from small numbers like 2 or 3, up to several thousand.

Of course you will appreciate that this is only a sketch of what has been done, and that much fine detail has had to be suppressed. Thus we may ask, what happens if we use non-Gaussian distributions, why do we assume circular statistical symmetry, why do we use this value for $\epsilon$, why do we not consider a range of $\epsilon$s, and so on. All these and many other questions are answered in the paper [4] by W. S. Kendall and myself. The answers are too technical to be appropriate for today's occasion.

There has been some controversy about the way in which these arguments should be extended to larger values of $k$. There is no difficulty in describing the geometry of the shape-manifold, and the nature of the shape-measures.

What has been in dispute is how we are to assess near-collinearity when there are more than three points. Fortunately the general theory I have been sketching provides a natural and attractive answer. Let us call the subset of *exactly* collinear shapes, 'Coll'. Then Coll is the subset $\mathbb{R}P^{k-2}$ of the shape-manifold $\mathbb{C}P^{k-2}$, and the shape of the $k$-ad to be tested is a point P lying in $\mathbb{C}P^{k-2}$. The Riemannian geometry provides us with the shortest geodesic distance in $\mathbb{C}P^{k-2}$ from P to Coll, and it is hard to conceive of a more natural measure of non-collinearity than that. If we call the above shortest distance $L$, then the shape-measure theory in the Gaussian case provides us with the statistical distribution of $L$, and indeed also with its statistical distribution when the generating Gaussian distribution is 'stretched' ($\sigma_1 \neq \sigma_2$). This means that (a) we can assess the significance of an observed effect, and (b) we can say how much more likely we are to get such an effect by chance, if the 'stretched' nature of the overall layout is allowed for.

The cosmos provides us with a neat example. Arp and Hazard [1] noticed a curious configuration of six QSOs involving collinearities. As they had quite different individual redshifts, it became important to know whether the observed collinearities could be dismissed as a chance effect. Treatment along the lines suggested here (for a detailed account of which see [5]) indicates that the collinearities can at any rate not be dismissed out of hand, the significance level being about 1/400, although it is clear that a firm decision on the matter must await a large-scale survey.

The results of such surveys are now becoming available in work by M. G. Edmunds and G. H. George (Cardiff), and by S. V. M. Clube, A. Savage and A. S. Trew (Edinburgh).

The ideas described above have been extended to the general $m$-dimensional case, and new features arise when $m$ is 3 or more. The representation-space for shapes then acquires homological torsion, and differential-geometric singularities also appear although for statistical (but not stochastic) purposes there is a 'big' smooth subspace on which most calculations can be carried out. Obvious further generalizations such as (1) the case of infinite $k$, and (2) the study of the 'shapes' arising when the product of $k$ copies of a homogeneous Riemannian manifold $M$ is quotiented by the action of a transitive group on $M$, remain problems for the future.

Another recent geometric-statistical study (Kendall and Young [6]) arose as a response to a remarkable paper by Birch [2] noting what he considered might be a large-scale asymmetry of the Universe. His paper has been criticized in three ways by Phinney and Webster [10], who pointed out that (1) the data could have been more judiciously selected, (2) the statistical significance of the effect was not obvious, (3) the effect, if it exists, could be local rather than global, super-galactically speaking (say as a consequence of incomplete correction for Faraday rotation). We met (1) by working with a revised data-set supplied by Dr Webster, and to investigate (2) we devised what we think is an appropriate and natural significance test. As non-professional astronomers we offer at present no comment on (3). I now describe some of the ideas used in the calculation, which shows that we have here to deal with an observation which, considered on its own, is surprising at a significance level of about 1/2000. This indicates that Birch indeed made a substantial point, whatever the correct astronomical interpretation may be.

Birch was concerned with 100 or so high luminosity classical double radio sources for which it is possible to determine (a) a geometrical elongation axis and (b) a polarization axis, and to measure the acute angle $\Delta$ between them, this angle being given an algebraic sign according to the direction in which one must move from (a) to (b) in order to describe that angle. Thus $-90° \leqslant \Delta \leqslant +90°$, and $\Delta = -90°$ must be identified with $\Delta = +90°$, because there is only one way in which the axis (a) can be perpendicular to the axis (b). If we think of the sources as living on the celestial sphere (e.g. on the surface $S^2$ of an astronomical globe) then we can write such a single observation in the form $(p, \Delta)$, where $p$ is the unit position vector from the centre of the globe (so $p^2 = 1$), and $\Delta$ is as above. Evidently we can think of $\Delta$ as a point in 1-D real projective space $\mathbb{RP}^1$, so that whenever it is convenient we can work instead with $2\Delta$ in $S^1$. Our set of observations will then be $((p_j, 2\Delta_j) : j = 1, \ldots, n)$ and a statistical analysis must proceed from the postulation of two probability measures for $(p, 2\Delta)$, one corresponding to a null Birch effect ($\Delta$ independent of $p$), and the other to a positive Birch effect ($\Delta$ statistically linked to $p$).

It is instructive to examine the topological character of the full sample-space in this problem. A basic data-point is of the form $(p, l, m)$, where the unsensed tangent line $l$ at $p$ locates the geometric axis of the source, and $m$ similarly locates the polarization axis. When $p$ and $l$ are given, then $\Delta$ is the angle by which we must rotate $l$ about $p$ to get $m$. Thus the sample-space for $(p, l, m)$ is a product with the total space $T$ of the $\mathbb{RP}^{-1}$-bundle over $S^2$ as first factor (for $p$ and $l$) and $\mathbb{RP}^1$ as second factor (for $\Delta$). We can replace the second factor by $S^1$ if we use $2\Delta$ instead of $\Delta$. It will be seen from the Appendix that this full sample-space for $(p, l, m)$ is homeomorphic to $L(4, 1) \times S^1$, where $L(4, 1)$ is a particular one of the so-called lens spaces. The calculations which follow take place in the marginal space for the reduced observations $(p, 2\Delta)$, and so require only the reduced sample-space $S^2 \times S^1$, but in further work it is likely that the lens space will play an important rôle.

The effect described by Birch amounts to a topographic regression of $\Delta$ (or $2\Delta$) on $p$, and in everyday terms it can loosely be described as follows. It is *as if*, when observing the heavens in one hemisphere of directions-of-viewing, we see a large number of capital roman letters N, together with a few capital cyrillic letters И, while in the opposite hemisphere of directions-of-viewing there are many cyrillic Иs and a few roman Ns. (Here the Ns correspond to say $0 < \Delta < \frac{1}{2}\pi$, and then the cyrillic Иs correspond to $-\frac{1}{2}\pi < \Delta < 0$; of course to make the simile more appropriate we must envisage a mixture of Ns, cyrtillic Иs, and 'intermediate' letter-shapes.) Such a phenomenon would be observed if the objects in question were maximally distant from us in Dante's $S^3$-universe, so that what we see in one direction as N, we also see again (from behind) in the 'opposite' direction as cyrillic И. This was not Birch's interpretation, of course, but it might have been the one that would have been put forward 50 years ago.

As $p$ specifies a point on the celestial sphere $S^2$, it represents a direction in the ordinary sense. As $\Delta$ (to be coded in the calculations as $2\Delta$) specifies a point on the projective space $\mathbb{R}\mathrm{P}^1$, we speak of it as an 'indirection'. The terminology here is borrowed from Shakespeare, and what he makes Polonius say is in fact remarkably appropriate to the present investigation:

*And thus doe we of wisedome and of reach*
*With windlesses, and with assaies of Bias,*
*By indirections finde directions out.*

We are thus concerned with a regression of an indirection on a direction.

While dealing with literary matters I had better add a few words on 'Dante's $S^3$-universe'. That Dante correctly described $S^3$, and perhaps seriously thought of it as a model for the universe, was pointed out by M. A. Peterson [9], who assembled evidence from the *Paradiso* indicating a representation of the Universe as the union of two copies of a solid sphere $B^3$. The first has the Earth at its centre, surrounded by the concentric shells of the Aristotelian universe, and ultimately by the Primum Mobile as its boundary $S^2$. The second copy of $B^3$ has a similar structure of concentric spherical shells, this time with God at its centre, and again the Primum Mobile as an $S^2$-boundary. It seems that Dante intends us to view the whole Universe as composed of these two copies of $B^3$ joined together by a 1:1 glueing together of their two boundaries $S^2$. And this is, of course, the standard construction of $S^3$. I am indebted to my friend Professor Manfred Gordon for this information.

At the heart of any mathematical formulation for a regression of the kind we have in mind, one will expect to find an expression of the form

$$\alpha \cos 2\Delta + \beta(q.p) \sin 2\Delta,$$

where $p$ is the observed direction, $q$ the directional pole of the regression, $\Delta$ records the observed indirection, $\alpha$ controls the null effect, and $\beta$ is the regression coefficient. It is natural to build up the model by making use of the Fisher and von Mises distributions familiar in directional statistics (for which see Mardia [7]), and of course we must incorporate a factor $f(p)dp$ to describe the distribution of source-positions $p$ on $S^2$. It would be quite inappropriate to postulate any simple form for the function $f(p)$, which describes the sociology of observatory-building rather than an astronomical phenomenon. These considerations led us to write

$$\frac{\exp(\alpha \cos 2\Delta + (\lambda \cdot p) \sin 2\Delta)}{2\pi\, I_0 (\sqrt{(\alpha^2 + (\lambda \cdot p)^2)})} f(p)\, dp\, d\,(2\Delta) \qquad (1)$$

for the joint distribution of $p$ and $2\Delta$, where now we have collapsed the parameters $\beta$ and $q$ into the vector $\lambda = \beta q$. We need to test the null hypothesis $|\lambda| = 0$ against the alternative hypothesis $|\lambda| > 0$, in the presence of the nuisance parameters $\alpha$ and $q$. (The expression $I_0(\cdot)$ in (1) is the zero-order modified Bessel function of the first kind.)

The standard procedure in such cases is to multiply together the factors at (1) above for each observed pair $(p_j, 2\Delta_j)$ so as to form what is known as the likelihood of the observations, and then to maximize this as $L_1$ when all parameters are free, and as $L_0$ when $\beta = |\lambda|$ is forced to zero. The conventional Neyman–Pearson test statistic is then $U = 2 \ln (L_1/L_0)$, and it is important to notice that there will be no contributions from $(f(p)_j : j = 1, \ldots, n)$ to $U$, so that we are not committed to any speculation about the form of the function $f$. It is known that asymptotically, when $n$ is large, then $U$ will have approximately the chi-square distribution with 3 degrees of freedom, but one would be hesitant to rely too much on the relevance of that fact in the present instance, where $n$ is only 134.

The maximized value $L_0$ is easily found. The calculation of $L_1$ is more difficult as it involves a global search over a 4-D space, but as it happens this is simplified by convexity considerations. It thus turns out that $U = 16.04$. The asymptotic arguments mentioned above indicate significance at a level of about 1/1000, but we do not wish to proceed in that way with such a small number of sources, so instead we use what is called a 'data-based' simulation test.

The idea behind this is to keep the partitioned sets of observations $(p_1, p_2, \ldots, p_n)$ and $(2\Delta_1, 2\Delta_2, \ldots, 2\Delta_n)$ unchanged, but to pair them up in random order, so as to erode the effects of any real regression that may be present. After each such randomization we compute the corresponding $U$-value, carrying out this operation some conveniently large number of times, and we then compare the observed $U = 16.04$ with the large collection of simulated test-statistics which are taken to represent the $U$-distribution that would have been obtained in similar circumstances if in fact there were no real Birch effect. The time-consuming global optimization involved in the calculation of $L_1$ makes it very costly to perform more than a few hundred such simulations, and so we make use of the fact that it is much quicker to calculate a first-order approximation $V$ to $U$. The value of $V$ for the actual data is $V = 14.64$, and this, when judged against 10 000 simulated $V$-values, turns out to be significant at a level of about 1/2000. Similar calculations with $U$, necessarily smaller in extent, confirm this result.

Accordingly we are driven to conclude that there is a strong indication of a real Birch effect which calls for explanation. Its estimated pole is

$$\text{RA } 13^{\text{h}}30^{\text{m}}, \text{Dec. } -37^{\circ}.4,$$

with an uncertainty of about 20°–30°, so that, to be acceptable, an explanation should imply the existence of a pole within that distance of the estimated pole.

In thanking you again for inviting me to deliver this lecture, and IBM UK Ltd for its generous patronage, I should like to say how pleasant it always is to visit the Mathematical Institute of which Milne was for many years the Curator.

## References


[1] Arp, H. and Hazard, C., 1980. Peculiar configurations of quasars in two adjacent areas of the sky, *Astrophys. J.*, **240**, 726–736.

[2] Birch, P., 1982. Is the universe rotating?, *Nature*, **298**, 451–454.

[3] Fisher, R. A., 1953. Dispersion on a sphere, *Proc. R. Soc. A*, **217**, 295–305.

[4] Kendall, D. G. and Kendall, W. S., 1980. Alignments in two-dimensional random sets of points, *Adv. appl. Probab.*, **12**, 380–424.

[5] Kendall, D. G., 1984. Shape manifolds, procrustean metrics, and complex projective spaces, *Bull. Lond. Math. Soc.*, **16**, 81–121.

[6] Kendall, D. G. and Young, G. A., 1984. Indirectional statistics, and the significance of an asymmetry discovered by Birch, *Mon. Not. R. astr. Soc.*, **207**, 637–647.

[7] Mardia, K. V., 1972. *Statistics of directional data*, Academic Press, New York.

[8] von Mises, R., 1918. Uber die 'Ganzzahligkeit' der Atomgewichte und verwandte Fragen, *Phys. Z.*, **19**, 490–500.

[9] Peterson, M. A., 1979. Dante and the 3-sphere, *Am. J. Phys.*, **47**, 1031–1035.

[10] Phinney, E. S. and Webster, R. L., 1983. Is there evidence for universal rotation?, *Nature*, **301**, 735–736.

[11] Rolfsen, D., 1976. *Knots and links*, Publish or Perish Inc., Berkeley.

## Appendix: The total space T of the $\mathbb{R}P^1$-bundle over $S^2$

A typical element $t$ of $T$ can be expressed in the form $(p, l)$, where as usual $p$ is a unit vector locating a point on $S^2$, and $l$ is an 'unsensed' tangent *line* for $S^2$ at that point. Suppose that $l = (\lambda v : \lambda \text{ real})$ where $v$ is a unit tangent *vector*; of course $l$ is not changed if we replace $v$ by $-v$. We can think of $p$ and $v$ in that order as the first two members of a right-handed mutually perpendicular triad $(p, v, w)$ of unit vectors got from a basic such triad $(\mathbf{i},\mathbf{j},\mathbf{k})$ by the quaternionic operation

$$(p, v, w) \to q\,(\mathbf{i},\mathbf{j},\mathbf{k})\,q^{-1},$$

where $q$ is a unit quaternion $x_0 + x_1\,\mathbf{i} + x_2\,\mathbf{j} + x_3\,\mathbf{k}$ $(\Sigma x^2 = 1)$. We can identify $q = (x_0, x_1, x_2, x_3)$ as a point of $S^3$, and we can also think of $q = (x_0 + ix_1, x_2 + ix_3)$ as a point of $\mathbb{C} \times \mathbb{C}$. Both interpretations will be useful here.

It is clear that we must identify $q^*$ with $q$ whenever both

$$q^*\,\mathbf{i}\,q^{*-1} = q\,\mathbf{i}\,q^{-1} \quad \text{and} \quad q^*\,\mathbf{j}\,q^{*-1} = \pm\, q\,\mathbf{j}\,q^{-1},$$

and in terms of $Q = q^{-1}\,q^*$ these relations are equivalent to

$$Q\,\mathbf{i} = \mathbf{i}\,Q \text{ and } Q\,\mathbf{j} = \pm\,\mathbf{j}\,Q,$$

so they assert exactly that $Q = y_0 + y_1\,\mathbf{i} + y_2\,\mathbf{j} + y_3\,\mathbf{k}$ satisfies

$$\text{(i) } y_2 = y_3 = 0,$$

and

$$\text{(ii) exactly one of } y_0 \text{ and } y_1 \text{ is } 0.$$

In terms of the original quaternion $q$, this just means that we are not to distinguish between $q$, $\mathbf{i}q$, $-q$, $-\mathbf{i}q$, and thus $T = S^3/(1, \mathbf{i}, -1, -\mathbf{i})$. A moment's algebra shows that when $q$ is considered as a point of $\mathbb{C} \times \mathbb{C}$ then the system of identifications is exactly such that each $(z_0, z_1)$ as a point of $\mathbb{C} \times \mathbb{C}$ is to be identified with $(iz_0, iz_1)$. But this last prescription is one of the several definitions of the lens place $L(4, 1)$ given in Rolfsen [11].

Another model of $L(4, 1)$, also described by Rolfsen, starts with the ball $B^3$ in $R^3$ with its boundary $S^2$ split into eight closed abutting pieces, these being then sewn together as follows. From the upper hemisphere we obtain four quadrants $H_0$, $H_1$, $H_2$ and $H_3$, successively related to one another by rotation about the positive $z$-axis through 90°. From the lower hemisphere in a similar way we obtain $H'_0$, $H'_1$, $H'_2$ and $H'_3$, but we arrange for $H'_0$ to lie directly below $H_1$, $H'_1$ below $H_2$, $H'_2$ below $H_3$ and $H'_3$ below $H_0$. The sewing programme is then: for each $j$, $H_j$ is to be sewn to $H'_j$ in a manner respecting the 90° rotations about $OZ+$ and the reflections in the plane $Z = 0$. What results from this operation is a closed connected (but not simply connected) orientable 3-manifold which is homeomorphic to $L(4, 1)$ as previously introduced, and so to $T$.

These representations show that the integral homology of $T$ is $Z/4$, 0, and $Z$ in dimensions 1, 2 and 3, and that its fundamental group is $Z/4$.

I am indebted to Professor J. F. Adams, FRS for assistance with this Appendix.

# The Earth's atmosphere: ideas old and new

Desmond G. King-Hele

Most previous Milne Lecturers have been either friends or pupils of E. A. Milne, but I cannot claim any such connection, although I did once hear him lecture on kinematic relativity when I was an undergraduate, and was much impressed. My link with Milne in this lecture is through the subject-matter, for Milne [1, 2] propounded several new ideas about the Earth's atmosphere —now transformed to old ideas by the passage of time.

My aim is to look at the Earth's atmosphere in its entirety, the Sphere of Air as the ancient Greeks called it. To set the scene, I shall outline modern views rapidly and superficially, and then offer some snapshots of past ideas. The air near the ground has always been familiar to people down the ages, so I shall not say much about the lower levels of the atmosphere but concentrate on the general picture.

I chose this subject because I feel that 'this most excellent canopy the air', as Hamlet called it, is unjustly underrated by those whose lives depend on it. Deprive us of air for even a few minutes and we should all be dead. Without it, life on Earth would either never have evolved or would have taken a quite different course. We are the creatures of air. Yet we just take it for granted. But why? The Moon has no air: how do we know that the Earth's air will not also escape into space? This is the very problem that occupied Milne.

Its power to keep us alive by letting us breathe is one great virtue of the atmosphere. But it has another virtue equally vital: it is transparent to sunlight, thereby allowing the photosynthesis on which our food depends. And even those astronomers whose thoughts are on higher things than breathing or eating should not curse the atmosphere for degrading their images but instead salute it for kindly allowing the stars to shine: otherwise astronomy would have been strangled before birth.

When our attitude towards the air is so offhand, it is not surprising to find that we are even more cavalier in our attitude towards the history of ideas about the atmosphere. Though historical scholarship on almost every conceivable subject has multiplied greatly in the past 30 years, there is, as far as I am aware, no book surveying the history of ideas about the atmosphere down the ages. This is in stark contrast with the many, many books about the history of ideas in astronomy, the ideas of Copernicus, Kepler and Galileo are all widely known. (Perhaps it is because the air controls our every-minute life, while astronomy has no effect on every-day life and is therefore a 'purer' subject?) Meteorologists might be expected to take some interest in the air, but they seem to have been too busy forecasting to bother much about looking backwards: in the UK it was not until 1983 that the Royal Meteorological Society set up a history section [3]. And even the meteorologists concentrate their attention on just a tiny fraction of the atmosphere, the lowest levels.

There is not even a name for the study of the atmosphere. 'Aeronomy' is available, and there is an International Association of Geomagnetism and Aeronomy, but in practice 'aeronomy' usually refers to the chemistry and electrodynamics of the upper atmosphere [4]. The Air would have a much better image today if there had been in the UK a Royal Aeronomical Society to match the Royal Astronomical Society. What happened was that in the nineteenth century the meteorologists were inevitably dominant among atmospheric scientists, and they pursued their specialism, leaving the Royal Astronomical Society to look after the higher reaches of the atmosphere, the realm of the aurora and (ironically) of meteors. The astronomers cannot be expected to be dedicated aeronomists. So, in the UK and many other countries, the science of the Air has provided a painful illustration of the maxim that organisms rarely function well when split into two parts.

## The modern view of the atmosphere

Because of its transparency I cannot give a useful picture of the atmosphere at optical wavelengths by day, but the situation is different at night. Looked at from space, the atmosphere then often shines quite brightly in regions near the magnetic poles, where charged particles of high energy excite fine displays of the aurora at heights of 100 km upwards. Plate I shows a moderate aurora over Canada, with the city lights of the USA below seeming quite puny by comparison: indeed the power in a strong aurora can reach $10^7$ MW, more than the total world electricity supply.

Our perception of the aurora has been much enhanced by these views from space: from the ground it can very rarely be seen at the low latitudes where most people live, and at high latitudes the weather is usually either too cloudy or too cold to encourage casual night sky-watching. A strong auroral display, changing in form and colour every few seconds and covering most of the sky, is a most beautiful and impressive sight. Several recent books [5, 6, 7] give splendid colour photographs but cannot capture the dynamic qualities. There are many poems about the aurora [5, 6] and I quote a few lines which concentrate on the rapid movements rather than the colours:

We watch the airy curtains flicker back and forth,
 See the sudden searchlights stab up and die,
 Column after ghostly column balanced in the sky.

Pale electric atom-streams shooting from the Sun
 Have felt the Earth's magnetic might
 And spiralled in to beautify the night.[8]

To return from poetry to science, the chief scientific parameters that describe the atmosphere are the temperature $T$, pressure $p$ and density $\varrho$, and they are connected by the well-known gas law,

$$\frac{p}{\varrho}=\frac{RT}{M} \tag{1}$$

where $R$ is the gas constant (8.31 $\mathrm{J\,K^{-1}mol^{-1}}$) and $M$ is the molecular weight of the gas. The decrease of pressure with height $y$ is given by the hydrostatic equation,

$$\frac{\mathrm{d}p}{\mathrm{d}y}=-\varrho g, \tag{2}$$

where $g$ is the acceleration due to gravity. Eliminating $\varrho$ between equations (1) and (2), we have

$$\frac{\mathrm{d}p}{p}=-\frac{Mg}{RT}\mathrm{d}y. \tag{3}$$

Thus, if we assume $RT/Mg$ is constant and denote it by $h$, which is called the 'scale height', equation (3) may be integrated to give the variation of $p$ with height as

$$p=p_0\exp\left(-\frac{y-y_0}{h}\right), \tag{4}$$

where $p_0$ is the pressure at a chosen reference level, $y = y_0$. If we further assume that the Earth is spherical of radius $r_E$, with an inverse-square gravity field, the variation of density $\varrho$ with height can be written

$$\varrho=\varrho_0\exp\left(-\frac{y-y_0}{H}\right), \tag{5}$$

where $H$, defined by the equation

$$\frac{1}{H}=\frac{1}{h}-\frac{2}{r_0}, \tag{6}$$

is called the 'density scale height'. In eqn (6), $r_0 = r_E + y_0$ and is the distance from the Earth's centre at height $y_0$. The assumption that $h$ is constant implies that $H$ is also constant; and since $h \ll r_0$, eqn (6) shows that the difference between $H$ and $h$ is never more than a few per cent. Equation (5) is not exact, but the error is negligible: the right side should be multiplied by a factor $(1 - (y - y_0)^2/r_0^2)$, which departs from 1 by less than $2.5 \times 10^{-4}$ if $(y - y_0) < 100$ km.

Equations (4) and (5) cease to apply when the height becomes so great that the mean free path of the air molecules exceeds $h$, and in practice the equations begin to lose accuracy at heights above about 500 km.

Although in reality $RT/Mg$ is not quite constant, eqns (4) and (5) provide powerful approximations, because we can usually divide the atmosphere into height bands thin enough to ensure that $RT/Mg$ is nearly constant. The equations tell us that if the temperature $T$ is higher, the pressure and density both fall off more slowly with height, because $h$ and $H$ are larger. The density falls off by a factor of 2.718 whenever the height increases by $H$, or by a factor of 20 for an increase of $3H$. For example, if $H = 33.3$ km and $y_0 = 200$ km, the density at $y = 300$ km would be $\varrho_0/20$ and at 400 km would be $\varrho_0/400$. If the temperature at heights above 200 km were to increase by a factor of 2, the value of $H$ would increase to 66.6 km and the density at 400 km would be $\varrho/20$. Thus if $\varrho_0$ stays nearly the same, doubling the temperature produces an increase in density at a height of 400 km by a factor of 20; and in general higher temperatures imply higher densities—usually much higher densities—for heights above 200 km.

At heights above about 100 km, diffusive equilibrium prevails, and equations (4) and (5) can be applied to the individual species of gases in the atmosphere. For a given temperature $T$, the gases of lower molecular weight $M$ have larger scale heights, because $h = RT/Mg$, and this confirms the intuitive belief that the lightest gases should rise to the highest reaches of the atmosphere:

> Where lighter gases, circumfused on high,
> Form the vast concave of exterior sky,

as Erasmus Darwin [9] expressed it in 1791.

As the temperature largely controls the rate at which pressure and density decrease, the variation of temperature with height [10, 11], shown in Fig. 1, is of crucial importance. The average temperature drops quite steadily from 290 K at sea-level to about 220 K at a height of about 10 km and then remains fairly constant in the stratosphere up to a height of about 25 km. Then the temperature rises to a maximum of about 280 K at a height of 50 km, because of the solar ultra-violet radiation being absorbed by ozone. Above that, the temperature falls again in the upper mesosphere to a mini-

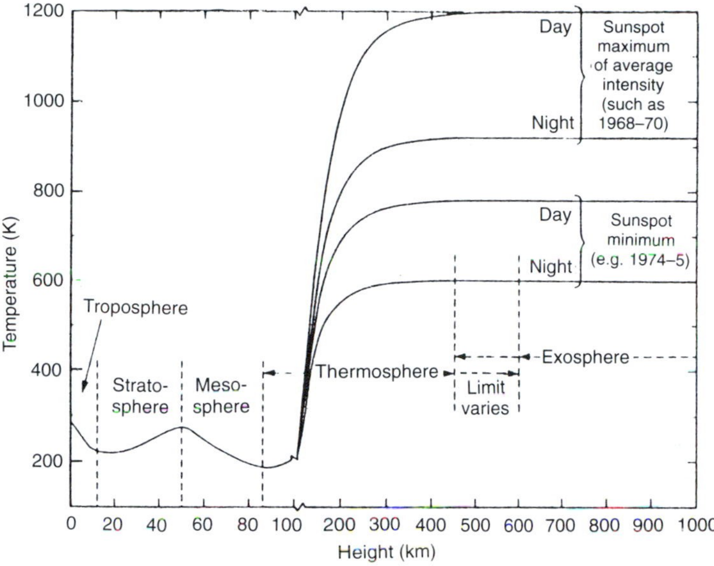


**Fig. 1** Average air temperature from sea-level to 1000 km, from *CIRA 1972*.

mum of about 180 K at a height of 85 km. Above 90 km, the temperature increases sharply as a result of the more extreme ultra-violet radiation from the Sun being absorbed, and this is the region known as the thermosphere, where at heights above 200 km the temperature becomes independent of height and has very high values. The temperature remains constant to heights above 500 km—indeed to as high aloft as the word 'temperature' remains meaningful. Above 500 km (for $T = 1000$ K) the mean free path of the atoms exceeds $h$, and many atoms pursue ballistic trajectories rather than continually colliding. Some escape, some collide, some fall back [4].

At heights above 200 km, the atmosphere is controlled mainly by the Sun, which has two quite different effects [10]. First, the thermospheric temperature is much higher during the afternoon than in the early hours of the morning, an effect not unfamiliar at ground level: the minimum temperature occurs at about 3 am and the maximum at about 3 pm (though the times vary with latitude and season). The variation is by a factor of about 1.3, so that if the minimum night-time temperature, $T_N$, is 1000 K, the temperature at 3 pm will be near 1300 K. The second effect is the control exercised by the Sun via its extreme ultra-violet radiation. This varies greatly during the 11-yr cycle of sunspot activity, and consequently the night-time temperature is very much lower in a year of sunspot minimum such as 1975 (when $T_N = 600$ K) than at sunspot maximum: in 1981, for example, $T_N = 1000$ K. These are the kinetic temperatures, but the air at such heights is so nearly a vacuum that the air temperature has no appreciable effect on a solid body like a satellite, the temperature of which is controlled by the radiant heat of the Sun, or its absence, and also by the reflectance of the satellite's surface.

As we have seen, the scale height depends on the molecular weight—in other words, on the atmospheric composition—as well as the temperature. There is mixing of the gases up to a height of about 100 km; above that the gases begin to sort themselves out according to their molecular weights [10]. Nitrogen remains dominant up to 170 km; then the main constituent is atomic oxygen, up to 500 km if the solar activity is low and the temperature is about 700 K. Helium takes over the leading rôle between about 500 and 900 km, and atomic hydrogen above that, as shown in Fig. 2. For a higher thermospheric temperature, of 900 K, atomic oxygen predominates up to 650 km and helium from there up to 1800 km.

A more tangible atmospheric parameter than the temperature is the density, which directly controls the air drag felt by the satellites passing through the upper atmosphere. Density can be determined by measuring the drag on orbiting satellites [12], and the results of thousands of such measurements are summarized in Fig. 3, which shows the variation of density with height for heights between 150 and 1000 km [10]. The density scale is logarithmic: it is conveniently centred on the value 1 nanogram per cubic metre (that is, 1 gram per cubic kilometre). The density decreases by a factor of 1 million between the right and left edges of the diagram.

As well as the million-fold variation with height, two other substantial vari-

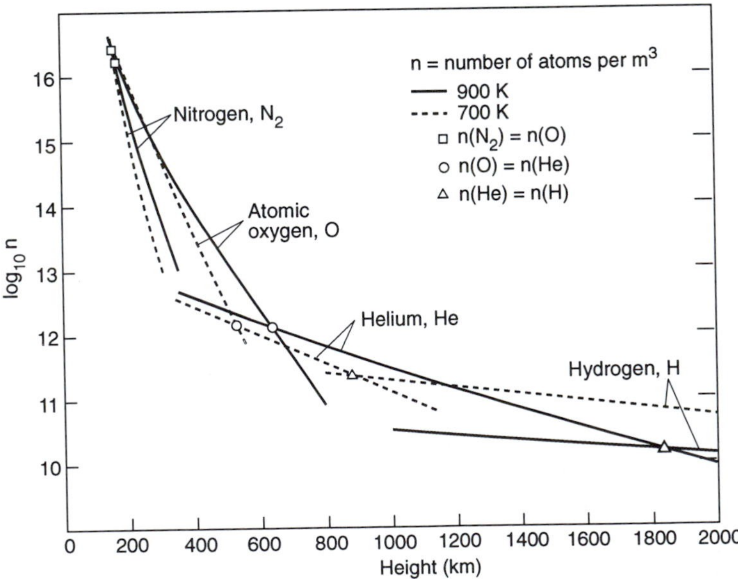


**Fig. 2** Concentrations of nitrogen, oxygen, helium and hydrogen, from *CIRA 1972*.

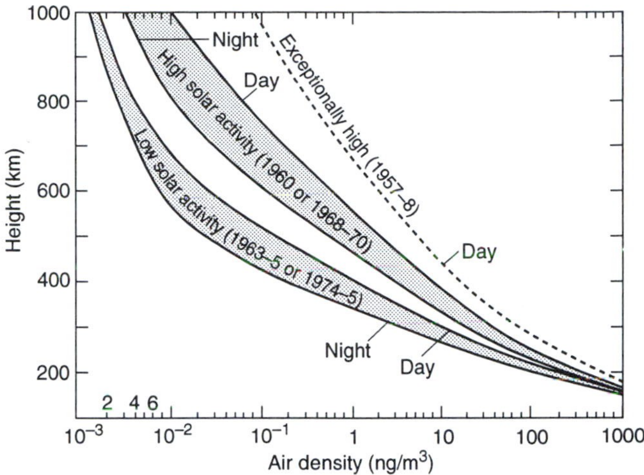


**Fig. 3** Variation of air density with height from 150 to 1000 km for high and low solar activity. Based on *CIRA 1972*.

ations are evident in Fig. 3. The first is a day-to-night variation in density, linked with the day to night temperature variation, though not exactly in phase with it. The density has a minimum at about 4 am and maximum at about 2 pm. This large variation occurs regularly each day, with the maximum density being about 5 times greater than the minimum at heights near 500 km. The factor of variation is smaller at lower heights, as Fig. 3 shows, but remains large at heights up to 1000 km when solar activity is high. The second major variation, even greater than the first, is that due to solar activity. This effect increases with height up to about 600 km, with the density at an average sunspot maximum (such as that of 1968–1970) exceeding the density at sunspot minimum by a factor of about 3 at 250 km height, a factor of about 8 at 400 km and about 20 at 600 km. However, these factors are even greater for a strong solar maximum—about 3.5, 12 and 60 respectively for 1957–1958, and about 3.2, 10 and 40 respectively for the 1981 maximum. Even larger factors arise when the two variations are combined: at a height of 600 km, the density by day at the end of 1957 was 250 times greater than in 1964 at night.

So there are immense regular variations in density. But there are also intense short-lived variations, and Fig. 4 shows a typical change in density in response to transient solar activity [13]. A solar storm occurred, disrupting the Earth's magnetic field, and the density at a height of 180 km (where density is relatively insensitive to solar activity) increased to nearly double in a few hours. At heights of 600 km the density can increase transiently by a factor of up to 8 in response to outbursts of solar activity [14].

In addition to these short-lived variations there is another important

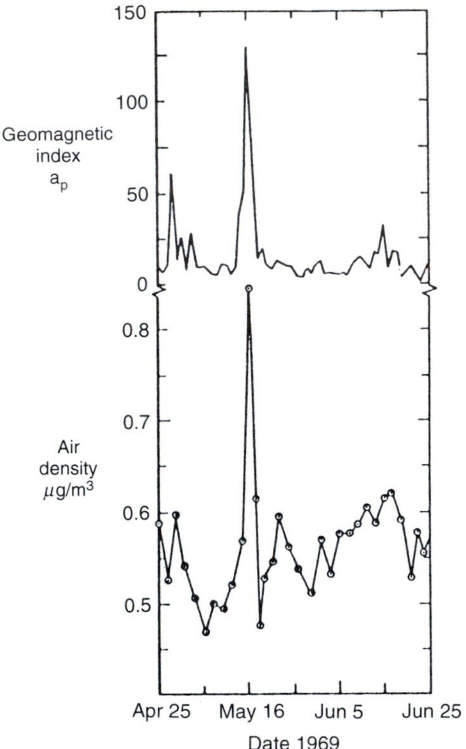


**Fig. 4** Increase in air density at 180 km height at the time of a magnetic storm [13].

longer-term irregular effect, the semi-annual variation. In a normal year the density has maxima during April and October, and minima during January and July: Figure 5 shows the variation of density in 1972, after correction for geomagnetic and day-to-night effects [15]. The density in October exceeded that in late July by a factor of about 1.5 in the example shown in Fig. 5, and this is at a height of 250 km. The factor increases to about 3 at heights near 500 km, and then decreases at greater heights [16]. The semi-annual variation probably originates from seasonal variations in the lower atmosphere, but full details have not yet been established. The strength and the timing of the oscillation vary appreciably from year to year [15, 16], but the example shown in Fig. 5 is fairly typical, and also illustrates the random-seeming variations of 5–10 per cent on time-scales of about a week, which characterize the behaviour of air density.

These are four important world-wide variations in upper-atmosphere density. I am ignoring many other effects that are confined to particular times, heights or localities, such as the 'winter helium bulge', propagating

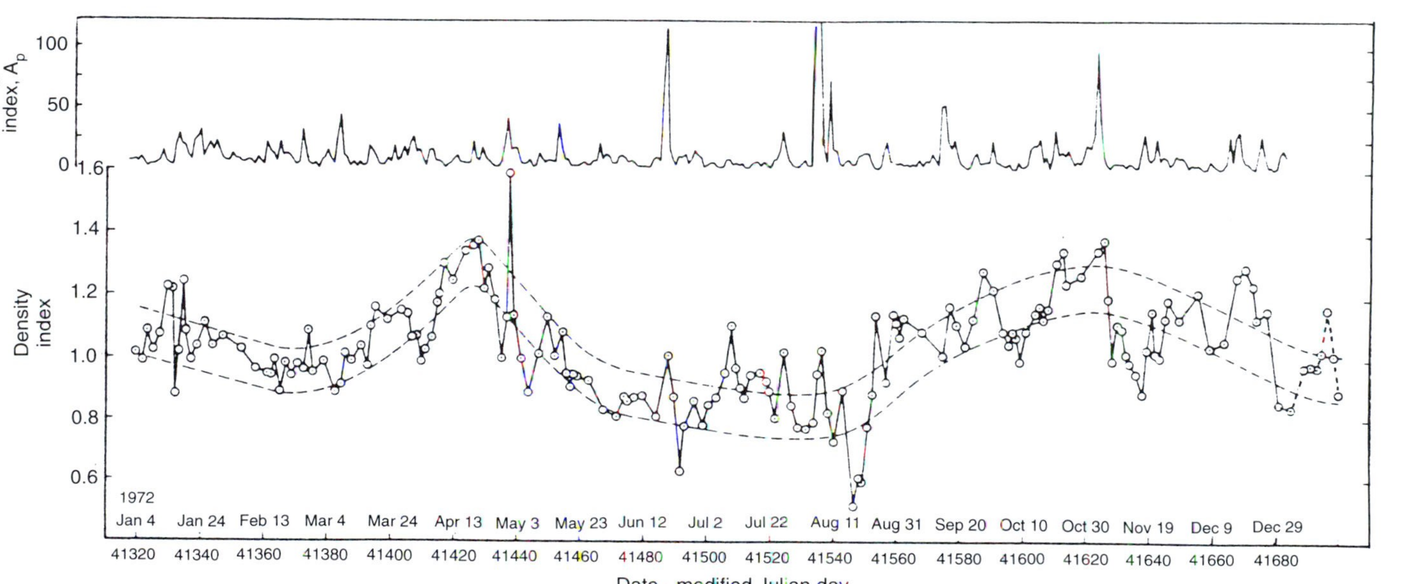


**Fig. 5** Variation of air density in 1972, at a height of 245 km, after removal of solar-activity and day-to-night effects, from analysis of the orbit of *Cosmos 462* [15].

gravity waves and dynamical effects in the auroral thermosphere. There are several recent atmospheric models [17–19] in which these effects are discussed or evaluated.

Most of the irregular variations in the upper atmosphere can be blamed on solar disturbances. To see why, we need to take a wider view [20]. The Sun is a hot ball of gas stuck in the midst of a vacuum, and an obvious response to this situation is that it might pour out its substance into interplanetary space. This is exactly what it is doing—just bleeding away really, though quite slowly, fortunately for us. Charged particles, mostly protons and electrons, stream out from the Sun at speeds of about 400 $\mathrm{km\,s^{-1}}$ when the Sun is quiet; but when the Sun suffers a convulsive eruption on its surface, the particles shoot out much faster in a plume. The solar wind, as this outpouring is called, sweeps across the Earth's orbit, and we are protected from its full effects by the outer regions of the atmosphere, the magnetosphere as it is called, which when looked at from the outside is like a huge tadpole-shaped cavity in the solar wind.

Figure 6 shows a rather outdated sketch of the magnetosphere which serves to indicate the main features. The magnetosphere acts as an obstacle in the strong outflow of particles of the solar wind, and a shock wave develops, quite like the shock wave arising when a spherical obstacle is placed in a supersonic wind tunnel—except that this shock wave at the boundary of the Earth's atmosphere is at ridiculously low pressures, and is a magneto-

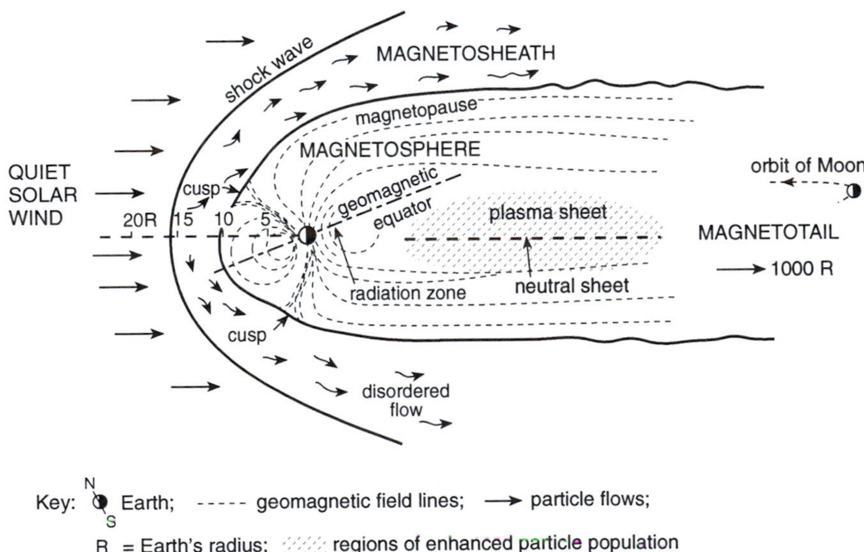


**Fig. 6** Sketch of the magnetosphere, reproduced from *Phil. Trans. A*, ***278***, 73 (1975).

hydrodynamic rather than an ordinary aerodynamic shock. When the Sun is quiet and the solar wind is blowing steadily, most of it flows round the boundary and we are protected from it. But when there is a solar flare or other strong disturbance, the higher-energy particles find their way into the magnetosphere. Some particles enter through the polar cusps, but most of them make their way in via the tail of the magnetosphere. From either point of entry they tend to follow the lines of magnetic force and therefore impinge on the Earth's atmosphere mainly at latitudes about 20–30° away from the magnetic poles, thus producing either visible aurorae or invisible but quite vigorous disturbances in the polar thermosphere. What I have said is of course a great over-simplification: magnetospheric physics is now virtually a branch of science of its own, and the 'meteorology' of the magnetosphere is an extremely complex subject [21, 22].

Any attempt to give a picture of the atmosphere is likely to leave the impression that it is static, like a picture: that is a false impression. The magnetosphere is not only rather like a tadpole in shape, but also wriggles vigorously, though more slowly than a tadpole, in response to the continual fluctuations in the flow of particles from the Sun. The air at lower levels in the atmosphere is also restless and dynamic: there are winds of more than hurricane force, driven by the great variations in pressure and density which I have already described. In the thermosphere the wind speeds can exceed 500 $\mathrm{m\,s^{-1}}$, especially in the auroral zone during disruptions due to solar storms [23]. And we find regular daily variations in the winds at lower latitudes by up to 200 $\mathrm{m\,s^{-1}}$: for example [24], at heights near 300 km, there are west-to-east winds of up to 150 $\mathrm{m\,s^{-1}}$: for example [24], at heights near 300 km, there are west-to-east winds of up to 150 $\mathrm{m\,s^{-1}}$ in the evening, and east-to-west winds (though not so strong) in the morning.

That concludes my quick survey of modern ideas on the atmosphere: as well as being superficial, the survey is also biased because it is designed so as to throw light on the historical topics ahead.

## The influence of Aristotle

Now I jump back more than 2000 years to look at some of the old ideas of the atmosphere. First, a word of apology: I shall limit myself to European culture. It would be interesting to make comparisons with early ideas in other cultures, but that would lead to far more material than would fit into a single lecture.

I begin with Aristotle, whose ideas had so much influence over this University during its first 400 years. [The lecture was given at the University of Oxford.] Aristotle's book called *Meteorologica*, written about 350 BC,

Περὶ μὲν οὖν τῶν πρώτων αἰτίων τῆς φύσεως καὶ περὶ πάσης κινήσεως φυσικῆς, ἔτι δὲ περὶ τῶν κατὰ τὴν ἄνω φορὰν διακεκοσμημένων ἄστρων καὶ περὶ τῶν στοιχείων τῶν σωματικῶν, πόσα τε καὶ ποῖα, καὶ τῆς εἰς ἄλληλα μεταβολῆς, καὶ περὶ γενέσεως καὶ φθορᾶς τῆς κοινῆς εἴρηται πρότερον. λοιπὸν δ' ἐστὶ μέρος τῆς μεθόδου ταύτης ἔτι θεωρητέον, ὃ πάντες οἱ πρότεροι μετεωρολογίαν ἐκάλουν· ταῦτα δ' ἐστὶν ὅσα συμβαίνει κατὰ φύσιν μέν, ἀτακτοτέραν μέντοι τῆς τοῦ πρώτου στοιχείου τῶν σωμάτων, περὶ τὸν γειτνιῶντα μάλιστα τόπον τῇ φορᾷ τῇ τῶν ἄστρων, οἷον περί τε γάλακτος καὶ κομητῶν καὶ τῶν ἐκπυρουμένων καὶ κινουμένων

**Fig. 7** The opening of Aristotle's *Meteorologica* in the Loeb edition.

dominated western European thought about the atmosphere until AD 1600; so we ought to stop and look at it. Your first thought may be to blame Aristotle for the disgustingly long word 'meteorological', that heptasyllabic millstone round the neck of its practitioners. Strangely enough, however, Aristotle does not seem to have been responsible for coining the word. Fig. 7 shows the first chapter of his book [25], and in the passage underlined he says he will discuss the subject 'which all our predecessors have called meteorologia'. He also says that the word covers 'everything that happens naturally', so that he deals with the entire realm of geophysics, all that is earthly rather than heavenly.

Aristotle's picture of the Earth and its atmosphere is in terms of the four spheres of Earth, Water, Air and Fire (Fig. 8). Classical scholars are often scornful about this picture. Indeed the translator of the *Meteorologica*, H. D. P. Lee, remarks in his preface that the book is little read because 'Aristotle is so far wrong in nearly all his conclusions that they can . . . have little more than a passing antiquarian interest'.

But is Aristotle wrong, or is he making a good first approximation? Aristotle is certainly correct in taking the solid Earth as the central sphere. He is also right to suggest that this is nearly covered by a sphere of Water, the hydrosphere as it is often called today. Though we now know that the oceans cover nearly 75 per cent of the Earth to an average depth of 5 km, Aristotle himself lived in a region where land was dominant: so he did well to avoid being misled. Above the water comes the sphere of Air: no one will quarrel with that. The air at the lower levels—the air we breathe—is usually quite cool and often humid. But, as we have already seen, the upper regions of the atmosphere, above 200 km, have dynamic temperatures ranging between 600 and 1500 K, much higher than any domestic oven; we call this the thermosphere, so why should be blame Aristotle for calling it the sphere of Fire? Beyond that in Aristotle's picture is the celestial region, which is of

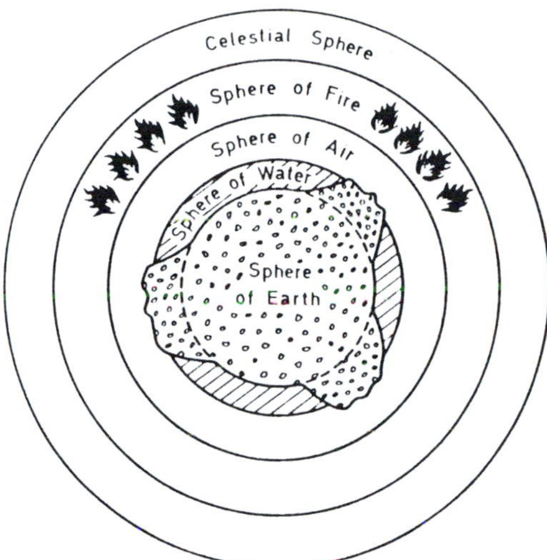


**Fig. 8** Aristotle's view of the Earth and its atmosphere.

course divided into further spheres belonging to the Sun, Moon and planets. However, if we take the outermost sphere as that of the Sun, the sphere of Fire melts easily into it and the boundary of the magnetosphere. Aristotle says that 'the celestial region as far down as the Moon is occupied by a substance which is different from air and fire, but which . . . is not uniform in quality'. By a slight stretch of the imagination we can identify that as the solar wind—and this is not really stretching the interpretation too far, because Aristotle does regard the sphere of Fire as being linked with the Sun's heat.

I must confess that I am hostile to Aristotelian physics in general, so it is rather unnerving for me to have to declare, misquoting Mark Antony,

> Friends, Oxonians, and countrypeople,
> I come to praise the Stagirite, not to bury him.

My 'conversion', even though strictly limited in scope, would no doubt have pleased the medieval scholars of this University.

Aristotle believed that the heat of the Sun drew up two sorts of 'exhalations' from the Earth, a hot dry exhalation, leading to thunder and lightning, shooting stars and the aurora; and, secondly, warm moist vapours, which cool and turn into clouds and rain and other 'watery meteors', as they were called. The aurora he regarded as having its home in the sphere of Fire, which is essentially correct if we equate that with the thermosphere. His division of the celestial sphere into shells housing the planets, Sun, Moon, etc., is shown in Fig. 9.

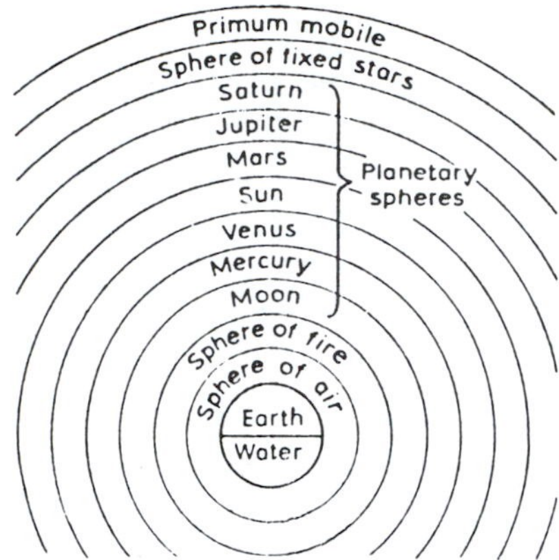


**Fig. 9** One version of Aristotle's model of the universe.

For my next dip into the past, I visit the medieval and Renaissance scholars, not only at Oxford but at any other of the seats of learning in Europe. And we find that Aristotle still rules: more than 125 editions of the *Meteorologica* were printed before 1600, and Aristotle's model is illustrated again and again [26]. Figure 10 shows a version from a book [27] published in Paris in 1551. It is just the same really, except that the captions are in French.

But there was one refinement of Aristotle's model which seems to have gained general approval in the sixteenth century. The sphere of Air was

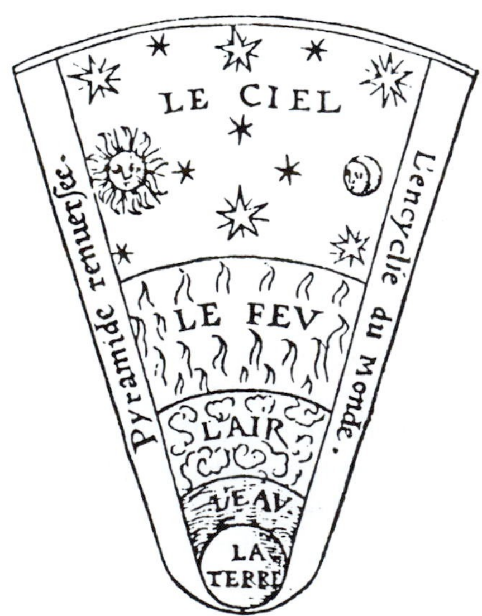


**Fig. 10** The spheres of Earth, Water, Air and Fire. After C. de Bouelles, *Géométrie practique* (1551).

divided into three regions. The lowest was regarded as being heated by the Earth and was usually shown as cloudless. The second layer was colder, and this was where the clouds formed. The third level, the 'Suprema regio aeris', was heated by the Sun and by its proximity to the sphere of Fire, and so it was free of watery vapours. This cloudless upper region would correspond in our terminology to the stratosphere and above, and would merge into the thermosphere or sphere of Fire at its upper boundary. Figure 11 shows this tripartite atmosphere, as illustrated in a book by Finé [28] published in 1532. To us, Fig. 11 seems unconvincing, because rain so often falls from the lowest layer. But there it is: they liked it. Many similar diagrams were published and have been collected by Heninger [29].

Figure 12 shows another picture of the three-layer atmosphere, which appears [30] in the later editions of Reisch's *Margarita philosophica* (first published in 1496). This diagram differs from Fig. 11 by showing the Sun as actively in command and 'breathing' on the outermost regions of the atmosphere. The sphere of Fire is now no longer a complete sphere, but is very much under solar control, and has become roughly equivalent to the solar wind, with the 'Suprema regio aeris' now corresponding to the entire thermosphere. The idea that the Sun can be regarded as 'breathing out' the solar wind is quite appealing: but the metaphor fails because the Sun never breathes in; so the anthropomorphized solar wind has to be the bleeding Sun, rather than the breathing Sun. Nevertheless, Fig. 12 is arrestingly

**Fig. 11** The three-fold atmosphere. After O. Finé, *Protomathesis* (1532).

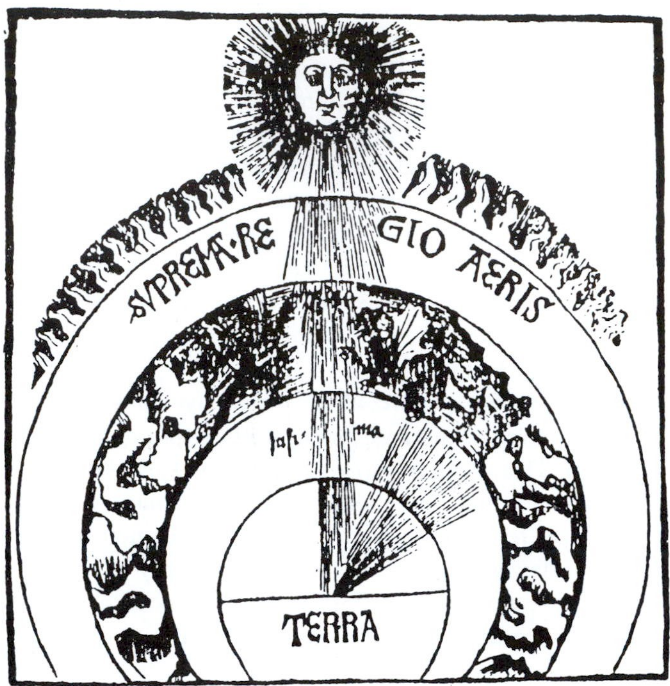


**Fig. 12** The Sun in command of the atmosphere. After G. Reisch, *Margarita Philosophica nova* (1512).

similar to some 'cartoon versions' of modern concepts of the Earth in the solar wind.

The atmosphere continued to be regarded as 'the triple region of the air' throughout the sixteenth century, a view that is often reflected in the imagery of the Elizabethan poets. Marlowe's *Tamburlaine the Great* is a good source, because Tamburlaine himself is for ever boasting about his prowess, and challenging the powers of the air, as when he treads on the defeated emperor Bajazeth:

> Now clear the triple region of the air,
> And let the majesty of heaven behold
> Their Scourge and Terror tread on Emperors.

And he seeks to rival even the most violent natural phenomena,

> As when a fiery exhalation
> Wrapt in the bowels of a freezing cloud,
> Fighting for passage, makes the Welkin crack,
> And casts a flash of lightning to the earth . . .
> So shall our swords, our lances and our shot,
> Fill all the air with fiery meteors.[31]

It is also worth remembering that earthquakes were regarded as largely meteorological, caused not by Earth movement but by air trying to escape from inside the Earth. A good example is provided by Shakespeare's lines:

> As when the wind, imprison'd in the ground,
> Struggling for passage, earth's foundation shakes,
> Which with cold terror doth men's minds confound.[32]

The imagery of the Elizabethan poets can fairly be called pre-scientific, although not necessarily wrong, of course. The enduring reality of their faith in Aristotle is nicely summarized by a slightly later poet, Cowley:

> Welcome, great *Stagirite*, and teach me now
> All I was born to know.[33]

This verse serves as a fitting finale to the Aristotelian epoch, for it was published in 1656, just as new ideas were sweeping over the dreaming spires of the Aristotelian stronghold, where today the poet's name, converted into a symbol of technology, is better known than the Stagirite's.

## The Invisible College, the Royal Society and Halley

The new scientific attitude to Nature, typified by the 'Invisible College' at Oxford in the 1650s and then by the Royal Society in the 1660s, bore its earliest fruits in the field of atmospheric physics. Robert Boyle came to Oxford in 1654, and recruited Robert Hooke to help him in 1655. The experiments made in the late 1650s in the house on the High Street—a site now occupied by the memorial to the aerial poet Shelley—led to the formulation of what is today known as Boyle's Law, which is a restricted version of equation (1), with $T$ and $M$ constant.

This was one decisive step towards a better understanding of the atmosphere. Another step, equally important, was the new emphasis on measurement of temperature, pressure, winds and rainfall, using newly-designed instruments [34]. Many such weather records began to be kept—enough to allow a detailed picture of the weather in central England from 1659 to the present day [35].

It was Edmond Halley (Plate II) who applied Boyle's Law to the atmosphere and determined the variation of air pressure with height, in a remarkable paper in the *Philosophical Transactions of the Royal Society* in 1686, entitled 'A Discourse of the Rule of the Decrease of the Height of the Mercury in the Barometer, according as Places are elevated above the Surface of the Earth' [36]. Although Halley conducts his annlysis with verbal argument, his procedure is equivalent to using equation (1) with $T$ constant, in conjunction with the hydrostatic equation (2). He then deduces that the decrease of

From thefe Rules I derived the following Tables.

| *A Table ſhewing the Altitude to given Heights of the* Mercury. | | *A Table ſhewing the Heights of the* Mercury *at given Altitudes.* | |
|---|---|---|---|
| *Inch.* | *Feet.* | *Feet.* | *Inch.* |
| 30 | 0 | 0 | 30, 00 |
| 29 | 915 | 1000 | 28, 91 |
| 28 | 1862 | 2000 | 27, 86 |
| 27 | 2844 | 3000 | 26, 85 |
| 26 | 3863 | 4000 | 25, 87 |
| 25 | 4922 | 5000 *feet* | 24, 93 |
| 20 | 10947 | 1 *mile* | 24, 67 |
| 15 | 18715 | 2 | 20, 29 |
| 10 | 29662 | 3 | 16, 68 |
| 5 | 48378 | 4 | 13, 72 |
| 1 | 91831 | 5 | 11, 28 |
| 0, 5 | 110547 | 10 | 4, 24 |
| 0, 25 | 129262 | 15 | 1, 60 |
| 0, 1 | 29 *mil. or* 154000 | 20 | 0, ~~95~~59 |
| 0, 01 | 41 *mil.* 216169 | 25 | 0, 23 |
| 0, 001 | 53 *mil.* 278338 | 30 | 0, 08 |
| | | 40 | 0, 012 |

**Fig. 13** Halley's tables of the decrease of atmospheric pressure with height (1686), with a misprint corrected.

pressure with height is exponential, as in equation (4). The table he gives, Fig. 13, goes up to 53 miles height, and when you compare his values with those in recent models, such as the *US Standard Atmosphere, 1976*[11], you find that he did very well indeed, as shown in Fig. 14. Although unaware of it, Halley was assuming that the temperature remains constant as the height increases. This assumption, though incorrect, is not too far from the truth up to about 120 km height, because the average temperature in that region is only about 10 per cent less than the sea-level temperature, as Fig. 1 shows. At heights above 120 km the temperature increases greatly; consequently the pressure in the real atmosphere comes nearer to Halley's model at heights above 120 km, and eventually the two curves cross over. If the standard thermospheric temperature is taken [11] as 1000 K, it turns out that Halley's model gives the correct pressure at a height of 160 km, where the pressure is 3 nanobars [11].

That takes us up to and above the heights of most auroral displays, and on this subject too Halley made a crucial contribution. He watched the great aurora of 1716 for several hours and wrote a long paper about it [37]. In this paper he not only gives a vivid description; he also deduces that the aurora is under the control of the Earth's magnetic field and he suggests measuring its height by triangulation. Halley also hinted that the actual 'luminous effluvia' of the aurora might be electrical in nature, and this idea became accepted

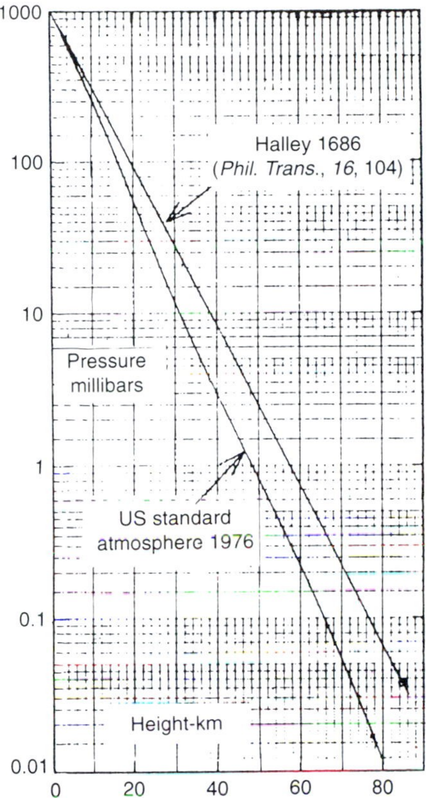


**Fig. 14** The decrease of atmospheric pressure with height, as given by Halley (1686) and by a modern standard.

during the eighteenth century. The 'electricians' of the eighteenth century were quite adept at producing glowing lights in what we would now call vacuum tubes, and the analogy with the aurora was recognized and accepted. In his treatise on the aurora [38] published in Paris in 1733, De Mairan went even further towards the modern view by suggesting that the aurora was an extension of the Sun's atmosphere. He also advocated measuring its height by triangulation, and a number of such height measurements by Bergman in the 1760s gave heights between 380 and 1300 km [6].

So, by 1790 the aurora was generally believed to be a glowing electrical discharge in the high atmosphere. As Erasmus Darwin put it in his poem, *The Botanic Garden*, the aurora

> Darts from the north on pale electric streams,
> Fringing Night's sable robe with transient beams. [39]

## Erasmus Darwin

That brings me to the subject of my next dip into ideas about the atmosphere, Erasmus Darwin (Plate III). In his classic paper [40] published in 1788, Fig. 15, Darwin establishes the principle of adiabatic expansion and explains the main mode of formation of clouds. First he shows by numerous experiments—with airguns, with the high-pressure air in the waterworks at Derby, and various other exotic examples— how air cools when allowed to expand from higher pressure to lower, as for example in air being let out of a car tyre. He then applies this principle to explain what he calls the 'devaporation of aerial moisture'. (*Devaporation* is a word coined by Darwin, meaning 'condense into droplets'—the opposite of evaporation. It would be preferable to our ambiguous modern word *condense*, as Knowles Middleton has remarked [41].)

'When large districts of air from the lower parts of the atmosphere are raised two or three miles high, they become so much expanded by the great diminution of the pressure over them', Darwin says, that the air 'robs the vapour which it contains of its heat, whence that vapour becomes condensed and is precipitated in showers'. (Darwin's use of the verb *precipitated* in this

IV. Frigorific Experiments on the mechanical expansion of Air, *explaining the Cauſe of the great Degree of Cold on the Summits of high Mountains, the ſudden Condenſation of aerial Vapour, and of the perpetual Mutability of atmoſpheric Heat. By* Eraſmus Darwin, *M. D. F. R. S.*; *communicated by the Right Honourable* Charles Greville, *F. R. S.*

Read December 13, 1787.

HAVING often revolved in my mind the great degree of cold producible by the well known experiments on evaporation; in which, by the expanſion of a few drops of ether into vapour, a thermometer may be ſunk much below the freezing point; and recollecting at the ſame time the great quantity of heat which is neceſſary to evaporate or convert into ſteam a few ounces of boiling water; I was led to ſuſpect, that elaſtic fluids, when they were mechanically expanded, would attract or abſorb heat from the bodies in their vicinity; and that, when they were mechanically condenſed, the fluid matter of heat would be preſſed out of them, and diffuſed among the adjacent bodies.

**Fig. 15** From the title page of Erasmus Darwin's paper explaining the formation of clouds by adiabatic expansion of moist rising air (*Phil. Trans.*, **78**, 43 (1788)).

sense is 75 years earlier than the first example given in the *Oxford English Dictionary*.) This insight on cloud formation is quite fundamental, of course, and Dalton built on it a few years later with his law of partial pressures and other ideas [42]. Darwin made several further contributions to meteorology, such as recognizing the existence and importance of what we now call cold and warm fronts [43].

Darwin also provided a picture of the complete atmosphere. That can be found in the notes to his poem *The Botanic Garden* (1791), which is a review of scientific knowledge in the Earth sciences, as well as a poem which gave him the highest reputation in the literary world at the time [44]. His description of the atmosphere, which I have converted into a diagram, Fig. 16, takes ideas from many predecessors [45] and adds several of his own. The result is quite a realistic three-layer model. Darwin's first stratum, with clouds and lightning, corresponds closely with what we call the troposphere, though Darwin's height of 6 km (actually he gives 4 miles) is much lower than the modern figure of about 10 km. However, his second and cloudless region does correspond closely to our stratosphere and lower mesosphere. This region ends, he says, 'where the air is 3000 times rarer than at the surface of the Earth' and his estimate of density, which closely follows Halley's, gives this as a height of 60 km. Above that, he says, 'The common air ends, and is surrounded by an atmosphere of inflammable gas [hydrogen] tenfold rarer than itself. In this region I believe fireballs sometimes to pass, and at other times the northern lights to exist' [46]. He was correct in thinking that fireballs appeared in this region, and he quotes the height of the great fireball meteor of August 1783, which he himself observed, as 'between 60 and 70 miles' which is nearly right, and its speed as 'about 20 miles in a second'. But

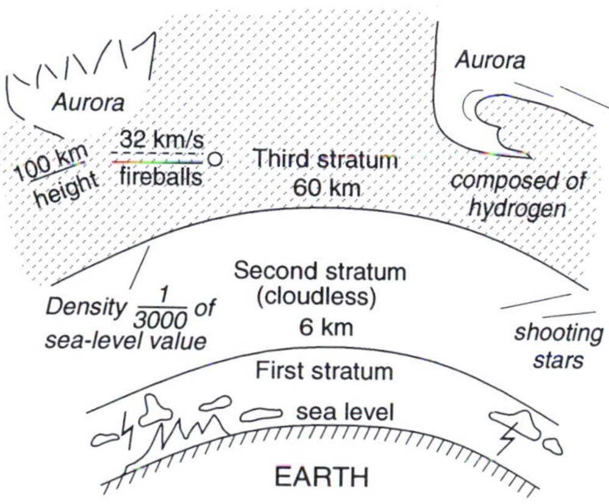


**Fig. 16** Erasmus Darwin's three-layer model of the atmosphere, as specified in *The Botanic Garden* (1791).

he wrongly thought that the smaller shooting stars were lower. He also placed the aurora in the correct region, as established some years later in measurements by Dalton [6].

An interesting feature of Darwin's picture is his belief that the outermost atmosphere is formed of hydrogen, the lightest gas. He reached this conclusion from the simple idea that the lighter gases will rise to greater heights, although his height for the base of the 'hydrogen exosphere' is far too low—it should be at least 600 km. He also thought the presence of hydrogen was confirmed by the red colouring often seen in the auroras. 'It was observed by Dr Priestley', he says, 'that the electric shock taken through inflammable air was red' [46]. This was good thinking, but actually hydrogen only gives a very small proportion of the red colour in the aurora, because most auroral displays are at heights near 100 km, where there are only traces of hydrogen.

Darwin's picture was in many ways better than anything for 100 years. For example, later ideas about hydrogen were very confused and contradictory, as we shall see shortly. As another example, 90 years later, in the first International Polar Year of 1882, the stations for observing the aurora were set up too close together, because it was assumed that the aurora was about 8 km high [47]. As the actual heights are more than 80 km, these stations were wrongly situated for triangulation and the measurements were much less accurate than they would have been if Darwin's (or Dalton's) heights had been adopted.

## Jeans, Chapman and Milne

I shall now exercise my right to skip, by skipping right over the nineteenth century and coming down early in the twentieth for my last snapshots of atmospheric ideas.

I begin with *The dynamical theory of gases* by Sir James Jeans [48], first published in 1904, with a second edition in 1916, and my quotations are from the fourth edition (1925). It is a classic text, and he shows quite clearly how in an isothermal atmosphere 'the heavier gases tend to sink . . . while the lighter ones rise to the top', because their decrease of density with height is slower. His theory is excellent and, regarded as a textbook, his work is superb. But when he ventures into the real world, he does not do so well. His table of densities in the atmosphere at a height of 170 km—near the height where Halley was correct—has errors so huge as almost to defy belief, as Table 1 shows. Certainly Jeans was extremely unlucky; even if he had picked four numbers at random he would probably have done better.

Why was he so much in error? The first reason is his assumption about upper-atmosphere temperature. He says: 'We shall obtain a fair approxima-

**Table 1** Number densities (per cubic centimetre at 170 km height, from Jeans's *Dynamical theory of gases* and the *US Standard Atmosphere, 1976*

| Component: | Nitrogen | Oxygen | Helium | Hydrogen | |
|---|---|---|---|---|---|
| Jeans's value | 350 | 3 | 1 300 000 | 182 000 000 | all × $10^6$ |
| US Standard Atmosphere: | 10 700 | 9800 | 17 | ≃0.2 | all × $10^6$ |
| Approximate error factor: | 30 | 3000 | 80 000 | $10^9$ | |

tion to average conditions by assuming that the temperature [at] a height of 10½ km . . . is −54 °C (219° absolute), and that beyond this the atmosphere is in isothermal equilibrium'.

So he assumed that the temperature was constant at 219 K, which is not nearly so good as Halley's tacit assumption that the temperature was the same as at sea-level. Because of the low temperature, Jeans had much too small a scale height, and this explains why his value for the nitrogen concentration is too low by a factor of 30. The same error arises with oxygen, but the discrepancy is much greater because Jeans naturally assumed that the oxygen was diatomic, whereas in reality it is largely monatomic by 170 km height, although he could not have known this. Jeans's very high values for the concentrations of helium and hydrogen arise purely from taking much too large a sea-level value for both. In Jeans's atmosphere, hydrogen is dominant at heights above 70 km, which, like Erasmus Darwin's 60 km, is far too low. (However, it should be remembered that the concentration of hydrogen at 170 km height is not securely established and is subject to many variations: see Chapter 16 of the book by Banks and Kockarts [4].)

Jeans, although unlucky in applying his ideas to the real atmosphere, was one of the greatest applied mathematicians of his day, and he pioneered the theory of the escape of planetary atmospheres. He appreciated that at great heights the mean free path of the molecules becomes great enough for some of them to go into orbit or escape. And he developed a self-consistent theory for the escape rates of various species. He concluded that the escape rate even of hydrogen would be very low: at his assumed temperature of 219 K, hydrogen would take $10^{24}$ yr to escape. But he does comment that if the temperature were 550 K, the escape rate would be $10^{18}$ times faster; so with a temperature of 1000 K all the hydrogen would rapidly vanish. It was not long before Jean's assumptions about hydrogen were challenged, and unfortunately this led to the wrong idea that there was no hydrogen at all in the upper atmosphere. What was wanted—but was not forthcoming—was a compromise between the extremes of zero and Jeans.

My last example is the important paper [1] by Sydney Chapman and E. A. Milne about 'The atmosphere at great heights', published in the *Quarterly Journal of the Royal Meteorological Society* in 1920. The theory in this paper

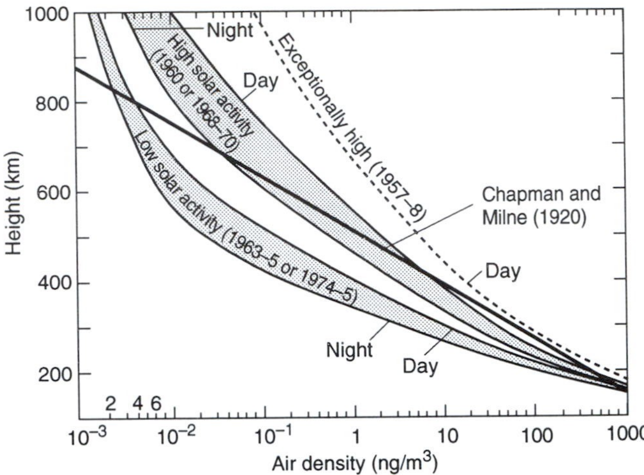


**Fig. 17** Variation of air density with height as given by Fig. 3 and by Chapman and Milne (1920) in their '$H_1$ = 30 km' model.

is excellent, and in applying the theory to the real atmosphere Chapman and Milne insured themselves against error by giving results for four possible values of the height above which there is diffusive equilibrium—and below which there is mixing and constant composition. The four possible values give densities at 200 km and above which differ by a factor of up to 100, so it is possible to choose the best of the values. Since this is the Milne lecture, I will be indulgent and do just that, choosing the value 30 km (although the actual height where diffusive equilibrium begins is near 100 km). The Chapman–Milne curve of density versus height then goes right through the middle of the modern diagram, as shown in Fig. 17, which is merely Fig. 3 with an additional line. The sight of this may well provoke applause for Chapman and Milne from all sides.

But their correct result, far from being a stroke of genius, was only luck, because they made four serious errors.

First, they assumed there was no hydrogen at all. Their comments are:

> Hydrogen is not indicated by the auroral spectrum, though this alone does not prove its absence . . . The case when hydrogen is present has been sufficiently discussed by Jeans, and accordingly for the purposes of the greater part of this paper the absence of hydrogen will be assumed.

The absence of hydrogen explains the low density at heights near 1000 km in their model, but it is only a minor constituent below 500 km, so at lower altitudes their assumption of zero hydrogen was much better than Jeans's assumption of about $10^9$ times too much hydrogen.

Their second error was the wrong choice for the height where diffusive equilibrium begins. Despite their 'insurance policy' of giving results for four different possible heights (12, 20, 30 and 50 km), all four were much lower than the 100 km which would now be recommended.

Their third error was to follow Jeans in assuming a constant temperature of 219 K, when we now know that the temperature ranges between 600 and 1500 K in the thermosphere, with an average of about 900 K.

Their fourth error was to take the concentration of helium in the stratosphere too high by a factor of about $10^4$. As a result of this error, helium, with molecular weight 4, becomes dominant in their atmosphere at heights of about 150 km, whereas it is only a minor constituent up to 400 km in the real atmosphere.

It may seem surprising that after all these errors Chapman and Milne emerge with reasonable values for air density and its variation with height for heights between 200 and 800 km, as Fig. 17 shows. Their luck came in because they arrived at nearly correct values for the scale height through making two errors, each by a factor of over 4, which happened to cancel out. At a height of 300 km they have an atmosphere with a molecular weight of 4.0 (almost entirely helium) and a temperature of 219 K. Thus they have $T/M = 55$, and a scale height $h = RT/Mg = 51$ km. At 300 km height in the *US Standard Atmosphere, 1976* the molecular weight is 17.7 and the temperature 976 K, which also gives $T/M = 55$ and $h = 51$ km. Their good luck is perhaps even more improbable than Jeans's bad luck.

Milne returned to the subject in 1923 in his paper on the escape of molecules from an atmosphere [2]. Though he again used the low temperature of 219 K, the heights he calculated for the level where molecules can escape—the base of the exosphere as we should now call it—are reasonably realistic: 630 km for helium and 1400 km for hydrogen. The latter is too high, but chiefly because he takes Jeans's high value for the hydrogen concentration.

## Retrospect

That completes my set of snapshots of past ideas of the atmosphere. It is up to you to conclude what you will from what I have said: I can only give my own views. I think the history of science sharpens our perceptions of the present in a revealing way, by making us see that presently-fashionable views about matters on the frontiers of knowledge in a particular subject may be partly or completely wrong—even when they have the highest authority, as with the results of Jeans, Chapman and Milne which I have mentioned. We also have to face the hard fact that old ideas may sometimes prove to have more truth than more modern ones. When I was an undergraduate, it was widely believed [49] that 'captured interstellar gas' was falling into the Sun at high speed—a concept now completely reversed in the *out*flowing solar wind. Also it should be remembered that 'the general reader' is not much interested in well-established science, so that writers of popular-science books need a continual flow of imaginative new concepts, particularly in astronomy, to provide material for their new books. Many imaginative astronomers gladly create these concepts, which are then popularized by the popularizers and earn credit for the creators. The history of science is like an X-ray that cuts through meretricious trappings and lets us see the bare bones of the modern ideas, which may look much less impressive if perceived as 500-year-old bones re-dressed.

The history of science teaches us to be humble, and also to be very sceptical of theories currently in fashion on subjects that are still in a state of flux. I would not wish to be accused of undermining all science: there are many areas that are securely established, and others where errors may be rampant. For example, no one is going to prove that the Earth is flat; its sea-level shape is now known correct to 2 m all round. But current ideas on the composition and character of the Earth's deep interior, which no one has yet seen, might be overturned completely in the next 30 years, just as the ideas of Jeans, Chapman and Milne on the outer atmosphere were overturned within 30 years. The same scepticism may be needed in some areas of astronomy.

Science can be defined as the thought-system favoured by the majority of current scientists. If you think differently, you are an independent thinker, not a scientist, and papers that you write will probably be rejected by scientific journals, which have a censorship system euphemistically called 'refereeing'. As in football, the referees are not necessarily good players, but they do know the rules. Having said that against the system, I must also say that most independent thinkers are wrong and some are quite nutty. But some are right, and they provide the framework for the orthodoxy of the next generation, as for example with Wegener and continental drift, once heretical but now orthodox. After these subversive thoughts, it is only fair to end by undermining myself with the warning, 'Don't believe all that your lecturers tell you.'

## References

[1] Chapman, S. and Milne, E. A., 1920. The composition, ionization and viscosity of the atmosphere at great heights, *Q. Jl R. met. Soc.*, **46,** 357–396.

[2] Milne, E. A., 1923. The escape of molecules from an atmosphere, with special reference to the boundary of a gaseous star, *Trans. Camb. phil. Soc.*, **22,** 483–517.

[3] Ogden, R. J., 1983. *Weather*, **38,** 250–251.

[4] Banks, P. M. and Kockarts, G., 1973. *Aeronomy*, two volumes. Academic Press, New York.
[5] Eather, R. H., 1980. *Majestic lights*, American Geophysical Union, Washington, DC.
[6] Brekke, A. and Egeland, A., 1983. *The northern light*, Springer-Verlag, Berlin.
[7] Meinel, A. and Meinel, M., 1983. *Sunsets, twilights and evening skies*, chapter 11, Cambridge University Press.
[8] King-Hele, D. G., 1972. *Poems and trixies*, Mitre Press, London.
[9] Darwin, E., 1791. *The Botanic Garden*, Part 1, Canto 1, lines 123–124, Johnson, London.
[10] *CIRA 1972 (COSPAR International Reference Atmosphere 1972)*, Akademie Verlag, Berlin.
[11] *US Standard Atmosphere, 1976*. US Government Printing Office, Washington, DC.
[12] King-Hele, D. G., 1987. *Satellite orbits in an atmosphere: theory and applications*. Blackie, Glasgow.
[13] King-Hele, D. G. and Walker, D. M. C., 1971. Air density at heights near 180 km in 1968 and 1969, from the orbit of 1967–31A, *Planet. Space Sci.*, **19,** 297–311.
[14] Jacchia, L. G., 1961. Satellite drag during the events of November 1960, *Space Research II*, pp. 747–750, North-Holland, Amsterdam.
[15] Walker, D. M. C., 1978. Variations in air density from January 1972 to April 1975 at heights near 200 km, *Planet. Space Sci.*, **26,** 291–309.
[16] Cook, G. E., 1969. The semi-annual variation in the upper atmosphere: a review, *Annls Géophys.*, **25,** 451–469.
[17] Jacchia, L. G., 1977. Thermospheric temperature, density and composition: new models, *Smithson, astrophys. Obs., Spec. Rept 375*.
[18] Barlier, F., Berger, C., Falin, J. L., Kockarts, G. and Thuillier, G., 1979. Comparisons between various semi-empirical thermospheric models of the terrestrial atmosphere, *J. atmos. terr. Phys.*, **41,** 527–541.
[19] Hedin, A. E., 1983. A revised thermospheric model based on mass spectrometer and incoherent scatter data: MSIS-83, *J. geophys. Res.*, **88,** 10 170–10 188.
[20] King-Hele, D. G., 1975. A view of Earth and air, *Phil. Trans. R. Soc. A.*, **278,** 67–109.
[21] Akasofu, S. I. (ed.), 1980. *Dynamics of the magnetosphere*, Reidel, Dordrecht, Holland.
[22] Carovillano, R. L., and Forbes, J. M. (eds), 1983. *Solar-terrestrial physics*, Reidel, Dordrecht, Holland.
[23] Rees, D., 1971. Ionospheric winds in the auroral zone. *J. Br. interplanet, Soc.*, **24,** 233–246.

[24] King-Hele, D. G. and Walker, D. M. C., 1983. Upper-atmosphere zonal winds from satellite orbite analysis, *Planet. Space Sci.*, **31,** 509–535.

[25] Aristotle, *c.* 350 BC. *Meteorologica*, with translation by H. D. P. Lee, Heinemann, London (1952).

[26] Heninger, S. K., 1960. *A handbook of Renaissance meteorology*, Duke University Press, Durham, North Carolina.

[27] De Bouelles, C., 1551. *Géométrie practique*, folio 53, Regnaud Chaudière, Paris.

[28] Finé, O., 1532. *Protomathesis*, folio 103, Paris.

[29] Heninger, S. K., 1977. *The cosmographical glass*, Huntington Library, San Marino, California.

[30] Reisch, G., 1512. *Margarita philosophica nova*, Liber VII, Strasbourg.

[31] Marlowe, C., 1592. *Tamburlaine the Great*, Part I, Act IV, sc. ii, lines 30–32, 43–45, 51–52, London.

[32] Shakespeare, W., 1593. *Venus and Adonis*, lines 1046–1048, London.

[33] Cowley, A., 1656. The Motto, in *Poems*, Vol. I, p. 15, Cambridge University Press, 1905.

[34] Middleton, W. E. K., 1966. *A history of the thermometer and its use in meteorology*, Johns Hopkins Press, Baltimore.

[35] Manley, G., 1974. Central England temperatures: monthly means 1659 to 1973. *Q. Jl R. met Soc.*, **100,** 389–405.

[36] Halley, E., 1686. A discourse of the rule of the decrease of the height of the mercury in the barometer, according as places are elevated above the surface of the Earth, *Phil. Trans. R. Soc.*, **16,** 104–120.

[37] Halley, E., 1716. An account of the late surprizing appearance of the lights seen in the air on the sixth of March last, *Phil. Trans. R. Soc.*, **29,** 406–428.

[38] De Mairan, J. J. D., 1733. *Traité physique et historique de l'Aurore Boréale*, Paris.

[39] Darwin, E., 1791. *The Botanic Garden*, Part 1, Canto 1, lines 129–130, Johnson, London.

[40] Darwin, E., 1788. Frigorific experiments on the mechanical expansion of air, *Phil. Trans. R. Soc.*, **78,** 43–52.

[41] Middleton, W. E. K., 1965. *A history of the theories of rain*, Oldbourne, London.

[42] Dalton, J., 1793. *Meteorological observations and essays*, Manchester.

[43] King-Hele, D. G., 1973. Erasmus Darwin, grandfather of meteorology?, *Weather*, **28,** 240–250.

[44] King-Hele, D. G., 1986. *Erasmus Darwin and the Romantic poets*, Macmillan, London.

[45] Wolf, A., 1952. *A history of science, technology and philosophy in the 18th century*, two volumes, Allen and Unwin, London.

[46] Darwin, E., 1791. *The Botanic Garden*, Part I, Note I, Johnson, London.
[47] Wilson, J. T., 1961. *IGY, the Year of the New Moons*, p. 28, Joseph, London.
[48] Jeans, J. H., 1925. *The dynamical theory of gases*, 4th edn, Cambridge University Press.
[49] Hoyle, F., 1950. *The Nature of the Universe*, p. 56, Blackwell, Oxford.

# Time, vacuum and cosmos

Sir William McCrea

I thank Professor Lyttleton and his colleagues for the honour of being invited to give this lecture. As a scientist and a personal friend, E. A. Milne meant a great deal to me. I am much moved by having my remembrance of him revivified by this invitation being conveyed to me by his granddaughter, Miranda Weston-Smith, who has achieved so much in keeping his memory fresh. In my recollection, Milne, more than anyone else, epitomizes those heroic days when quantum physics, astrophysics and—as it seemed— even the cosmos itself exulted in the joy of youth.

You will find in *Who's Who* an entry from which I extract: 'Lyttleton, Raymond Arthur . . . *Recreations* . . .; wondering about it all'. I trust that Professor Lyttleton will permit us to share his recreation for an hour or so—although the topics of our wonderment may be not precisely the same as his.

My topics are, in fact, some concepts of the vacuum and the manner in which they underlie certain aspects of physics and cosmology. You might think that there can be nothing in this—we shall see.

The topics come out of everyday experience. For instance: Why does your modern wrist watch tell the same time as your old grandfather clock ticking away at home? I am not asking merely about adjusting them to be in agreement; I ask why should there be *any* intelligible relation between them?

Or, what about athletic records? In 1954 Roger Bannister in Oxford was the first man on record to run a mile in less than four minutes. It is claimed that successive records have been broken almost every year since them. But what does that mean? You might try to preserve the Oxford measured mile and insist that you will recognize only records made on that identical track. What, though; about the minute? How can you preserve four Oxford minutes for thirty-odd years? If you cannot, how do you compare Steven Cram's running in 1985 with Roger Bannister's in 1954?

Or again, you could discover that the Earth has a centre and you could measure the Earth's radius, without stepping outside this room. Just hang up as many plumlines as you like and—provided you possess the very accurate apparatus required—you will find that they all aim at a single point some 6000-odd km below your feet. You have observed nothing outside the room; just what have you observed in it?

## Summary

We shall consider quantum transitions and their significance; special relativity in this context; the Casimir effect as a fulfilled prediction from these ideas. Then we shall extend such ideas to particles other than photons, and thence to the physics of time-keeping. General relativity must be mentioned as a prototype field theory. Then we have to try to understand photon waves and electron waves in the vacuum. Finally we look at the significance of all this for physics in the large—for the cosmos.

## Spontaneous quantum jumps

In any one simple experiment on a quantized system some two particular quantum states are usually all that matter much. So an atom that is supposed to possess only two possible quantum states is a not unduly mythical beast. Let these have energies $E_1$, $E_2$ where $E_2 - E_1 = h\nu$ where $h$ is Planck's constant. Then $\nu$ is the frequency of the single spectrum line given by quantum transitions between these states. We take the simplest case in which the states have equal so-called quantum weights.

Suppose the atom to be immersed in isotropic radiation having intensity $I_\nu$ at frequency $\nu$.

Standard radiation theory asserts that in a given time interval:

(a) if the atom is in state 1 it may absorb a quantum of energy, i.e. a 'photon', $h\nu$ which takes it into state 2; the probability of this happening is proportional to the radiation-intensity, say $B_{12} I_\nu$;

(b) if the atom is in state 2 it may emit a photon $h\nu$ and fall back into state 1; there is probability $A_{21}$, say, that this occurs 'spontaneously'—whatever that may mean; but there is also probability $B_{21} I_\nu$, say, that such a jump is 'stimulated' by the radiation field. So the total probability of the $2 \rightarrow 1$ jump is $B_{21} I_\nu + A_{21}$.

$B_{12}$, $B_{21}$, $A_{21}$ are constants of the sorts first defined by Einstein some 70 years ago. Incidentally, stimulated emission is the phenomenon responsible for LASER action—*L*ight *A*mplification by *S*timulated *E*mission of *R*adiation.

The usual theory proceeds to show that

$$B_{12} = B_{21} \tag{1}$$
$$A_{21} = 2h\nu^3 B_{21}/c^2 \quad (c = \text{light-speed}) \tag{2}$$

so that we may write

$$B_{21}\, I_\nu + A_{21} = B_{21}\,(I_\nu + I_\nu^*) \tag{3}$$

where
$$I_\nu^* = 2h\nu^3/c^2. \tag{4}$$

We know that the $1 \to 2$ transition is caused by the encounter of the atom with a photon $h\nu$, so the simple relation (1) implies that the stimulated $2 \to 1$ transition is also caused by such an encounter.

Relation (3) then implies that emission proceeds as though caused by the applied radiation $I_\nu$ and by some universal radiation $I_\nu^*$.

Relation (4) makes no mention of the particular atom considered; so $I_\nu^*$ has the spectrum expressed by (4) for all $\nu$.

An atom therefore recognizes the existence of *virtual photons* described by the field $I_\nu^*$. Thus every radiative quantum transition has to be regarded as *caused*—either by a 'real' photon that we have deliberately put there by switching on the light, or by a 'virtual' photon that is there whether we like it or not. Presumably, all that an atom in state 2 'knows' is that a photon comes along and tips it out of that state—the *atom* cannot tell whether this is what *we* call a 'real' photon or a 'virtual' photon.

The notion of virtual photons goes back to about the 1920s, but this particular way of inferring their existence may be not generally familiar.

Quantum field theory treats a virtual photon as a real photon that comes into fleeting existence before annihilating (generally) with a virtual anti-photon—any individual such photon $h\nu$ having existence for time about $\nu^{-1}$ (of order $10^{-15}$ s for an optical photon). Presumably this is in general too brief an existence for the virtual photon to be *absorbed* by an atom. Such absorption would be the extraction of energy from the vacuum, and this must in general be forbidden.

The appearance of virtual photons is a manifestation of *vacuum fluctuations*. The intensity $I_\nu^*$ is the statistical outcome of all such photons present at any instant. It implies an energy density of amount

$$\begin{aligned} dE_\nu &= 4\pi\, I_\nu^* d\nu/c \\ &= 8\pi h\nu^3 d\nu/c^3 \end{aligned} \tag{5}$$

per unit volume, in the frequency interval $\nu$, $\nu+d\nu$. By putting in numerical values, this provides one way of seeing how in ordinary experience so-called 'spontaneous' transitions predominate over stimulated emission transitions. In the case of optical atomic transitions, for instance, in order to produce temperature radiation that would cause stimulated transitions comparable in number to 'spontaneous' transitions, a temperature of the order of $10^4$ K is required; in the case of nuclear transitions a temperature of the order of $10^9$ K, or more, is required.

We have inferred that matter behaves as though the 'zero-point' energy density (5) is present in the vacuum. Here I call $I_\nu^*$ *substratum radiation*. Milne had a 'substratum' in his cosmology, and the term appealed to me. So I take the liberty of using his word, although his own use had a quite different application.

Quantum field theory is usually interpreted as giving one-half the value of $dE_\nu$ obtained in (5). I believe that my interpretation of zero-point energy is different but that my treatment is self-consistent.

## Vacuum time

We have spoken in terms of frequency, probability, and so forth. These involve the notion of *time*. We have not yet said how time is to be measured.

We accept that vacuum fluctuations occur. If we have not said how time is measured, it is meaningless to say that these fluctuations occur at random intervals in time. But we can, and do, now say that *time is to be measured so that these fluctuations are random in time.* That is to say, we take time from these fluctuations. We claim that we then have a self-consistent account thus far.

*Time* so determined will be called *vacuum time.*

*Distance* will be measured so that light-speed in the vacuum is a universal constant $c$. The $c$ in our formulae is to be this quantity, and this preserves self-consistency.

(Some further discussion of time-keeping will come later.)

## Special relativity (SR)

The radiation field $I^*_\nu$ possesses certain remarkable properties; one of them is this:

*An isotropic radiation field having the spectrum*

$$I_\nu^* = \text{universal constant} \times \nu^3 \tag{6}$$

*is the one and only radiation field that is seen as the same by every freely-moving (i.e. inertial) observer therein.* This is a known theorem—maybe not all that well known! It is true only by virtue of non-convergent properties to which we return below.

This room contains a lot of ordinary light that appears to be about equally bright in all directions. But if we could streak across the room at speed comparable to light-speed, owing to aberration and doppler effects, looking forward the radiation would appear to us brighter than if we were looking backward. If, however, the radiation in the room could have the spectrum specified by (6) then aberration and doppler effects for such radiation would conspire to make it appear exactly the same in all directions, however fast we could possibly go. The proof of the theorem consists in showing in detail that this is true for this spectrum and no other.

To assert that a particular excited atom has a certain probability of making a particular 'spontaneous' quantum jump means that this is true in its rest-frame, if it is moving freely. We can now assert that this is true because according to the theorem the atom sees always the same substratum radiation, and because according to the ideas being presented that radiation actually causes the jump. This is beautiful!

Another way of saying it is simply that these ideas are thus discovered to be fully compatible with the theory of special relativity.

Further investigation makes it all even more interesting. Such investigation was carried out a few years ago by P. C. W. Davies and W. G. Unruh. They showed (independently) that an accelerated observer, i.e. one not moving freely (non-inertial) in the radiation field would indeed see a different field, the modification being, however, inappreciable unless his acceleration is enormous.

This clarifies much in atomic physics. Such physics might seem to require an atom to carry a clock with it. However, we now see that we may contemplate the atom simply as picking up time from the substratum radiation, i.e. from the vacuum wherever the atom happens to be and in whatever way it is moving. Or perhaps we ought simply to say that an atom does what the vacuum in its neighbourhood tells it to do.

## Casimir effect

In what has been said so far the notion of substratum radiation has supplied a neat means of relating known phenomena to each other, but it has predicted nothing new.

In order to see whether a vacuum property can yield something observable, it is natural to look for a situation in which we can vary what we may call the 'amount of vacuum' available to some feature. If we vary it, does anything happen?

The simplest case that can be proposed is that of two flat plates held parallel to each other in what may be regarded as otherwise unbounded vacuum. The distance between the plates is supposed to be adjustable. The plates are to be made of solid conducting material and 'earthed' so that they cannot exhibit electrostatic effects. Let

$L$ = separation between plates,
$A$ = area of each plate, so that
$V = AL$, volume of space between the plates.

Suppose $L$ is small compared with the linear dimensions of the plates, so that edge effects are negligible.

Expressing the substratum energy density in terms of wavelength $\lambda = c/v$, from eqn (5)

$$dE_\lambda = 8\pi hc d\lambda/\lambda^5 \tag{7}$$

gives the energy per unit volume in the range $\lambda$, $\lambda + d\lambda$.

Within the gap between the plates, let us for simplicity think of the radiation as composed of one-third moving across the gap, and two-thirds moving parallel to the plates. Intuitively it is hard to see that radiation could be established moving across the gap if its wavelength is more than about the width of the gap. So for this component of the radiation we shall require

$$\lambda < \lambda_L \equiv kL \tag{8}$$

where $k$ is some number of order unity.

If then the plates be brought closer by amount $dL$ a further quantity of substratum energy is lost, namely that having $\lambda_L > \lambda > \lambda_L - d\lambda_L$, where from (8) $d\lambda_L = kdL$, in the component moving across the gap. From (7) this is

$$\tfrac{1}{3}VdE_\lambda = 8\pi hcAdL/3k^4L^4. \tag{9}$$

This implies that there is per unit area an *attractive force of one plate on the other of amount.*

$$\frac{8\pi}{3k^4}\frac{hc}{L^4}. \tag{10}$$

Using standard quantum field theory, Casimir (1948) calculated the value

$$\frac{\pi}{480}\frac{hc}{L^4}. \tag{11}$$

for this force.

In short, some fraction of normal substratum radiation cannot exist in the gap because its wavelength would be too great compared to the width of the gap. So if the plates are brought closer a further amount of energy is lost. Such a loss implies that there is a force of attraction between the plates. A crude argument yields the formula (10) where $k$ is a number not greatly different from unity. Accepted theory gives formula (11).

### Comments

(a) It is now generally accepted that the prediction (11) has been well confirmed by experiment. This includes the confirmation of the $L^{-4}$ dependence, which in turn confirms the $\nu^{-3}$ dependence of the substratum radiation in (5).

(b) Formulae (10), (11) do have the same form; they agree if $k \approx 6$ which, by being rather different from unity, shows how crude is the derivation of (10). Nevertheless this derivation helps to elucidate the physics in (11).

(c) The force (11) is minute unless $L$ is very small; by way of illustration, if $L = 10^{-5}$ cm, this force $\approx$ 130 dyne $\text{cm}^{-2}$.

(d) Casimir made a *positive prediction of a new phenomenon depending upon vacuum properties.*

(e) It is a quantum effect in the sense that it vanishes if we put $h = 0$.

(f) It is not electromagnetic since the quantity $e$ is not involved.

(g) It depends upon no particular property of the solid material of the plates—only upon its presence.

(h) The same basic ideas are now well known to account for a whole range of observed phenomena: the Lamb shift in the energy levels of the hydrogen atom; certain intermolecular forces, e.g. in paints having the character of gels; and so forth. Perhaps mention should be made also of Hawking radiation from a black hole, as another prediction of a vacuum effect, albeit one that has not yet been observed.

## Virtual particles

Quantum physics demands the existence of vacuum fluctuations that give rise to virtual neutrinos, virtual electrons, and their antiparticles, as well as to virtual photons.

We shall take it as axiomatic that *fluctuations of the various sorts are randomly distributed amongst each other*. This implies:

(1) Any recognizable species of fluctuation determines a vacuum time.

(2) Any other such species determines the same vacuum time, that is

*vacuum time is unique.*

This is the 'natural time' implicit in the formulation of quantum physics.

An example is physics of the solid state. So vacuum time would be the time by a clock depending upon the resulting properties of matter, e.g. a spring-driven clock, a quartz oscillator, . . .

We now state:

*Proposition*

*Every quantum transition called 'spontaneous' is stimulated by a virtual particle* —a photon, or electron, or . . . *associated with a vacuum fluctuation.*

This is the natural generalization of what we inferred about an atomic transition; it appears to be implicit in most of our discussion.

## Radioactivity

A radioactive atomic nucleus is a system that has been got into an 'excited' quantum state—state 2 of our two-state atom was the one possible excited state of *that* system. So radioactive decay is a quantum transition. As normally experienced it is what is described as 'spontaneous'. According to the present discussion, therefore, it is caused by a vacuum fluctuation—in general in radioactive processes this would be one involving a virtual particle other than a virtual photon.

A body of radioactive material of some particular species serves therefore as a clock measuring time by counting decays.

This we may call an 'atomic clock' or 'radium clock'—using 'radium' in a generic sense.

Our foregoing discussion implies that all atomic clocks keep the same time—vacuum time, the same as all the other clocks we have mentioned.

## Clock problem

A radium clock tells the same time as a wrist watch, a cuckoo clock, a sundial . . . Why?

A radium atom is said to decay at random. We cannot predict when the decay will happen. Nevertheless all experience shows that all these clocks (when properly serviced) agree with each other with, to all intents and purposes, perfect accuracy.

Common sense leads us to infer:

There has to be a physical connection between all these clocks. In particular, this must cause the radioactive decays—subject to irreducible uncertainties in details. The randomness of which we are aware in the details must reside in the universal physical connection. Thus from common experience we infer the existence of something having the nature of universal vacuum fluctuations.

Almost all my life I wondered about this clock problem; apparently I lacked courage to draw these commonsense inferences.

## Master-clock: time's arrow: Problems

We have inferred that vacuum affects the behaviour of matter. So we infer that matter affects the behaviour of vacuum.

This is confirmed by the reality of the Casimir effect and all related effects.

So vacuum furnishes the *ideal* master-clock only if there is meaning in our being able to use it apart from effects produced by the presence of matter, or in the limit when these effects may be regarded as negligible. I do not know whether or how this problem can be solved. Nevertheless, 'ideal' or not, it seems that the vacuum is the one master-clock available to physics.

Any clock that needs winding, or a new battery, or whatever, has an arrow of time pointing in the running down sense.

All clocks take time from vacuum fluctuations—according to the view advanced here.

It is hard to see why all clocks should have the *same* arrow unless the vacuum itself has this arrow. Has it? We shall be persuaded, I believe, that vacuum time does not possess such an arrow—a situation that is decidedly perplexing.

## General relativity (GR)

GR is the prototype field theory in physics; its field is everything with an existence recognized by the theory.

GR contemplates a single field in a topological four-space; it is specified by a tensor $g_{pq}$ and it is taken to include any tensors that can be derived from this.

The field yields exactly the number of features needed to depict matter, sustaining stress, in space–time that is itself depicted by the field.

Any such system is a *universe*, i.e. anything that affects any part of the system belongs to the system.

It is found that matter described in the manner thus proposed inevitably exhibits a property that we are constrained to term *self-gravitation.*

The situation is: GR yields the barest necessities for a universe; this exhibits what we want to call gravitation; we hope that it indicates broadly the basic (macroscopic) nature of gravitation; in this, experience suggests that it has proved to be successful so far as it has been possible to test it.

Probably most physicists would agree that GR should not be expected properly to treat gravitational effects of electromagnetic (EM) phenomena because the GR field has no features that are available to depict electromagnetism.

*GR vacuum.* In a GR universe there may be a region $\mathscr{V}$ that we recognize as *vacuum*—or, more correctly, *vacuum history* since it is in four-space. Without needing additional postulates, it can be inferred:

The history of any test-particle in $\mathscr{V}$ is a geodesic: an 'ordinary' geodesic if the particle has small non-zero rest mass; a 'null' geodesic if every observer regards the particle as throughout having fundamental speed *c*—it is then called a *photon*.

Going over to the language of three-space, these inferences are expressed by saying that in the vacuum there exists an aggregate of trajectories of free particles, including photons. This we call a *vacuum state*.

*This state is observable*; we may release as many free particles—ordinary particles and photons—as we please; by seeing how they behave, we 'see' as much of the state as we wish.

The plumblines mentioned earlier would be part of the observation of the vacuum state in this room. If they were not responding to the vacuum state, to what would they be responding?

Also in SR the straight lines in four dimensions, which are histories of free particles, are features of the vacuum state of that theory.

## Optical interference

Consider a line source LL of monochromatic light of wavelength $\lambda$ parallel to and equidistant from two narrow slits, S, S in an opaque screen. Let there be a photographic plate P behind and parallel to the screen. Using the concepts of elementary physical optics it is inferred that S, S behave as coherent line sources of $\lambda$ light. So an interference pattern is formed on P, with illumination-maxima along lines having distances to S, S differing by integral multiples of $\lambda$, and minima close to mid-way between the maxima—a very familiar experiment in physics.

Electromagnetic theory (EM) gives a good account of this in terms of vectors satisfying EM field equations together with boundary conditions which serve to specify the particular apparatus in the experiment.

In this way the graph of the plate-darkening on P may be predicted with what is deemed 'complete' accuracy.

The equations are linear in the field-vectors; hence if the source power of LL is reduced by a factor 100, say, and the exposure time is increased 100 fold, the picture on the (supposedly perfect) plate remains the same. This seems to be common sense.

The light in the experiment consists actually of photons. This also seems common sense, for otherwise the photographic emulsion composed of actual atoms and molecules could not function. This ultimately is why the actual photograph is 'granular'.

For this reason we interpret the EM field as in reality yielding the *probability distribution of photons* arriving on the plate.

Now if the passage length for a photon traversing the apparatus is, say, about 30 cm, the passage time would be about $10^{-9}$ s. If then we had fewer than about a billion photons per second traversing the system, on average there would be no more than a single photon going through at any instant. It would be easy to adjust the apparatus so that this is in fact the case. Then *no photon can interfere with any other.* Nevertheless *the interference pattern remains the same.* This seems to be *not* common sense!

## EM vacuum

We seek a commonsense resolution of the situation just described; it seems to be forthcoming in the following manner:

The EM vacuum is a charge-free, matter-free region. The matter present in the interference experiment is described merely by conditions at the boundary of this region. We assert that the *vacuum state* in the region is described by the EM field that satisfies the EM vacuum field-equations together with these boundary conditions. We envisage this vacuum state as being there—as subsisting—for as long as the apparatus is set up, whether or not any photons are traversing it. We regard this state as *'determining' the trajectories of whatever free photons are present.* We claim that the state is observable—that it is observed by releasing free photons and noting what happens. In other words, the experiment itself is to be regarded precisely as the observation of the EM vacuum state produced by the apparatus.

### Comments

There is general similarity to the interpretation of GR as yielding a vacuum state that is observable by appropriate means; this is why I mentioned GR.

There are differences, however, because the GR case is claimed to be purely and simply *macroscopic*, whereas we hope to use the EM case for not only macroscopic optics but also for quantum aspects, at any rate to the extent of describing *statistical features* of photon optics. So we claim to evaluate only *probabilities* regarding the behaviour of individual photons; their trajectories are to be considered as 'determined' in only this sense.

'Observation' has also to be carefully interpreted. In the given experiment we could not observe a photon *during* its passage through the apparatus. Any attempt to do this would produce a different experiment.

Should anyone wish to claim that a photon traversing the apparatus itself causes the vacuum state that we claim to observe, I cannot see that we can do an experiment directly to falsify this—any more than we can do one to falsify the view advocated here. Mainly the situation appears to be that it is easier to think of the photon behaving *as if* the vacuum state is there all the time.

This attitude is, however, reinforced by considering the two-slit experiment with a very massive body in the vicinity. Photons would be deflected by its gravitation. We envisage this gravitation as determining the vacuum state, so far as gravitation is concerned, for as long as the body is present. So it seems simplest to think of the vacuum state, so far as electromagnetism is concerned, as being determined by the bodies providing the slits, etc., so long as those bodies are present.

These latter bodies need to be regarded as affecting the EM vacuum state simply in their immediate vicinity—this being the significance of the boundary conditions employed in the mathematical account. Unlike the gravitational case, the presence of the bodies does not appreciably affect the vacuum state elsewhere throughout the region.

A rough analogy is that a photon traversing the vacuum of the experiment is like a pin-ball negotiating the pins on a pin-table—in a case where the pins themselves are jittery. The outcome is predictable, but *only* statistically.

## Electron 'optics'

As is well known, for the statistical behaviour of actual electrons passing through the slits, etc., there are analogues of all optical phenomena such as those in the interference experiments we have been discussing.

In the electron case we have the wave function of quantum mechanics in place of the pattern of EM waves.

In a given experiment the relevant vacuum state is the pattern of all wavefunctions satisfying the appropriate boundary conditions. Once again, the predictions are of *probabilities*.

## Vacuum fluctuations and cosmos

Vacuum fluctuations have a fundamental rôle in physics—that is the burden of all I have been saying. All random processes in quantum physics are linked with these—that is my thesis. In everyday physical experience we always trace a random effect to a random cause—so it seems, and this may be little more than a truism. We seem now to be able to trace all such causes to random *vacuum fluctuations*. Thus it seems that *randomness is primitive*—put into the Universe at its creation.

In big-bang cosmology, the existence of *galaxies* is generally traced to primeval fluctuations in the distribution of contents of the Universe. These have been regarded as chance fluctuations. However, most cosmologists seem now to regard these as originating in actual quantum fluctuations in the *very* early Universe.

Unless we are driven to adopt some form of steady-state cosmology, I think this last inference in itself is an inevitable consequence of the trend of thought we have been following. But whether it is indeed possible to trace a credible development from quantum fluctuations right through to galaxies is a daunting problem. This does appear to be what some people are attempting in the theory of phenomena called 'strings', 'loops', 'superstrings', and so forth.

Some people may have thought the title of this lecture to imply that it was to be about those very ideas. If so, I can only assure them that the mere fact of their doing so suggests that they must know more than I do about this aspect! I wish to devote the rest of the lecture to broadly cosmical features that I think may be basic, but not so recondite as such innocent-sounding topics as 'strings' and the like.

## Vacuum time and cosmic time

Let us provisionally accept the big-bang cosmological model—call it GRC since it is cosmology provided by pure GR. The model is determined by the matter (energy) content with its equation of state. The treatment makes gravitational interaction inevitable. The use of GRC rests upon the hypothesis that this is the only interaction that is significant for the large-scale behaviour of the Universe.

This implies that time in GRC is measured by a gravitational clock, e.g. a planet going round a sun. Standard GRC yields *cosmic time* $t$ at any event as *proper time, since the big bang, for the co-moving observer who experiences that event.*

Earlier we had *vacuum time* as a universal time. If gravitation is to be included in grand universal theories (GUTs) then we should require gravitational time, like atomic time, to be the same as vacuum time. We should then conclude that cosmic time is measured in vacuum time. This would, of course, be subject to the provisos we make about the measurement of vacuum time. There is no known empirical objection to this conclusion. We now include it in our provisional acceptance of GRC.

## Substratum radiation in an expanding universe

We come now to another remarkable property of the substratum radiation.

From eqn (5) this radiation has energy density in the frequency interval $\nu_1$, $\nu_1 + d\nu_1$

$$dE_1 = (8\pi h/c^3)\, \nu_1^3 d\nu_1. \tag{12}$$

Dividing by $h\nu_1$ the photon number-density $dn_1$, say, is

$$dn_1 = (8\pi/c^3)\, \nu_1^2 d\nu_1. \tag{13}$$

This gives for the number $dN_1$ in a cube with edge $l_1$

$$dN_1 = l_1^3 dn_1. \tag{14}$$

In an expanding universe each free photon remains one and only one free photon. If two co-moving observers observe the photon to have frequencies $\nu_1$, $\nu_2$ then

$$\nu_2 = \nu_1/K,$$

where $K$ is the factor by which the Universe has expanded between the observations. This is the familiar cosmological redshift.

We make the hypothesis that in these regards *virtual photons behave in the same way as free photons.*

In the expansion contemplated $l_1$ expands to $l_2 = Kl_1$. Let $dn_2$ be the number density in $\nu_2$, $\nu_2 + d\nu_2$.

We have
$$\begin{aligned} &\nu_1 = K\nu_2 \qquad d\nu_1 = Kd\nu_2 \\ &l_1 = K^{-1} l_2 \\ &dN_1 = l_1^3 dn_1 \qquad dN_2 = l_2^3 dn_2. \end{aligned} \tag{15}$$

Conservation of photon number in accordance with the hypothesis is

$$dN_2 = dN_1,$$

so that
$$\begin{aligned} dn_2 &= K^{-3}\, dn_1 \\ &= K^{-3}\, (8\pi/c^3)\, \nu_1^2 d\nu_1 \end{aligned}$$

giving, from (15),

$$dn_2 = (8\pi/c^3)\, \nu_2^2 d\nu_2.$$

Multiplying by $h\nu_2$, the energy of each photon, the energy density in the interval $\nu_2$. $\nu_2 + d\nu_2$ is

$$dE_2 = (8\pi/c^3)\, \nu_2^3 d\nu_2. \tag{16}$$

In short, if the substratum photons behave like any other photons *vis-à-vis* the expansion of the Universe, then *the substratum radiation is invariant throughout the expansion*—as shown by the identical forms (12) and (16).

## Other vacuum properties

We now have to presume that all zero-point (vacuum) properties have the same invariance as the substratum radiation.

## Cosmological consequences and queries

The invariance revealed in this manner is obviously a remarkable property. It follows from the infinity in the substratum. As photons are red-shifted to lower frequencies in space that is expanding they maintain the same number density as before at those lower frequencies. All the time there is continual drawing upon higher frequencies from an inexhaustible supply. The Universe can never overdraw its account in substratum photons.

This constancy of the substratum makes the assumption of standard clocks consistent with our relating them to vacuum fluctuations. In fact it justifies the very idea of there existing such things as standard clocks. Of course, we know that in the last resort something has simply to be deemed standard. But we have to be able to defend our particular choice. Physics would be in sorry confusion were we unable to defend our chosen time-reckoning. And it is rather appealing that we should find a need to go to cosmology in order to be satisfied that athletic records make sense.

Apparently we have to think of space–time, and this immutable substratum, as being inseparable. If so, this must be an expression of the underlying homogeneity of the Universe. In particular, if galaxies arise from fluctuations of the substratum, the 'universe of galaxies' would of necessity be homogeneous on some scale. But this in turn reminds us that we have to think of the substratum as being endowed with a basic randomness.

We are left with some big questions:

1. Is the concept of an unchanging substratum reconcilable with the concept of a unique big bang?
2. Where does the basic (or primitive) randomness come into the picture?
3. How are we to regard the infinity of the substratum?

Must we perhaps accept that physicists and astronomers are always dealing with only some parts of an infinite universe that we can never apprehend as a whole? We must recognize that we never do get away from infinities of one sort or another. The simplest big-bang model starts with a singularity—an 'infinity'. If we take the extreme means of avoiding this by postulating a steady-state universe then the Universe is a perpetual infinity. Or, in quantum theory, H. Everett's many-worlds interpretation is another, but far more abstruse, exploitation of an infinity.

Once again, what becomes of the arrow of time? The invariance of the substratum seems to imply that vacuum time does not have an arrow—because, roughly speaking, this invariance means that the vacuum is not running down. But cosmic time does have an arrow—in the direction of the expansion of the Universe. How is the paradox to be resolved?

## Constants of physics

In what we are calling the substratum we have come upon a basic constancy in physics—or, what comes to the same thing, we have discovered that physics is so constructed that we are bound to regard it as constant. Inevitably, we then ask: Does this tell us something about the more commonly recognized constants of physics?

The number density of substratum photons (13) was written

$$dn = (8\pi/c^3)\, v^2 dv \tag{17}$$

Here a photon is just an entity with a label $v$ that has dimension the reciprocal of time, so in order to express a number per unit volume we need the conversion constant $c^3$. The purely numerical factor $8\pi$ has no particular significance since it could be absorbed into the other quantities if desired. The distribution (17) has the interest we have been discussing precisely because it is an infinite system. It then turns out that it increases the interest if we say that a photon labelled $v$ is said to carry an amount $hv$ of a quantity we choose to label 'energy', where *h is a new conversion constant.*

In GR both the fundamental speed $c$ and the gravitation constant $G$ are conversion constants and nothing more.

As regards electric charge, anything like a cosmological principle requires a particle substratum to have zero net charge. For non-zero charge implies a non-zero electromagnetic field that cannot be everywhere the same. Hence any virtual charged particles must come in pairs of equal and opposite charge $\pm Q$, say. A virtual charge $Q$ must be able to annihilate completely with any charge, actual or virtual, of opposite sign, or at any rate with a finite number of them. I believe this requires every $|Q|$ to be the same or a finite integral multiple of a single value $e$. It is thus seen that vacuum properties require there to exist no more than one single basic quantum of charge.

Thus we seem to have some clue to the origin of the constants $c$, $G$, $h$, $e$.

## Cosmical constant $\Lambda$

The so-called 'cosmical' terms in GR may be interpreted as zero-point values of energy and stress. They contain a constant factor $\Lambda$. A non-zero value of $\Lambda$ might therefore be regarded as a recognition of possible gravitational effects of zero-point properties. However, the substratum properties that we have been considering possess the infinities that appear to be such essential features. So A could not arise directly from such properties. In fact it is plausible to accept that an immutable universal substratum should have no gravitational effect anywhere. However, anything like a Casimir disturbance, that renders the substratum not perfectly uniform, might have such an effect. Properties of this general category have been discussed by others; they can apparently lead to a non-zero $\Lambda$, but I think that the cosmological status of the result is still unclear.

## Laws of physics

If we accept a traditional view of laws of physics then big-bang cosmology provokes awkward questions:

I. At what stage of the big bang did the laws come into existence?

II. If the laws came into existence along with the Universe, why do they not evolve along with the evolution of the Universe?

Such questions have been asked ever since anything like the big bang was first proposed. Possible escapes are:

(1) The Universe is in a steady state.

(2) There are no laws.

As regards (2), in GR we need no laws of motion or gravitation; once the field is given the world-lines of all free particles simply are there as features of the vacuum state. We have recognized a generally similar situation regarding the behaviour of photons and electrons in the vacuum states of electromagnetism and quantum mechanics.

Accordingly in big-bang cosmology we may dispense with laws of motion and the rest. Instead we may regard vacuum states, along with all matter, as features of the Universe emerging from the big bang. The vacuum state concept tends thus to mitigate some of the difficulties about laws in physics. This may tend further to weaken the case for steady-state cosmology.

On the other hand the conserved substratum properties are reminiscent of steady state cosmology. So we are tempted to flirt with a concept that seems to have, or have had, attractions for Sir Fred Hoyle, of 'bangs—big or otherwise—happening in an overall steady-state setting. 'Flirt' seems indeed to be the word for playing with two so different notions at the same time.

## Conclusions

The discussion may be claimed to go some way to demonstrate how the concept of vacuum effects:

> rationalizes 'spontaneous' quantum processes—correlates apparently unrelated phenomena—unifies time-keeping—predicts phenomena not previously

> recognized—accords well with special relativity and with basic notions of general relativity—fits satisfactorily with some main aspects of cosmology—sheds some light upon the status of the constants of physics and of laws of physics.

Many deep problems only too obviously call for discussion, for example:

> interpretation of infinities—meaning and origin of randomness—possible relevance of steady-state cosmology, as well as clarification of the whole discussion.

## References

Probably every individual conclusion in this lecture has long ago been presented in due professional style by specialists in the fields concerned. I hope there may be interest in seeing how some of their essentials may be reached in less sophisticated ways. Since this is in fact how I did reach most of them, it would be difficult to provide references to original publications.

It should at least be stated that the Casimir effect was first described in H. B. G. Casimir (1948) *Proc. K. ned. Akad. Wet.*, **60,** 793. One valuable account of some of the ideas is Dennis Sciama, 'The ether transmogrified', *New Scient.*, 1978 February 2, p. 298. Much of the physics is expounded in the well-known books of recent years by Paul C. W. Davies. Also I believe that, were some of my ideas on quantum theory formulated more fully, they might be found to be along the lines of some long advocated in writings of David Bohm. In the early stages of thinking about some of the material here presented I had much benefit from discussions with Dr John D. Barrow who has, however, no responsibility for the form it has eventually taken.

# The age of the observable universe

William A. Fowler

## Personal notes

Preparation of the Tenth Milne Lecture and this article involving cosmological chronology proved to be a very heady experience for me. Therefore, I hope it is not inappropriate to begin on a somewhat personal note. I first learned about cosmology from Richard Chase Tolman in his graduate student course from the spring quarter of 1934 through the fall quarter of 1935 at Caltech. Then I eventually became a colleague and good friend of Howard Percy Robertson who tried to teach me the finer points of cosmology *à la* Einstein's general relativity. Then I came under the influence of Fred Hoyle who insisted that I learn a little about the steady-state cosmology of Hoyle, Bondi and Gold. I soon came to know Hermann Bondi and Thomas Gold. But I also met George Gamow who served as a counter balance and who interested me in nucleosynthesis in the big bang. Then because I had two sabbaticals in England and abroad and spent a number of years at Hoyle's Institute of Theoretical Astronomy, I eventually came to know George Lemaître, William McCrea, George McVittie, G. J. Whitrow, Roger Penrose and Dennis Sciama. At Caltech I later learned about cosmology in a very special way from Kip Thorne. But I never met Milne or Eddington or Walker. What a pity! It does show however that a nuclear physicist on the periphery of astronomy and cosmology could touch base with some of the great men and I am grateful for knowing them personally. In particular I'll never forget spending a week in 1957 on the beach at Sorrento, Italy with Hoyle and Lemaître, listening to them discuss the relative merits of the steady state versus the primeval atom. I listened and never said a word, believe it or not. What an experience!

But enough of name-dropping, although I think it is really more than that to recount what was possible for an outsider during the golden years in cosmology. The heady experience I referred to came when I began to read, in preparing this article, the papers of Milne (1934), McCrea and Milne (1934) and Walker (1940) on Newtonian expanding universes in kinematical relativity and the critical papers of McVittie (1940) with responses by Milne (1941) and Walker (1941). Those were the days! Tolman very gently, Robertson very vehemently and Thorne very carefully told me that Einstein's general relativity is most probably correct. I must say however that Milne's classical Newtonian treatment of the universal expansion, which I learned about many years ago from other sources than his own papers, appeals to me, as it must to many, as a neat way to understand the universal expansion without recourse, as Bondi (1952) put it, to 'the cumbrous tensor calculus'. My main point in this article is that nuclear chronologies are compatible with a ratio of the age of the observable universe to the Hubble time equal to two-thirds. Imagine my delight in finding in Milne's 1934 paper for a universe with zero curvature, zero cosmological constant [his equation (5)], $v = 2r/3t$, which for the present time (subscript zero) reads $v_0/r_0 = H_0 = T_0^{-1} = \frac{2}{3}t_0^{-1}$ or $t_0 = \frac{2}{3}T_0$, where $H_0$ is the present value of the Hubble constant, $T_0 = H_0^{-1}$ is the Hubble time and $t_0$ is the present age of the observable universe. For that delight I repay Edward Arthur Milne with homage in this tenth lecture in his memory.

I know that the model with zero curvature and zero cosmological constant and critical mean density is called the Einstein–de Sitter Model (1932). (Milne's name is usually applied to a model with negative space curvature, zero cosmological constant and zero mean density.) It is interesting to read the Einstein–de Sitter paper (1932). They justify $H_0^2 = \frac{8}{3}\pi G\rho_0$, an equation derived in their paper, on the then current value of $H_0 = 500\ \mathrm{km\,s^{-1}\,Mpc^{-1}}$ and de Sitter's estimate for the mean density of the universe, $\rho_0 = 4 \times 10^{-28}\ \mathrm{g\,cm^{-3}}$. Their $H_0$ was high by $\sim 10$ and their $\rho_0$ by $\sim 10^2$. In some way, Milne (1934) knew enough to side with Newton and did not try to compare his equation with observations. Of course, in the end, Einstein triumphed but I must say Milne put up a good fight for Newton.

## Introduction

In a paper published in *Nature* in 1929 Lord Rutherford established the use of radioactive nuclei as chronometers in geochronology. Geologists, geochemists and geophysicists use his idea to date terrestrial rocks, meteorites and even the Moon. But Rutherford did more than that—he also established the use of radioactive nuclei as chronometers in cosmochronology. Almost 30 years later Burbidge *et al.* (1957) were able to calculate the production of $^{232}$Th, $^{235}$U and $^{238}$U in the rapid ($r$)-process of neutron capture. They realized that they could then calculate the duration of nucleosynthesis in the Galaxy using Rutherford's ideas and found values of the order of 10 Gyr. My interest in nuclear cosmochronology has continued with the publications of Fowler and Hoyle (1960), Fowler (1961, 1967, 1972, 1977, 1978), Hoyle and Fowler (1963) and Fowler and Meisl (1986*a, b*). All of this work follows from my strong conviction that the only way to determine the ages or times in cosmochronology accurately is through the use of nucleosynthesis and nuclear radioactivity.

The recent work of Fowler and Meisl (1986*a, b*) has been stimulated by the work of Guth (1981) and others on the Inflationary Model of cosmology. An excellent account of this new cosmology is given in Guth and Steinhardt (1984) and informative discussions are given in Bludman (1984) and Peebles (1986). The claims made by the proponents of the Inflationary Model are not accepted in some quarters. For differing points of view in this regard the reader is referred to Barrow and Turner (1982), Gibbons, Hawking and Siklos (1983) and Penrose (1986). In this article one consequence of the Inflationary Model is accepted—Einstein's curvature parameter is taken to be zero. The main conclusion is that the 'best' or most probable value for the cosmological constant is very close to zero. This is possible in supersymmetric particle theory but has not been definitely shown to be the case. For a less restrictive approach to cosmochronology than that taken in this article the reader is referred to Tayler (1986).

## The inflationary model

In general the universal expansion rate is governed by the Friedmann equation

$$H^2 = T^{-2} = \left(\frac{1}{a}\frac{\mathrm{d}a}{\mathrm{d}t}\right)^2 = \frac{8\pi G}{3}(\rho + \rho_v) - k\frac{c^2}{a^2}. \tag{1}$$

In this equation $t$ is the time coordinate in general relativity, $H(t)$ is the Hubble expansion parameter, $T(t) = H^{-1}$ is the Hubble time, $a(t)$ is any arbitrary distance measure, $G$ is the gravitational constant, $c$ is the velocity of light, $k$ is Einstein's curvature parameter, $\rho$ is the mass–energy density of matter and radiation and $\rho_v$ is the constant vacuum mass–energy density for which the corresponding pressure is $p_v = -\rho_v c^2$. This constant vacuum density corresponds to the cosmological constant $\Lambda$, which appears in much of the literature. The correspondence is given by $\Lambda = 8\pi G\rho_v$. For non-relativistic matter $\rho$ is proportional to $a^{-3}$ while for relativistic matter and radiation $\rho$ is proportional to $a^{-4}$. When the time variables are evaluated at the present time, $t = t_0$, a subscript zero is used, e.g. $H_0$. In this article $a_0 = a(t_0)$ will be taken as the present radius of the observable universe with the meaning of 'observable' to be discussed eventually in what follows.

The inflationary model postulates that, in the early universe from $\sim 10^{-34}$ s to $\sim 10^{-32}$ s, $\rho_v$ was very large with the constant value $\sim 10^{74}$ g cm$^{-3}$ and dominated the right-hand side of eqn (1). Thus during this period $H$ was constant with the value $\sim 10^{34}$ s so that the period exhibited an exponential expansion or inflation given by

$$a(t) = \text{constant} \times \exp(Ht). \tag{2}$$

The exponential factor equalled $\sim$ exp 65 at the end of inflation (Brandenberger 1985), and the constant is chosen to yield $a$ equal to $\sim 1000$ cm given by the standard big bang models of cosmology at $t = 10^{-32}$ s. Inflation was ended by a phase transition from vacuum energy to matter–radiation energy. For the purposes of this article the main consequence of inflation was that the term $kc^2/a^2$ was reduced by $\sim$ exp $(-\,130)$ during the exponential expansion and thus after inflation $k$ can be taken equal to zero and eqn (1) reduces to a universe which is flat and Euclidean in three dimensions, namely

$$H^2 = \left(\frac{1}{a}\frac{\mathrm{d}a}{\mathrm{d}t}\right)^2 = \frac{8\pi G}{3}(\rho + \rho_v). \tag{3}$$

Depending on one's point of view, eqn (3) may or may not result from inflation. In this article it will be assumed that (3) is correct. If $\rho_v$ vanishes as a result of elementary particle processes during inflation then $H^2 = 8\pi G\rho/3$ which describes the Einstein–de Sitter (1932) universe. It has been noted that Milne (1934) derived this equation using Newtonian mechanics.

Whether or not $\rho_v$ decreases during inflation from its initial enormous value of $10^{74}$ g cm$^{-3}$ to exactly zero thereafter is an intriguing problem. For the nuclear physicist the phase transition from vacuum energy to matter–radiation energy can be likened to radioactive decay. In this case $\rho_v$ decreased exponentially on a very short time-scale and was essentially zero after inflation. But the analogy with radioactive decay may not be appropriate. This article attacks the problem of the value of $\rho_v$ or $\Lambda$ after inflation using nuclear cosmochronology.

Proponents of the inflationary model emphasize that the model solves the horizon problem which plagues the standard big-bang models. The horizon is the limit on the vision of an observer given by the finite speed of light. It can be shown, e.g. Weinberg (1972), that the proper distance corresponding to the horizon is given by

$$d_{\mathrm{H}}(t) = ca(t)\int_0^t \frac{d\tau}{a(\tau)}, \tag{4}$$

where the factor $c$, omitted by Weinberg (1972), has been inserted on dimensional grounds. In the standard big bang models the cosmological constant is taken equal to zero so $\rho_v = 0$. In the early universe the temperature was high, so the density was dominated by radiation and relativistic particles for which $\rho \propto a^{-4}$, so that the $k$-term $\propto a^{-2}$ can be neglected when $a$ is small. Thus eqn (1) yields $a\, da/dt$ = constant and on integration $a(t) \propto t^{1/2}$. Thus

$$d_H = 2ct = 6 \times 10^{10}\, t\, \text{cm}. \tag{5}$$

The argument to follow is independent of the numerical factors used but it is useful to take $a \sim 1000$ cm at $t = 10^{-32}$ s as indicated previously so that $a(t) = 10^{19}\, t^{1/2}$ cm and

$$\frac{d_H(t)}{a(t)} = 6 \times 10^{-9}\, t^{1/2}. \tag{6}$$

This relation holds until baryonic matter becomes non-relativistic and the density of such matter $\propto a^{-3}$ dominates over the density of radiation $\propto a^{-4}$ as $a$ grows larger. The time at which this occurs depends on the matter density assumed but is in the order of $10^{12}$–$10^{13}$ s. This is also the time when matter de-ionizes with neutral hydrogen forming from protons and electrons, for example. When this happens radiation and matter decouple. (The possible significance of the coincidence in these two times is discussed by Schramm and Vishniac 1986.) Thus when radiation decoupled from matter $d_H/a \sim 0.01$ and what is now the observable universe could be divided into $10^6$ regions which were not in causal contact with each other. This is a problem because today the universal microwave background radiation is observed to have the same value, 2.75 K, in all directions to one part in $10^4$ (Wilkinson 1986). The temperature of the background radiation was established at decoupling and so it must have also been extremely uniform at that time. How could this have been the case when what is now our observable universe then consisted of many small volumes not in causal contact? The only answer in the standard model is to assume it by *fiat*.

In the inflationary model $d_H/a \sim 2 \times 10^{26}$ at the beginning of inflation and from eqn (4) this ratio was only reduced by a factor of two during inflation to $\sim 10^{26}$ and has remained so through decoupling to the present time. All parts of what is now the observable universe were in causal contact before and at decoupling and heat transfer assured a uniform temperature throughout.

There are those who do not accept the argument that the inflationary model solves the horizon problem as well as the argument that it yields a flat universe. The matter will not be debated here but a suggestion will be made concerning what to take for the radius of the observable universe now, since defining this radius is difficult in flat cosmologies with $k = 0$. The value $d_H/a \sim 10^{26}$ is not very useful. As a matter of fact as we look out in space we look back in time. But before decoupling the universe was opaque to all wavelengths of light just because the Thomson cross section for the scattering of light by electrons is so large. Thus the practical limit on the radius of the observable universe is the increase in the horizon since decoupling. As noted above, since decoupling the density of non-relativistic matter has dominated over radiation and so $\rho \propto a^{-3}$. Neglecting $\rho_v$ and the $k$-term in eqn (1) then yields $a^{1/2}\mathrm{d}a/\mathrm{d}t$ = constant and on integration $a(t) \propto t^{2/3}$. Inserting this relation into eqn (4), and neglecting the time of matter domination relative to the present time so that the limits on the integral are zero and $t_0$ one finds

$$a_0 \equiv d_{\mathrm{H}}(t_0) = 3ct_0 = 9 \times 10^{10}\, t_0 \,\mathrm{cm}. \tag{7}$$

The result $3ct_0$ rather than $ct_0$ is not difficult to understand. The lengths traversed by light signals prior to reaching us at the present time have all expanded by $a(t_0)/a(t) \propto (t_0/t)^{2/3}$ in the intervening time from $t$ to $t_0 \gg t$.

As the conclusion of this section it is useful to return to eqn (3) and express it in terms of dimensionless quantities defined by

$$\Omega = \frac{8\pi G\rho}{3H^2} \tag{8}$$

and

$$\lambda = \frac{8\pi G\rho_v}{3H^2} = \frac{\Lambda}{3H^2} \tag{9}$$

Substitution of these quantities into eqn (3) yields

$$\Omega + \lambda = 1. \tag{10}$$

For all times and in particular at the present time

$$\Omega_0 + \lambda_0 = 1. \tag{11}$$

It will be clear that a critical density, $\rho_{c0}$, can be defined at the present time by

$$\rho_{c0} = \frac{3H_0^2}{8\pi G\rho} = \frac{3}{8\pi G\rho T_0^2}$$

$$= 1.879 \times 10^{-29}\, h_0^2 = 1.793 \times 10^{-27}/T_0^2 \,\mathrm{g\,cm^{-3}}, \tag{12}$$

where $h_0 = H_0/100$ for $H_0$ in $\mathrm{km\,s^{-1}\,Mpc^{-1}}$ and $T_0 = 9.769/h_0$ Gyr ($10^9$ yr).

In what follows it will be convenient to refer from time to time to the cosmological deceleration parameter given in general by

$$q \equiv -\frac{(\ddot{a}/a)}{(\dot{a}/a)^2} = \tfrac{1}{2}\Omega - \lambda. \tag{13}$$

so that for the inflationary model with $\Omega + \lambda = 1$ one has

$$q = \tfrac{3}{2}\Omega - 1 = \tfrac{1}{2} - \tfrac{3}{2}\lambda. \tag{14}$$

## Cosmological evolution in a flat universe

The time variation or evolution of cosmological parameters in a flat universe has been studied for many years (e.g. Bondi 1952) and very recently by Bludman (1984) and Peebles (1984*a*). In all cases the solutions we are interested in apply to the time after the matter mass–energy density became greater than the radiation energy density. Since this occurred when the universe was about $10^5$–$10^6$ yr old we are safe in neglecting this short interval of time. Thus, for all but a negligible time period it is assumed that $\rho/\rho_0 = (a_0/a)^3$. It is not a subject for discussion in this account but we gratuitously note that big bang nucleosynthesis was completed by $t$ = 300 s when the universe was radiation dominated and $\rho$ was proportional to $a^{-4}$.

Three cases must be treated to yield the variation of the distance measure with time and the ratio $t_0/T_0$.

*Case I*: $\lambda_0 < 0, \Omega_0 > 1, q_0 > ½$

$$\left.\begin{aligned}\left(\frac{a}{a_0}\right)^3 &= \left(\frac{\Omega_0}{\Omega_0 - 1}\right)\sin^2\left[3(\Omega_0 - 1)^{1/2}\frac{t}{2T_0}\right],\\ &\Rightarrow \left(3\Omega_0^{1/2}\frac{t}{2T_0}\right)^2 \quad \text{for} \quad t \ll T_0,\end{aligned}\right\} \tag{15}$$

$$\left.\begin{aligned}\frac{t_0}{T_0} &= \frac{2}{3(\Omega_0 - 1)^{1/2}}\tan^{-1}(\Omega_0 - 1)^{1/2},\\ &\Rightarrow ⅔ \quad \text{for } \Omega_0 \approx 1.\end{aligned}\right\} \tag{16}$$

In general $$0 < t_0 < ⅔T_0.$$

*Case II*: $\lambda_0 = 0, \Omega_0 = 1, q_0 = ½$

$$\left(\frac{a}{a_0}\right)^3 = \left(\frac{3t}{2T_0}\right)^2 \tag{17}$$

$$t_0 = ⅔T_0. \tag{18}$$

*Case III*: $1 \geq \lambda_0 > 0, \quad 0 \leq \Omega_0 < 1, \quad -1 \leq q_0 < ½$

$$\left.\begin{aligned}(a/a_0)^3 &= \left(\frac{\Omega_0}{1 - \Omega_0}\right)\sinh^2\left[3(1 - \Omega_0)^{1/2}\frac{t}{2T_0}\right],\\ &\Rightarrow \left(\frac{3\Omega_0^{1/2}t}{2T_0}\right)^2 \quad \text{for} \quad t \ll T_0.\end{aligned}\right\} \tag{19}$$

$$\left.\begin{aligned}t_0/T_0 &= \frac{2}{3(1 - \Omega_0)^{1/2}}\tanh^{-1}(1 - \Omega_0)^{1/2},\\ &\Rightarrow ⅔ \quad \text{for} \quad \Omega_0 \approx 1.\end{aligned}\right\} \tag{20}$$

In general $$\infty > t_0 > ⅔T_0.$$

Note that the distance measure can be arbitrarily chosen since $a$ appears only in the ratio $a/a_0$, where $a_0 = a(t_0)$ is the present value of $a$.

Case I represents a recontracting universe with $|\lambda_0|$ equivalent to an attractive force. The distance measure reaches a maximum given by

$$a_{\max}/a_0 = \left(\frac{\Omega_0}{\Omega_0 - 1}\right)^{1/3} \tag{21}$$

at

$$t/T_0 = \frac{\pi}{3(\Omega_0 - 1)^{1/2}} \tag{22}$$

and returns to zero at

$$t/T_0 = \frac{2\pi}{3(\Omega_0 - 1)^{1/2}}. \tag{23}$$

Bludman (1984) discusses this final 'crunch' in detail. The canonical condition for Case I, $q_0 > ½$, follows from eqn (14).

Case II represents an expanding universe in which the distance measure expands forever with $a \propto t^{2/3}$, but the expansion rate $a \propto t^{-1/3}$ approaches zero. Again eqn (14) yields the canonical condition for Case II, $q_0 = ½$.

Case III represents an ever expanding universe with $\lambda_0$ equivalent to a repulsive force. The distance measure eventually expands with time according to

$$(a/a_0) \propto \exp[(1 - \Omega_0)^{1/2}\, t/T_0] = \exp[\lambda_0^{1/2}\, t/T_0]. \tag{24}$$

Once again eqn (14) yields the canonical condition for Case III, $q_0 < ½$.

Using eqns (16), (18) and (20) it is possible to calculate $t_0/T_0$ for a range of values for $\Omega$ and $\lambda$ subject to the condition $\lambda_0 + \Omega_0 = 1$ for a flat, $k = 0$, universe. The results taken from Fowler and Meisl (1986) are shown in Fig. 1. The lowest value for $\Omega_0$ is taken to be 0.1. The production of the isotopes of hydrogen and helium in the big bang at $t \sim 300$ s is found to agree with solar system abundances for a present dimensionless baryon density given by $\Omega_{b0} = 0.1$ according to Wagoner, Fowler and Hoyle (1967) and Boesgaard and Steigman (1985) so $\Omega_0 > 0.1$ only if non-baryonic matter exists. Scenarios in which $\Omega_0 = \Omega_{b0} = 1$ will be discussed later. Also shown in Fig. 1 is the ratio $t_0/T_0$ calculated for the standard big bang model, $\lambda_0 = 0$ and $k = 0, \pm 1$ calculated by Sandage (1961). Figure 1 was given for a much larger range in the coordinates in Fig. 4 of Robertson (1956).

It will be clear from Fig. 1 that, even for the flat universe, $t_0/T_0$ varies slowly with $\Omega_0$. For example, differentiation of eqns (16) or (20) yields $d\Omega_0/d(t_0/T_0) = -9/2$ near $\Omega_0 = 1$. Thus a determination of $t_0/T_0$ must have a small uncertainty in order to yield a precise value of $\Omega_0$. The final results obtained in this article yield $1.8 > \Omega_0 > 0.7$ for $0.54 < t_0/T_0 < 0.76$ which illustrates how reducing the uncertainty in $t_0/T_0$ to $\approx 17$ per cent still yields a considerable spread in the resulting values of $\Omega_0$. It is interesting to note that for the standard model,

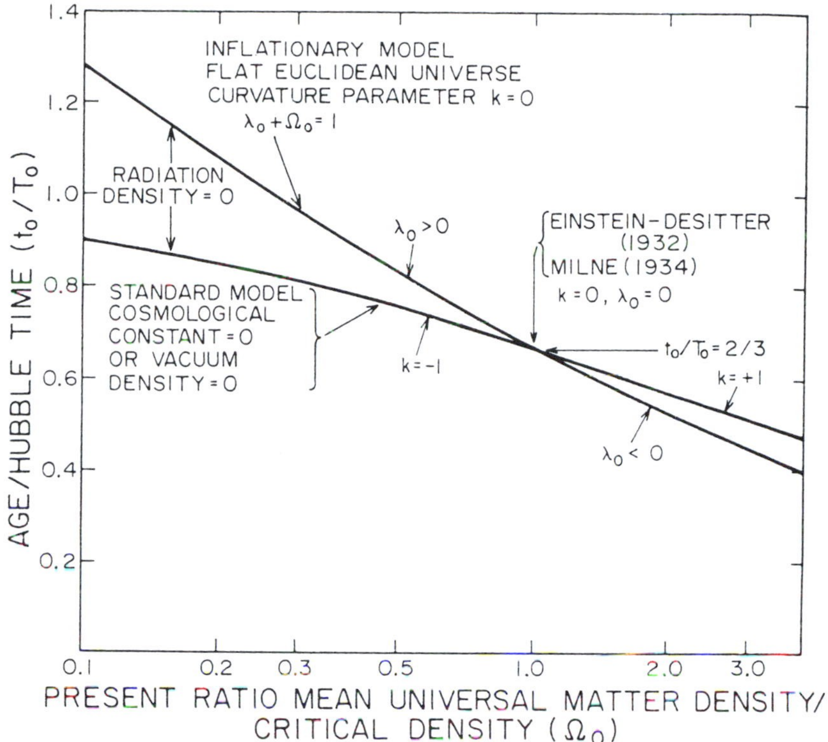


**Fig. 1** The ratio of the age of the observable universe, $t_0$, to the Hubble time, $T_0$, versus the dimensionless ratio, $\Omega_0$, of the present universal mean density to the critical value given in eqn (12). The curve labelled the inflationary model or flat Euclidean universe holds for $k = 0$, $\lambda_0 + \Omega_0 = 1$ and $q_0 = 3\Omega_0/2 - 1$ from eqn (14). The curve labelled the standard model holds for $\lambda_0 = 0$, $k = -1, 0, +1$ and $q_0 = \Omega_0/2$ from eqn (13). The point $t_0/T_0 = ⅔$ is shown for the case $k = 0$, $\lambda_0 = 0$.

with $\lambda_0 = 0$, one finds $d\Omega_0/d(t_0/T_0) = -15/2$. It is even more difficult than in the case of the flat universe to obtain a precise value for $\Omega_0$ by determining $t_0/T_0$.

## Supersymmetric particle theory and the cosmological constant

This section is based on articles by Freedman and Das (1977), Deser and Zumino (1977) and Pagels (1984). In elementary particle theory, rigorous supersymmetry implies the existence of supermultiplets made up of pairs of fermions and bosons with equal masses. Corresponding to the zero mass photon, a boson with spin one, there should exist a photino, a fermion of spin one-half, with zero mass. Corresponding to the zero mass graviton, a boson with spin two, there should exist a gravitino, a fermion of spin three-halves, with zero mass, and so on for quarks and leptons.

Rigorous supersymmetry yields a cosmological constant equal to zero but there is no observational support of rigorous supersymmetry. A boson partner of the fermion electron with equal mass should have been detected quite easily. There is considerable evidence that the mass of the photino and gravitino are not zero. Using the generally accepted results of light element isotopic abundances produced in the big bang and other considerations, Pagels (1984) finds that the photino and gravitino masses must fall in the range $10^{-2\text{–}103}$ eV or must be greater than $10^{22}$ eV. In the real world supersymmetry is broken in the vacuum. However, there are no direct observations on the mass of the photino or gravitino.

In the simplest forms of broken supersymmetry with supergravity (the graviton plus the gravitino) a relation can be derived for the square of the cosmological constant ($\Lambda$), involving a positive term from the breaking of the supersymmetry and a negative term from a possible gravitino mass ($M_g$). This relation is

$$\Lambda^2 = \frac{a^2}{2} - \frac{3M_g^2}{8\pi G}, \tag{25}$$

where $a$ is the strength of the breaking of supersymmetry. The condition for $\Lambda = 0$ in broken supersymmetry is then

$$M_g(\Lambda = 0) = (4\pi G a^2/3)^{1/2} \tag{26}$$

Note that $\Lambda$ is imaginary if $M_g$ exceeds the right-hand side of eqn (26). There is no direct observational evidence for the value of $a$ just as there is no observational evidence for the value of $M_g$ and there are no theoretical arguments for eqn (26). In broken supersymmetry the cosmological constant may be equal to zero but the proof of this remains for the future. Peebles (1984*a*) has pointed out that if the constant vacuum density $\rho_v = 3H^2\lambda/8\pi G = \Lambda/8\pi G$, is the same order of magnitude as $\rho$ just during our epoch then $\rho_v \sim 10^{-103}\,\rho$ just after inflation. A reduction by this enormous factor but not to zero requires fine-tuning which is not understood theoretically. This type of argument cannot be taken as definitive. Thus there remains a problem on which the determination of the Hubble time and of the age of the universe may shed some light.

## The Hubble time

In *The realm of the nebulae*, Hubble (1936) gave the value for his constant, $H_0 = 530\ \mathrm{km\,s^{-1}\,Mpc^{-1}}$, corresponding to $T_0 = 1.84$ Gyr. Hubble's value for $t_0$ was taken to be the maximum age of the universe but was soon challenged by geologists who were convinced that there existed terrestrial rocks some 3–4 Gyr in age.

Hubble's analysis of his red-shift observations are in principle still correct but his expression $H_0 = (c/r)\ (\Delta\lambda/\lambda_0)$, where $\Delta\lambda$ is the red-shift in light from a galaxy at proper distance, $r$, has always been plagued by uncertainties in the distance scale for $r$. The wavelength of the light from a source at zero velocity with respect to the observer is $\lambda_0$. Baade, Humason, Mayall, Sandage and Tammann have corrected Hubble's distance scale, always using observational techniques and analysis which improved with the years, so that Tammann and Sandage (1985) claim $H_0 = 50 \pm 7\ \mathrm{km\,s^{-1}\,Mpc^{-1}}$ on the basis of their latest observations or $T_0 = 19.5 \pm 2.7$ Gyr.

That is not all of the story. There is considerable controversy even today about the value of the Hubble constant as determined by classical redshift-distance observations. The values $H_0 = 95 \sim 10\ \mathrm{km\,s^{-1}\,Mpc^{-1}}$, $T_0 = 10.3 \sim 1$ Gyr are given by de Vaucouleurs (1982). These widely different values of $H_0$ arise from controversies about the distance scale and about the velocity component of the local group of galaxies towards the Virgo cluster which is superimposed on the cosmological Hubble flow and about the velocity component of the local supercluster towards the supercluster in Hydra-Centaurus. Our Galaxy, the Milky Way, is a member of the local group which in turn is a member of the local supercluster. For the extent of the disagreement about the Virgocentric velocity see Davis and Peebles (1983). For a general discussion see Chapter 6 of Rowan-Robinson (1985).

A promising new technique as a distance indicator for spiral galaxies is that developed by Tully and Fisher (1977). This technique relates the velocity width of the 21-cm line emission of neutral hydrogen from galaxies (i.e. essentially twice their maximum rotational velocity) to their absolute magnitude. Comparison of absolute magnitude with the observed apparent magnitude yields the distance to the galaxy. With the distance the observed redshift yields the Hubble constant. Aaronson *et al.* (1986) used infrared magnitudes to obtain $H_0 = 90\ \mathrm{km\,s^{-1}\,Mpc^{-1}}$. Calibration of the Tully–Fisher technique is discussed in detail by van den Bergh (1984) using observations in blue light and in the infrared. Van den Bergh concludes that $H_0 = 68 \pm 17\ \mathrm{km\,s^{-1}}$ $\mathrm{Mpc^{-1}}$, $T_0 = 14.4 \pm 3.6$ Gyr from the Tully–Fisher technique using observations in blue light and $H_0 = 82 \pm 18\ \mathrm{km\,s^{-1}\,Mpc^{-1}}$, $T_0 = 11.9 \pm 2.6$ Gyr fron infrared observation.

The differences in the optical determinations of the Hubble constant and the Hubble time necessitate the application of nuclear physics to cosmochronology. Nuclear physics is capable of determining an independent measurement of the Hubble time. There is little doubt that the light curve of Type 1a Supernovae (SN) is powered by the radioactive decay of $^{56}$Ni through $^{56}$Co to $^{56}$Fe. The arguments are summarized in van den Bergh (1985). The basic papers are Hoyle and Fowler (1960), Colgate, Petschek and Kriese (1980), Woosley, Axelrod and Weaver (1984), Arnett, Branch and

Wheeler (1985), Wheeler and Harkness (1986), Hillebrandt *et al.* (1986) and Nomoto (1986). Type 1a SN originate as a white dwarf accreting matter from a giant binary companion. Normal stellar and nuclear evolution leads to a white dwarf consisting mainly of carbon and oxygen. When the accretion leads to a total white dwarf mass exceeding the Chandrasekhar limit of $\sim 1.4$ $M$ explosive deflagration of the carbon and oxygen occurs. The rapid deflagration results in the synthesis of nuclei with equal numbers of protons and neutrons all the way to $^{56}$Ni. In the references previously cited the amount of $^{56}$Ni produced covers the range from $0.3 \leqslant M(^{56}\mathrm{Ni})/M \leqslant 0.9$ with the most probable value being $M(^{56}\mathrm{Ni})/M = 0.6$. The lower limit is substantiated by the observations of Graham *et al.* (1986) who discovered $0.3 \pm 0.2$ $M$ of iron in Type 1a SN 1983n. This was accomplished by the observation of strong (Fe II) line emission in the infrared at 1.644 μm from the optically thin ejecta. These observations provide strong support for the $^{56}$Ni–$^{56}$Co–$^{56}$Fe radioactive decay scenario. However, the authors note that SN 1983n showed unusual UV and optical spectra and was an atypical Type 1a. They also note that the considerable uncertainty quoted is due to uncertainty in the distance of SN 1983n and in the relevant atomic data. The upper limit follows from the fact that the maximum mass involved, the Chandrasekhar limit, is 1.4 $M_\odot$. The supernova event converts $^{12}$C and $^{16}$O predominantly into nuclei with $Z = N$ from $^{20}$Ne to $^{56}$Ni with a small production of $^{60}$Zn and beyond. It is unreasonable to expect the total mass to be converted into $^{56}$Ni and 0.9 $M_\odot$ is a firm upper limit. The references cited in this paragraph have been studied carefully and have led to the conviction that the range 0.3–0.9 $M_\odot$ corresponds to a $3\sigma$-uncertainty (confidence limit = 0.997) so that $M(^{56}\mathrm{Ni})/M_\odot = 0.6 \pm 0.1$ $(1\sigma)$. [Professor Craig Wheeler has since informed me that SN 1983n is a Type 1b Supernova so that the observations by Graham *et al.* (1986) are irrelevant to the determination of the lower limit for $M(^{56}\mathrm{Ni})/M_\odot$ produced in Type 1a Supernovae. The lower limit thus follows only from theoretical arguments.]

Arnett, Branch and Wheeler (1985) calculate the maximum absolute luminosity of Type 1a Supernovae and find $L_{\mathrm{max}} = 2.2 \times 10^{43}(M_{\mathrm{Ni}}/M_\odot)\ \mathrm{erg\,s^{-1}}$. In detail they find that the maximum occurs 17 days after the initial event. This is in good agreement with those cases where the initial event or the rise to the maximum luminosity has been observed.

The apparent luminosity is related to the absolute luminosity by $L_{\mathrm{app}} \propto r^{-2}L_{\mathrm{abs}} \propto r^{-2}M_{\mathrm{Ni}}/M_\odot$, where $r$ is the distance to the supernova. Thus measurement of the apparent luminosity at maximum permits determination of $r$ and thus of the Hubble constant, $H_0 = v/r = cz/r$, if the redshift, $z$, of the galaxy in which the supernova occurs is known. Observations on six Type 1a Supernovae in clusters of galaxies with known redshifts greater than $cz = 3000\ \mathrm{km\,s^{-1}}$, are used to determine$H_0$. With redshifts this large, departures from the ideal Hubble flow, discussed previously in this section, introduce only small uncertainties. This is reinforced by Lynden-Bell (1986) who shows that all galaxies out to $cz = 12\,000$ km s$^{-1}$ have the same motion superimposed on the Hubble flow. The result is $H_0 = 45[\mathrm{M}_\odot/\mathrm{M}(^{56}\mathrm{Ni})]^{1/2}$ km s$^{-1}$ Mpc$^{-1}$. With this equation $M(^{56}\mathrm{Ni})/M_\odot = 0.6 \pm 0.1$ ($1\sigma$) yields $H_0 = 58 \pm 5(1\sigma)$ km s$^{-1}$ Mpc$^{-1}$ and

$$T_0 = 16.8 \pm 1.4\,(1\sigma)\ \mathrm{Gyr}. \qquad (27)$$

These results are in excellent agreement with those of Branch (1979) who found $H_0 = 56 \pm 15$ km s$^{-1}$ Mpc$^{-1}$ from composite photometric and spectroscopic data on Type 1 Supernovae.

## The age of the universe from astronomy

The age of the universe, $t_0$, the time from the big-bang singularity to the present, is calculated according to the equation

$$t_0 = t_F + t_G, \qquad (28)$$

where $t_F$ is the time of formation of galaxies in general and of the Milky Way in particular and $t_G$ is the age of galaxies in general and of the Milky Way in particular. We use Milky Way loosely to describe our Galaxy in general and not just the disk.

The time of formation of galaxies is a matter of some debate and is discussed by Peebles (1984*b*). Fowler and Meisl (1986*a*, *b*) adopt the following value

$$t_F = 1.0 \pm 0.4\,(1\sigma)\ \mathrm{Gyr}. \qquad (29)$$

The large $1\sigma$-uncertainty allows for the considerable uncertainty in the time in the early universe after decoupling at which appropriate conditions in density and temperature for galaxy formation occurred. If those conditions could be calculated more precisely, then $t_F$ could be expressed more precisely. There is also the very real possibility that galaxies formed at different times under varying local conditions. The $t_G$ to be used in what follows will be the age of the Milky Way so $t_F$ for the Milky Way is required in eqn (28). This introduces additional uncertainty. Finally in spiral galaxies there may be a difference in the time of formation of the halo, the nuclear core and the disk. More uncertainty in $t_F$ results. In spite of the large uncertainty in $t_F$ it will be found in what follows that it contributes little uncertainty to $t_0$.

The age of the Milky Way is taken to be that of the oldest systems in the Galaxy. For the halo the oldest systems are the metal-poor globular clusters and in the disk the oldest systems are the cool, faint white dwarfs.

Iben and Renzini (1984) take M92 as representative of globular clusters in our Galaxy and find

$$\log_{10} t_G = 1.071 - 1.88\,(Y_I - 0.3), \tag{30}$$

where $t_G$ is in Gyr and $Y_I$ is usually taken equal to $Y_P$, the primordial abundance of helium by mass fraction produced in big-bang nucleosynthesis. Since Peebles (1966) and Wagoner, Fowler and Hoyle (1967) many attempts have been made to determine $Y_P$ from the abundances of $^2$D, $^3$He, $^4$He and $^7$Li produced in the big bang. Iben and Renzini (1984) state that these attempts yield $Y_P = 0.23 \pm 0.04$ with a rather large uncertainty as indicated. In addition pregalactic and protogalactic nucleosynthesis can increase $Y_I$ over $Y_P$. Thus it is preferable to turn to observations such as those of Kunth and Sargent (1983) who find what they term $Y_P$ and which we term $Y_I = 0.235 \pm 0.003$ from high-precision spectroscopy on He/H II regions in 12 emission-line galaxies. Kunth and Sargent (1983) call their $Y_P$ the primordial or pregalactic helium abundance and show that other estimates which lead to a lower value of $Y_P$ do not stand up to critical study, usually because of lack of attention to errors. For contrary results see, for example, James and Demarque (1983).

Pagel, Terlevich and Melnick (1986) re-analyse determinations of $Y_P$ from emission-line spectra of extragalactic H II regions and conclude that

$$Y_P = 0.232 \pm 0.004. \tag{31}$$

Using this value for $Y_I$ in eqn (30) one finds

$$t_G = 15.8 \pm 2.4\,(1\sigma)\ \text{Gyr} \tag{32}$$

where the uncertainty is reduced from the $\pm$ 3.5 Gyr given by Iben and Renzini (1984) because the uncertainty in $Y_P$ given by Pagel, Terlevich and Melnick (1986) is so small. The other sources of uncertainty in eqn (30) are unchanged. Using eqns (29) and (32), eqn (28) yields

$$t_0 = 16.5 \pm 2.4\,(1\sigma)\ \text{Gyr}. \tag{33}$$

[It has since come to my attention that Noerdlinger and Arigo (1980) introduced helium diffusion and found a reduction in cluster ages of 22 per cent from conventional values. If this correction is made, eqn (32) yields 12.3 Gyr and eqn (33) yields 13.3 Gyr.]

The preceding discussion has been based on the conventional belief that significant mass loss does not occur during the main sequence stage of stellar evolution. This belief has been recently challenged by Willson, Bowen and Struck-Marcell (1987). They investigate the effect of reasonable mass loss rates during the early main sequence life of globular cluster stars. They are able to explain the observed increase in *apparent* ages of globular clusters with decreasing metallicity found by Ciardullo and Demarque (1977). Their final result is expressed in this quotation from their paper: 'We conclude that it is possible that no stars in our galaxy are older than 7–10 × $10^9$ yr old. A new method for determining the ages of globular clusters will need to be found, if indeed it is mass loss, not nuclear evolution, that is determining the "turnoff" mass.' The nuclear physicist is saved from contradictory results using nuclear physics as will be seen in the next section.

White dwarfs provide an independent measure of the age of the Galactic disk. Winget (1987) has analysed observations of the number density of white dwarfs as a function of luminosity. The observed density of white dwarfs increases rapidly with decreasing luminosity and then shows a sharp cut-off at $\log (L_{\text{wd}}/L_{\odot})$ equal to $-$ 4.5. Winget's conclusion is that the disk is not old enough for cooling white dwarfs to become fainter than this luminosity. His final result is expressed in this quotation from his paper: 'Thus, our current best value from all these considerations, including the uncertainty in the exact luminosity of the downturn, is probably about 9.3 ± 2 Gyr for the age of the oldest population of white dwarfs.' It can only be concluded that the astronomical evidence for the age of the Galaxy and the universe is very uncertain indeed. We turn once again to our bias for chronology determined by the methods of nuclear physics.

## The age of the universe from nuclear cosmochronology

Nucleocosmochronology uses the production and decay of radioactive nuclei as chronometers to determine $\Delta$, the duration of nucleosynthesis in the Galaxy before the formation of the solar system. Addition of $t_s$, the age of the solar system, to this duration yields the age of the Galaxy

$$t_G = \Delta + t_S \tag{34}$$

and then, through eqns (28) and (29), the age of the universe. Here, we will include in $t_S$ small additions ($\lesssim$ 0.1 Gyr) for the time of the formation and evolution of the stars which first synthesized the chronometers in the Galaxy and for a possible interval of free decay between the end of nucleosynthesis and the formation of the solar system. From the many very close values in the literature we round off that of Wasserburg *et al.* (1977) to

$$t_S = 4.6 \pm 0.1\,(1\sigma)\ \text{Gyr}. \tag{35}$$

The most suitable chronometers are $^{232}$Th($^{208}$Pb) with mean lifetime $\tau$ = 20.270 Gyr, $^{235}$U($^{207}$Pb) with $\tau$ = 1.0154 Gyr and $^{238}$U($^{206}$Pb) with $\tau$ = 6.4464 Gyr. The mean lifetimes have been calculated from the half-lives given in

Tuli (1985). The final decay products, all isotopes of Pb, are given in parentheses. Other chronometers, their decay products and their mean lifetimes include:

$^{26}$Al($^{26}$Mg) 1.039 Myr, $^{87}$Rb($^{87}$Sr) 69.2 Gyr,
$^{129}$I($^{129}$Xe) 22.7 Myr, $^{187}$Re($^{187}$Os) 62.8 Gyr
$^{244}$Pu($^{240}$U and fission products $\sim$ 0.1 per cent) 0.117 Gyr.

Of these, we believe that the three short-lived chronometers can tell us little about the long duration of nucleosynthesis in the Galaxy and that there are many observational and theoretical problems involved in the use of the two long-lived ones.

The history of nuclear cosmochronology and the current situation are discussed in Fowler (1984) and in Fowler and Meisl (1986*a*). Here we will discuss nucleosynthesis in the Galaxy using observational input on the abundances, $N_A$, of $^{232}$Th, $^{235}$U and $^{238}$U plus theoretical calculations of their abundances, $P_A$, produced in the *r*-process involving rapid neutron capture and beta decay. $^{238}$U is taken as the standard of comparison and the mass number, $A$, will be designated by the last digit (2, 5, or 8) in the number when used as a subscript

$$K_{28} = \frac{N_2/N_8}{P_2/P_8} = \frac{n_2}{p_2}, \tag{36}$$

$$K_{58} = \frac{N_5/N_8}{P_5/P_8} = \frac{n_5}{p_5}. \tag{37}$$

The relative abundances, $n_A$, are evaluated at the origin of the solar system 4.6 Gyr ago taken as the end of nucleosynthesis for solar system elements. These relative abundances are calculated from the observed values, $n_A^P$, at the present time using the relative mean lifetimes in equations for free decay. The relative production rates, $p_A$, are taken as the average over all *r*-process events, presumably supernovae, which occurred in the Galaxy from its formation to that of the solar system. It will be clear that early in the Galaxy before appreciable radioactive decay took place, $n_2^F = p_2$ and $n_5^F = p_5$, where F as a superscript designates the time of formation of the Galaxy. It is taken that the *r*-process occurs in supernovae of fairly large mass so that the time of formation and evolution of the stars becoming supernovae is small, $\lesssim 0.1$ Gyr.

The input data are shown in Table 1. The absolute values of the relative *mean* lifetimes, $\tau_{A8}$, are calculated from the half-lives $t_5 = 0.7038$ Gyr, $t_2 =$ 14.05 Gyr and $t_8 = 4.4683$ Gyr given in Tuli (1985). The ratio $n^P{}_5$ at the present time is also taken from this reference and the uncertainty is taken to be negligible since the ratio involves isotopes of the same element. The ratio $n_2^P$ is taken from Anders and Ebihara (1982) who analysed measurements on C1 chondritic meteorites. These authors give a 5.7 per cent standard deviation

**Table 1** Relative abundance and production input data for $^{235}U$ and $^{232}Th$ with $^{238}U$ as standard

| A | $\tau_{A8}$(Gyr) | $n^P_A$ | $n_A$ | $p_A$ | $K_{A8}$ |
|---|---|---|---|---|---|
| 235 | 1.2052 | $7.2526 \times 10^{-3}$ | 0.330 | $1.34 \pm 0.19$ | $0.246 \pm 0.035$ |
| 232 | 9.4526 | $3.75 \pm 0.23$ | 2.305 | $1.71 \pm 0.07$ | $1.351 \pm 0.100$ |

(Numbers following the $\pm$ sign indicate $1\sigma$-uncertainties.)

for their assigned abundance of Th and an 8.4 per cent standard deviation for their assigned abundance of U. Since these two actinides have similar chemical properties the abundance ratios have been studied for individual cases with the resulting standard deviation for $n^P_2$ given in Table 1. Using the relative mean lifetimes, $\tau_{58}$ and $\tau_{28}$ given in the table, it is then possible to calculate $n_5$ and $n_8$ as tabulated.

The *r*-process production of $^{232}Th$, $^{235}U$ and $^{238}U$ has been re-calculated with results slightly different from those given in Fowler (1977). The changes result mainly from the fact that $^{250}Cm$, one of the progenitors of $^{238}U$ at the termination of the *r*-process, is now known to decay by spontaneous fission only 65 per cent of the time as indicated in Tuli (1985) rather than 90 per cent of the time as previously believed. Thus $^{238}U$ has 3.35 progenitors rather than 3.10 while the number of progenitors remains at 5.8 for $^{232}Th$ and at 6.0 for $^{235}U$. If *r*-process production were uniform for all of the progenitors the *a priori* ratios would be $p_2 = 5.8/3.35 = 1.73$ and $p_5 = 6.0/3.35 = 1.79$. The detailed calculations, with careful attention paid to the uncertain factors involved, yield $p_2 = 1.71 \pm 0.07(1\sigma)$ and $p_5 = 1.34 \pm 0.19(1\sigma)$. Both $^{232}Th$ and $^{238}U$ are even *A*, even *Z* and even *N* nuclei and all of their progenitors during the *r*-process and at the termination of the *r*-process should show very similar behaviour. On the other hand $^{235}U$ is an odd *A* nucleus and all of its progenitors will also have *A* odd. On very general grounds it can be expected that the odd nucleon will have a binding energy less than that of a paired nucleon so that in the *r*-process this will lead to increased destruction rates and lower abundances for odd *A* nuclear species. It is quite reasonable that the *a priori* ratio, $p_5 = 1.79$ is reduced in detailed calculations to 1.34. Having evaluated the relevant $n_A$ and $p_A$ it is possible to calculate the final column in Table 1 showing $K_{28} = 1.35 \pm 0.100$ and $K_{58} = 0.246 \pm 0.035$, which become the basic input data. More significant figures than are justified are retained until round-off in the final results.

Since we restrict ourselves to two input data, models for calculating $\Delta$ can involve only one more parameter to be evaluated. Fowler and Meisl (1986*a*) discussed three models which express the time dependence of nucleosynthesis in the Galaxy over its duration $\Delta$. The first model involved exponential nucleosynthesis ($\exp - t/\tau_R$) with time constant $\tau_R$ for the *r*-process.

The second model assumed an early spike of nucleosynthesis with relative magnitude $S_E$, followed by uniform synthesis ($\tau_R = \infty$) with relative magnitude $1 - S_E$. The third model assumed a late spike of nucleosynthesis with relative magnitude $S_E$, preceded by uniform synthesis ($\tau_R = \infty$) with relative magnitude $1 - S_E$. It has long been known that the value of $\Delta$ obtained is relatively insensitive to the model employed. For example, if one asks for the average age of stable elements for $S_E$ or $S_L = 1$ it will be obvious that the average will just be $\Delta$ while for $S_E$ or $S_L = 0$ the average age will be $\Delta/2$. Thus, for not too short-lived nuclei, nucleosynthesis in a single spike will yield a $\Delta$ approximately one-half that of the uniform nucleosynthesis. Detailed calculations always yield a spread in $\Delta$ considerably less than this maximum factor of two.

In recent years it has become increasingly apparent that the metallicity of Galactic disk material was somewhat greater than zero when the disk formed and increased more or less uniformly with time thereafter. Standard infall models for the history of the disk are consistent with this result as shown by Twarog and Wheeler (1982) who provide detailed references to the relevant literature. Twarog (1980), in the curve in his Fig. 12, shows the derived enrichment history of the Galactic disk assuming a constant star formation rate and constant infall rate. This figure is derived from his four-colour and $H\beta$ photometry of a large sample of southern *F* dwarfs in the neighbourhood of the Sun. We find that Twarog's curve can be fitted prior to the formation of the solar system by the linear relation $0.16 + 0.84t/\Delta$, in the notation of this article in which $t = 0$ when nucleosynthesis begins and $t = \Delta$ when it ends.

It is thus clear that a time-dependent model for nucleosynthesis in the Galaxy with an early 'spike' followed by uniform synthesis is to be preferred and only this model will be considered in what follows. For this model Fowler and Meisl (1986*a*) show that

$$K_{AS} = \frac{S_E \exp(-\Delta/\tau_A) + (1 - S_E)(\tau_A/\Delta)[1 - \exp(-\Delta/\tau_A)]}{S_E \exp(-\Delta/\tau_S) + (1 - S_E)(\tau_S/\Delta)[1 - \exp(-\Delta/\tau_S)]}, \tag{38}$$

where for the sake of generality we use a subscript *S* for any standard nucleus. In this discussion $S = 8$ for $^{238}U$.

Computer calculations to determine $\Delta$ and $S_E$ have been performed using the values for $K_{58}$ and $K_{28}$ and their $1\sigma$-uncertainties given in Table 1. The final results are $S_E = 0.17 \pm 0.02$ ($1\sigma$) and $\Delta = 5.4 \pm 1.5$ ($1\sigma$) Gyr. It is somewhat surprising that the $1\sigma$ uncertainty in $S_E$ is rather small but that in $\Delta$ is in line with previous calculations using differing values of $K_{58}$ and $K_{28}$.

The $\Delta$ and $S_E$ we have obtained can be used to obtain the mean age of stable elements. It is $\bar{\Delta} = 0.83 \times 2.7$ plus $0.17 \times 5.4 = 3.2$ Gyr. This can be compared with the mean age of our two long-lived chronometers, $^{232}Th$ and

$^{238}$U, using the prescription of Schramm and Wasserburg (1970). In our notation this can be expressed as $\bar{\Delta}_{28} = \tau_{28} \ln K_{28} = 2.8$ Gyr. The agreement between these two mean ages is very satisfactory since they should and do differ slightly in the right direction.

Before concluding this section it must in all candour be noted that considerably larger values for $\Delta$ have been obtained by the introduction of beta-delayed fission into the *r*-process network of reactions. By including substantial beta-delayed fission Thielemann, Metzinger and Klapdor (1983) found $\Delta = 16.2\ (+2, -4)$ Gyr, Thielemann and Truran (1986) found $7.0 < \Delta < 15.0$ Gyre while Cowan, Thielemann and Truran (1986) found $7.8 < \Delta < 10.1$ Gyre. All of these results require increasing *r*-process during $\Delta$ or a late 'spike' just before the formation of the solar system in contradiction with the results of Twarog (1980) and others.

It is well known that the probability of any fission process depends critically on the height of the fission barrier used in calculating the probability. All of the references cited in the preceding paragraph used the fission barrier heights calculated in a systematic way for many neutron-rich heavy nuclei by Howard and Möller (1980). These authors used a modified oscillator potential model for heavy nuclei. On very general grounds Nix (1986) has suggested that this will lead to low barrier heights and increased beta-delayed fission rates. According to Nix (1986) the problem can only be resolved by using a more realistic diffuse edge potential for the heavy nuclei.

An indication that beta-delayed fission may not be important in the *r*-process has been emphasized by Hoff (1986). Tests involving thermonuclear explosions subject $^{238}$U target nuclei to intense neutron fluxes, and successive neutron captures produce isotopes of U and Pa up to $A = 257$ which beta-decay to relatively long-lived detectable nuclei after the thermonuclear explosion. The abundances of these neutron rich isotopes produced in the tests agree well with simple theoretical calculations based on neutron reactions without beta-delayed fission in the ultimate decay chain. Hoff (1986) has introduced the beta-delayed fission rates used by Thielemann, Metzinger and Klapdor (1983) and finds it difficult to reproduce theoretically the observed results. He concludes '. . . since there are alternate possibilities that explain the observations adequately, the presence of substantial amounts of beta-delayed fission is not required in order to understand the data'. It should also be noted that Meyer *et al.* (1985) calculated *r*-process abundances including beta-delayed fission but using new beta-strength distribution functions and concluded only that $\Delta > 4.2$ Gyr but specified no upper limit. Their result does not contradict the value $\Delta = 5.4 \pm 1.5$ obtained here. Fuller (1986), following Bloom and Fuller (1986), finds beta strength distribution functions in general agreement with Meyer *et al.* (1985). The beta-strength used by Thielemann, Metzinger and Klapdor (1983) peaked at much too high an energy in the daughter nucleus and consequently permitted beta-delayed fission over their low fission barrier heights.

The long-lived chronometers,

$$^{87}\mathrm{Rb}(\beta^-)^{87}\mathrm{Sr} + 0.283\ \mathrm{MeV},\ ^{187}\mathrm{Re}(\beta^-)^{187}\mathrm{Os} + 2.64\ \mathrm{keV}$$

have been used to obtain values for $\Delta$. Beer and Walter (1983) find $\Delta = 9 \pm 6$ Gyr for $^{87}$Rb($\beta^-$)$^{87}$Sr which is not in conflict with the value found here. Winters, Macklin and Hershberger (1986) have completed the necessary nuclear physics measurements needed in the case of $^{187}$Re($\beta^-$)$^{187}$Os. Using the terrestrial mean lifetime, 62.8 Gyr, for the $^{187}$Re decay they find $\Delta = 11.0 \pm 2.5$ Gyr in conflict with the value found here. However, there is a serious problem in the atomic physics of the $^{187}$Re-decay in astronomical environments as discussed by Takahashi and Yokoi (1983). When $^{187}$Re is subjected to temperatures of $10^7$ K or greater, it loses atomic electrons, becomes ionized, and rapid beta-decay to bound states in the daughter $^{187}$Os takes place. The lifetime is decreased by orders of magnitude so that the effective lifetime to be used in the $^{187}$Re-decay is reduced by astration in much the same way that the big bang abundance of deuterium has been reduced. The lifetime could well be reduced by a factor of two and the $\Delta$ of Winters, Macklin and Hershberger (1986) is certainly an upper limit and could be reduced to as low as $\sim$ 6 Gyr, again not in conflict with our determination.

Returning to the calculation of the age of the universe from nuclear cosmochronology, the final result is

$$\left.\begin{aligned} t_0 &= t_F + \Delta + t_S, \\ &= (1.0 \pm 0.4) + (5.4 \pm 1.5) + (4.6 \pm (4.6 \pm 0.1)\ \mathrm{Gyr}, \\ &= 11.0 \pm 1.6(1\sigma)\ \mathrm{Gyr}. \end{aligned}\right\} \tag{39}$$

It must be emphasized that the $1\sigma$-uncertainty arises primarily from statistical uncertainties in *r*-process abundances and does not include possible systematic uncertainties in $t_F$ and $\Delta$ if the accepted model for Galactic evolution and nucleosynthesis is in error.

## Conclusions

Nuclear cosmochronology yields determinations of the age of the universe and of the Hubble time given in Table 2. The value of the age comes from the updated computer calculations of Fowler and Meisl (1986*b*) discussed above. The value for the Hubble time comes primarily from the work of Wheeler and Harkness (1986) but the $1\sigma$-uncertainty is the best estimate made by the present author. The ratio of age to the Hubble time is

$$t_0/T_0 = 0.65 \pm 0.11(1\sigma). \tag{40}$$

**Table 2** Conclusions

| |
|---|
| Age (Fowler and Meisl 1986) $t_0 = 11.0 \pm 1.6(1\sigma)$ Gyr |
| Hubble time (Wheeler and Harkness 1986) $T_0 = 16.8 \pm 1.4(1\sigma)$ Gyr |
| Ratio: $t_0/T_0 = 0.65 \pm 0.11(1\sigma)$ |
| |
| Results: |
| Curvature parameter: $k = 0$ (inflationary model) |
| Cosmological constant: $\lambda_0 = -0.1$ (+0.3 to −0.8) |
| Critical density: $\Omega_0 = +1.1$ (+0.7 to +1.8) |
| Deceleration parameter: $q_0 = +0.6$ (+0.0 to +1.7) |

It is now possible to use Fig. 1 to evaluate the values of various cosmological parameters which correspond to the $1\sigma$-range $0.54 \leqslant t_0/T_0 \leqslant 0.76$. These values for a flat, $k = 0$, universe derived from the inflationary model are also given in Table 2. For example, the value of $\Omega_0$ corresponding to the most probable value of $t_0/T_0 = 0.65$ is $\Omega_0 = +1.1$ but the $1\sigma$-uncertainty permits the range, $+0.7 \leqslant \Omega_0 \leqslant +1.8$. It will be noted that a zero cosmological constant, $\lambda_0 = 0$, with $\Omega_0 = 1$ is not far off the mark.

For the purists who prefer $\lambda_0 = 0$, $k = 0 \pm 1$, it is found from Fig. 1 that the range $0.6 \leqslant \Omega_0 \leqslant 2.2$ is permitted by the $1\sigma$-uncertainty in $t_0/T_0$. There is not much in it! It must also be strongly emphasized that there is no support for the adoption by Klapdor and Grotz (1986) of the range of values for a non-vanishing cosmological constant given by $\Lambda = (4.7 \text{ to } 19) \times 10^{-57}\ \text{cm}^{-2}$ in their units or $\lambda_0 = +0.4$ to $+1.6$ for our dimensionless parameter using $T_0 = 16.8$ Gyr. This is to be contrasted with the range in Table 2, namely $\lambda_0 = -0.8$ to $+0.3$ which includes $\lambda_0 = 0$ while the results of Klapdor and Grotz (1986) exclude it.

There exists some supporting evidence for our general result that our universe could well be flat and Euclidean in three dimensions with zero cosmological constant, have a density equal to the closure value and a future of asymptotic expansion leading to a final expansion velocity approaching zero.

Applegate, Hogan and Scherrer (1987) and Alcock, Fuller and Mathews (1987) suggest that the abundances of the isotopes of the light elements produced in the big bang could agree with observations for an inhomogeneous universe with dimensionless baryon density given by $\Omega_{bo} = 1$ rather than the generally accepted value $\Omega_{bo} \approx 0.1$. Pronounced non-linear fluctuations in the neutron–proton ratio throughout the early universe may be produced during quark–hadron phase transitions. Neutron enriched regions produced abundances which when averaged over all regions simulate those produced in the standard big bang with $\Omega_{bo} = 0.1$. This suggestion is illustrative of the difficulty of ruling out $\Omega_0 = \Omega_{bo} = 1$ on the basis of currently accepted big bang calculations.

Loh and Spillar (1986) and Loh (1987) have measured the redshifts and fluxes of 1000 field galaxies with a median redshift of 0.5 up to a maximum redshift of 1.2. With these observations they determine $\Omega_0$ by measuring the volume element (to a scale factor $H^3$) as a function of redshift. Their measurement detects any kind of matter, luminous or dark. Loh (1987) finds that $\Omega_0 = 0.9(+0.4, -0.2)$ and $\lambda_0 = 0.1\ (-0.4, +0.3)$ with 95 per cent confidence and constrains $t_0/T_0$ to the range 0.60–0.88 which agrees very well with our range 0.54–0.76. It is gratifying that nuclear chronology and optical galaxy counts are in such good agreement.

Yahil, Walker and Rowan-Robinson use IRAS 60 $\mu$m sources to map the local ($\lesssim 200\ h_0^{-1}$ Mpc) gravitational field and to determine its dipole component. The gravitational dipole moment is found to point $26° \pm 10°$ away from the direction of the velocity of the Local Group relative to the microwave background radiation. Comparison of the two anisoptropies using linear perturbation theory yields the value $\Omega_0 = 0.85 \pm 0.16$ (statistical IRAS error only) with nonlinear effects increasing $\Omega_0$ by $\approx$ 15 per cent. This is additional strong supporting evidence for $\Omega_0 \approx 1$.

It clearly cannot be claimed that $\Omega_0 = 1$ when the range allowed by our results is $+0.7$ to $+1.8$. There is the possibility just discussed that $\Omega_{bo}$ can be unity but this may well be in conflict with astronomical observations on luminous ($l$) matter in galaxies on large scales of the order of 100 MPc which suggest $\Omega_{l0} \approx 0.1$–$0.3$. Biased galaxy formation in regions where density fluctuations were large in the early universe may provide the answer as discussed in Peebles (1986) and Dekel (1986). In this scenario most of the baryonic matter in the universe lies outside luminous galaxies. There have also been many suggestions that dark matter—hot, warm or cold—exists in the form of exotic, nonbaryonic particles such as neutrinos, photinos, gravitinos, other 'inos', axions and so on. The reader is once again referred to Pagels (1984) and references therein. The contents of the universe may add up to the critical density, but, to put it simply, we do not know what it is! The search for dark matter stimulates the invention of many new sophisticated detectors enumerated by Waldrop (1986).

Finally it is important to note that there are other combinations of $\lambda_0$ and $\Omega_0$ which yield $t_0/T_0 = 2/3$. For example McVittie (1962) in his table 1 finds $t_0 = 0.68$ for $\lambda_0 = -2$, $\Omega_0 = 0$ ($q_0 = 2$) and for $\lambda_0 = -1$, $\Omega_0 = +0.3$ ($q_0 = 1.15$). With appropriate choices of $a_0$ these correspond to $k = -1$. It is also clear from his table 1 (entry Ex. 8) that $t_0/T_0$ can also approximately equal $2/3$ for cases with $k = +1$. These are pointed out even though this article accepts $k = 0$ from the early universal inflation so that $\lambda_0 + \Omega_0 = 1$ and stresses that the case which yields $t_0/T_0 = 2/3$ is $\lambda_0 = 0$, $\Omega_0 = 1$ ($q_0 = 1/2$).

## Acknowledgements

I thank Roger Penrose and Wadham College for hospitality during the period of the Milne Lecture and Miranda Weston-Smith, Milne's granddaughter, and G. M. Harding of IBM UK, Ltd for the arrangements which made my visit to Oxford possible. I am greatly indebted to Chris C. Meisl for performing the computer calculations which yielded the numerical results given in the article. I acknowledge many discussions on cosmochronology with Friedel Thielemann who tolerated my prejudices and shared his with me. Support of the National Science Foundation through grant PHY85-05682 is acknowledged. The Milne Lectures are sponsored by IBM UK Ltd.

## References

Aaronson, M., Bothun, G., Mould, J., Huchra, J., Schemmer, R. A. and Cornell, M. E., 1986. *Astrophys. J.*, **302,** 536.

Alcock, C., Fuller, G. M. and Mathews, G. J., 1987. Preprint.

Anders, E. and Ebihara, M., 1982. *Geochim. Cosmochim. Acta*, **46,** 2363.

Applegate, J. H., Hogan, C. J. and Scherrer, R. J., 1987. Preprint.

Arnett, W. D., Branch, D. and Wheeler, J. C., 1985. *Nature*, **314,** 337.

Barrow, J. D. and Turner, M. S., 1982. *Nature*, **298,** 801.

Beer, H. and Walter, G., 1983. *Astrophys. Space Sci.*, **100,** 243.

Bloom, S. D. and Fuller, G. M., 1986. *Nucl. Phys.*, **A440,** 511.

Bludman, S. A., 1984. *Nature*, **308,** 319.

Boesgaard, A. M. and Steigman, G., 1985. *Ann. Rev. Astr. Astrophys.*, **23,** 319.

Bondi, H., 1952. *Cosmology*, p. 89, Cambridge University Press.

Branch, D., 1979. *Mon. Not. R. astr. Soc.*, **186,** 609.

Brandenberger, R. H., 1985. *Rev. Mod. Phys.*, **57,** 1.

Burbidge, E. M., Burbidge, G. R., Fowler, W. A. and Hoyle, F., 1957. *Rev. Mod. Phys.*, **29,** 547.

Ciardullo, R. and Demarque, P., 1977. *Trans. Yale Univ. Observ.*, **33.**

Colgate, S. A., Petschek, A. G. and Kriese, J. T., 1980. *Astrophys. J.*, **237,** L81.

Cowan, J. J., Thielemann, F. K. and Truran, J. W., 1986. In *Proc. Second IAP Rencontres (Nuclear Astrophysics), 7–11 July 1986, Paris, France.*

Davis, M. and Peebles, P. J. E., 1983. *Ann. Rev. Astr. Astrophys.*, **21,** 109.

Dekel, A., 1986. *Comments on Astrophysics*, **11,** 235.

Deser, S. and Zumino, B., 1977. *Phys. Rev. Lett.*, **38,** 1433.

de Vaucouleurs, G., 1982. *Nature*, **299,** 303.

Einstein, A. and de Sitter, W., 1932. *Proc. Natn. Acad. Sci. U. S. A.*, **18,** 213.

Fowler, W. A., 1961. In *Proc. Rutherford Jubilee International Conference*, p. 640, Heywood and Co. Ltd., England.

Fowler, W. A., 1967. *Nuclear astrophysics*, American Philosophical Society, Philadelphia.

Fowler, W. A., 1972. In *Cosmology, fusion, and other matters*, p. 67, ed. Reines, F., Colorado Associated University Press, Boulder.

Fowler, W. A., 1977. In *Proc. Welch Conferences on Chemical Research, XXI, Cosmochemistry*, p. 61, ed. Milligan, W. D., Welch Foundation, Houston.

Fowler, W. A., 1978. *Bull. Am. Acad. Arts and Sci.*, **32** (2), 32.

Fowler, W. A., 1984. In *Les Prix Nobel en 1983*, pp. 136–141, the Nobel Foundation, Stockholm.

Fowler, W. A. and Hoyle, F., 1960. *Ann. Phys., **10,*** 280.

Fowler, W. A. and Meisl, C. C., 1986*a*. In *Cosmogonical processes*, ed. Arnett, W. D., VNU, Netherlands. Ibid., 1986*b*. Updated calculations.

Freedman, D. A. and Das, A., 1977. *Nucl. Phys.*, **B120,** 221.

Fuller, G. M. 1986. Private communication.

Gibbons, G. W., Hawking, S. W. and Siklos, S. T. C., 1983. *The very early universe*, Cambridge University Press.

Graham, J. R., Meikle, W. P. S., Allen, D. A., Longmore, A. J. and Williams, P. M., 1986. *Mon, Not. R. astr. Soc.*, **218,** 93.

Guth, A. H., 1981. *Phys. Rev.*, **D23,** 347.

Guth, A. H. Steinhardt, P. J., 1984. *Sci. Am.*, **250,** 116.

Hillebrandt, W., Lechle, M., Ziegart, W., Nomoto, K. and Thieleman, F. K., 1986. In *Accretion processes in astrophysics*, XXIth Rencontre de Moriond Lew Arcs.

Hoff, R. W., 1986. In *Int. Symp. on Weak and Electromagnetic Interactions in Nuclei, Max-Planck Institute, Heidelberg, FRG, 1–5 July 1986.*

Howard, W. M. and Möller, P., 1980. *Atomic data and nuclear data tables*, **25,** 219.

Hoyle, F. and Fowler, W. A., 1960. *Astrophys. J.*, **132,** 565.

Hoyle, F. and Fowler, W. A., 1963. In *Isotopic and cosmic chemistry*, p. 516, eds Craig, H., Miller, S. and Wasserburg, G. J., North-Holland Publishing Co., Amsterdam.

Hubble, E., 1936. *The realm of the nebulae*, Yale University Press, New Haven, p. 170, Dover Publications, 1958.

Iben, I., Jr. and Renzini, A., 1984. *Phys. Repts*, **105,** 331.

James, K. and Demarque, P., 1983. *Astrophys. J.*, **264,** 206.

Klapdor, H. V. and Grotz, K., 1986. *Astrophys. J.*, **301,** L39.

Loh, E. D. and Spillar, E. J., 1986. *Astrophys. J.*, **307,** L1.

Loh, E. D., 1987. *Phys. Rev. Lett.*, in press.

Lynden-Bell, D., 1986. *Q. Jl R. astr. Soc., **27,*** 319.

McCrea, W. H. and Milne, E. A., 1934. *Q. Jl Math.* (Oxford series), **5,** 73.

McVittie, G. C., 1940. *The Observatory*, **63,** 273.
McVittie, G. C., 1962. *J. Soc. Ind. Appl. Math*, **10,** 737.
Meyer, B. S., Howard, W. M., Mathews, G. I. Möller, P and Takahashi, K., 1985. In *ACS Symp. on Recent Advances in the Study of Nuclei off the Line of Stability, Chicago, IL, 8–13 September 1985.*
Milne, E. A., 1934. *Q. Jl Math.*, **5,** 64.
Milne, E. A., 1941. *The Observatory*, **64,** 11.
Nix, R., 1986. Private communication.
Noerdlinger, P. D. and Arigo, R. J., 1980. *Astrophys. J.,* **237,** L15.
Nomoto, K., 1986. *Prog. Particle Nucl. Phys., **18.***
Pagel, B. E. J., Terlevich, R. J. and Melnick, J., 1986. *Publs Astr. Soc. Pacif,* **98,** 1005 and private communication.
Pagels, H. R., 1984. *Ann. N.Y. Acad. Sci.*, **442,** 15.
Peebles, P. J. E., 1966. *Astrophys. J.*, **146,** 542.
Peebles, P. J. E., 1984*a*. *Astrophys. J.*, **284,** 439.
Peebles, P. J. E., 1984*b*. *Science*, **224,** 1385.
Peebles, P. J. E., 1986. *Nature*, **321,** 27.
Penrose, R., 1986. *The Observatory*, **106,** 20.
Robertson, H. P., 1956. *Helv. Phys. Acta. Suppl.,* **4**, 128.
Rowan-Robinson, M., 1985. *The cosmological distance ladder*, W. H. Freeman and Company
Sale, K. E. and Mathews, G. J., 1986. *Astrophys. J.*, **309,** L1.
Sandage, A., 1961. *Astrophys. J., **133,*** 355.
Schramm, D. N. and Vishniac, E., 1986. *Comm. Nucl. and Part. Phys.,* **16,** 51.
Schramm, D. N. and Wasserburg, G. J., 1970. *Astrophys. J.*, **162,** 57.
Takahashi, K. and Yokoi, K., 1983. *Nucl. Phys.,* **A404,** 578.
Tammann, G. A. and Sandage, A., 1985. *Astrophys. J.*, **294,** 106.
Tayler, R. J., 1986. *Q. Jl R. Astr. Soc.*, **27,** 367.
Thielemann, F. K., Metzinger, J. and Klapdor, H. V., 1983. *Z. Phys. A.*, **309,** 301.
Thielemann, F. K., 1984. In *Stellar nucleosynthesis*, p. 389, eds Chiosi, C. and Renzini, A., Reidel, Dordrecht.
Thielemann, F. K. and Truran, J. W., 1986. In *The extragalactic distance scale and deviations from the Hubble expansion*, Reidel, Dordrecht.
Tuli, J. K., 1985. *Nuclear wallet cards*, Nat. Nucl. Data. Cen. Brookhaven, Nat. Lab., Upton, N.Y.
Tully, R. B. and Fisher, J. R., 1977. *Astr. Astrophys.*, **54,** 661.
Twarog, B. A., 1980. *Astrophys. J.*, **242,** 242.
Twarog, B. A. and Wheeler, J. C., 1982. *Astrophys. J.*, **261,** 636.
van den Bergh, S., 1984. *Q. Jl R. astr. Soc.*, **25,** 137.
van den Bergh, S., 1985. *Nature*, **314,** 320.
Wagoner, R. V., Fowler, W. A. and Hoyle, F., 1967. *Astrophys. J.*, **148,** 3.
Waldrop, M. M., 1986. *Science*, **234,** 152.

Walker, A. G., 1940. *Proc. Lond. Math. Soc.*, **46,** 113.

Walker, A. G., 1941. *The Observatory*, **64,** 17.

Wasserberg, G. J., Papanastassiou, D. A., Tera, F. and Huneke, J. C., 1977. *Phil. Trans. R. Soc. A.*, **285,** 7.

Weinberg, S., 1972. *Gravitation and cosmology*, p. 489, John Wiley & Sons, New York.

Wheeler, J. C. and Harkness, R. P., 1986. *Galaxy distances and deviations from universal expansion*, p. 45, eds. Madore, B. F. and Tully, R. B., Reidel, Dordrecht.

Wilkinson, D. T., 1986. *Science*, **232,** 1517.

Willson, L. A., Bowen, G. H. and Struck-Marcell, C., 1987. *Comments on Astrophysics*, in press.

Winget, D. E., 1987. *Astrophys. J.*, in press.

Winters, R. R., Macklin, R. L. and Hershberger, R. L., 1986. Preprint.

Woosley, S. E., Axelrod, T. S. and Weaver, T. A., 1984. In *Stellar nucleosynthesis*, p. 263, eds Chiosi, C. and Renzini, A., Reidel, Dordrecht.

# Geometry, topology and physics

Sir Michael Atiyah

## 1 Introduction

The relation between Geometry and Physics is a very ancient one going back to the earliest times, but in the modern period it is best exemplified by Einstein's Theory of General Relativity. As is well known, Einstein interpreted gravitational force in terms of the curvature of space–time and Maxwell's Theory of Electromagnetism was subsequently interpreted in a somewhat similar manner. At the present time these geometrical ideas have been extended to incorporate the other fundamental forces of nature, the 'weak' and 'strong' forces that are encountered at the nuclear level. As a result we are witnessing now a remarkable interaction between mathematicians and physicists which has been extremely fruitful.

One of the new features of the present interaction is that *quantum theory* enters in an essential manner in the physics while, correspondingly, *topology* enters into the relevant mathematics. An indication that this ought to happen is that both quantum theory and topology lead from the continuous to the discrete. The relationships that enter at this level are more 'global' and less local than in classical physics and differential geometry. Because of this novelty I will be emphasizing this global or topological aspect and ignoring the more established and better understood local theory.

The interaction between mathematicians and physicists is a two-way process with benefits for both sides. Mathematicians are, of course, pleased to see that theories which they have developed have now turned out to be useful to physicists, and are increasingly involved in the latest fundamental physical theories. How successful these theories will be is still uncertain and as a mathematician I am not qualified to judge. However, in the opposite direction, it is quite remarkable that ideas flowing from physics have yielded new and important results in mathematics. These results are much easier to assess, since fortunately their validity does not depend on a comparison with experiments! In this lecture, therefore, I will in the main concentrate on explaining the benefits mathematics has derived from its current association with theoretical physics.

## 2 Topology

Let me begin by reviewing briefly the topological ideas which will be relevant in connection with physics.

To a topologist any closed curve is the same as a circle but is different from a straight line. Similarly a sphere is the same as an ellipsoid but different from a plane. Moreover lines, planes, etc. are topologically 'uninteresting' whereas circles, spheres, etc. are the simplest examples with interesting topology.

A 'punctured' plane, i.e. a plane with a single point removed, is topologically equivalent to a cylinder. This is implicit in the use of polar coordinates $(r, \theta)$. Similarly removing a line (say the $z$-axis) from 3-space leaves something topologically equivalent to the product of a circle and a plane, as one sees by using coordinates $(r, \theta, z)$.

After these elementary but important basic examples, perhaps I should more formally recall that in topology size and distance are ignored. Two geometric objects are topologically equivalent if one can be deformed into the other (without cutting or tearing).

More subtle topological examples are provided by *knots*. By definition a knot is a closed curve in 3-space or more physically and conventionally a closed piece of string. A simple but interesting example is the *trefoil* knot:

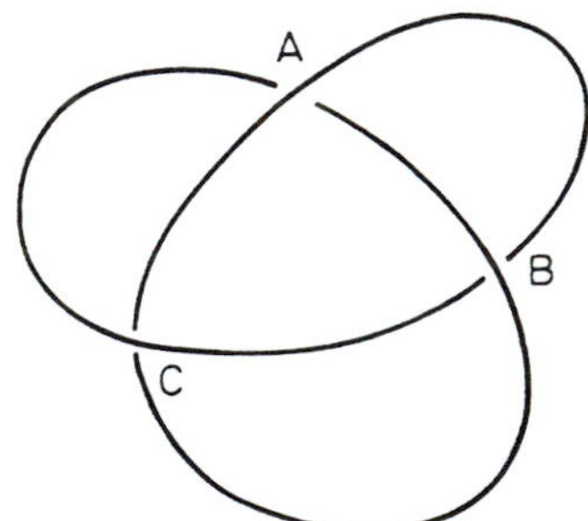


This picture is of course a plane projection of the knot and there are over/under crossings at A, B, C as indicated. It is best for this purpose to imagine the string lying on a table.

A different plane picture of the same knot is the following 'braid':

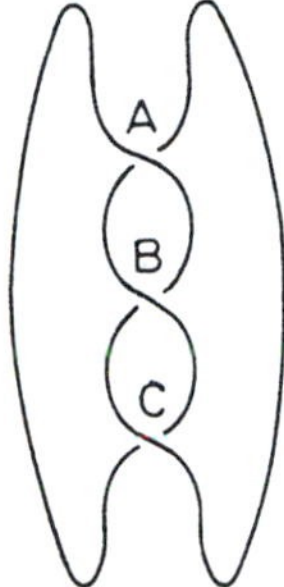


To see that this is the same as the previous trefoil picture lift up the Southeast portion of the string and lay it down above the Northwest corner. This braid is made up by iterating three times the basic 2-strand braid:

and then closing it up. It is a general fact that all knots can be described by appropriate braids, using sufficiently many strands.

Knots provide an excellent illustration of topology. It is intuitively clear and well known to boy scouts and sailors that they can be very complicated and subtle. It is also the case that the topics I will discuss in relation to physics are essentially concerned with knots. For all these reasons knots will figure prominently throughout my lecture.

## 3 Topology in physics

I shall next review some well known situations in physics where topological notions enter in an essential way. The examples will be presented in order of mathematical sophistication and this does not correspond to the historical order of discovery in physics.

### 3.1 Orientation of 3-space

The distinction between left-handed and right-handed systems is one of the most basic examples of a topological phenomenon. It is now well known, as proposed by Yang and Lee, that some physical processes distinguish between left-handedness and right-handedness. This is known as *parity violation*. Theories in which such processes can take place are called *chiral* theories and finding appropriate chiral theories is a delicate and important part of model building. Chirality has many topological consequences and it underlies all the topics I will discuss.

## 3.2 Spin

The fact that the rotation group SO(3) of 3-space has a double-covering group SU(2) is the topological fact which underlies the notion of spin (e.g. for the electron) and hence leads to the distinction between fermions and bosons.

## 3.3 Point source

If we have a point charge in 3-space, the total flux through any surface surrounding the point is constant. It does not depend on the shape or size of the surface, only on whether it surrounds the point-source or not. This is clearly a topological effect and represents the interesting topology of a punctured space.

## 3.4 Dirac quantization of electric charge

Perhaps the most important application of topological ideas in physics is due to Dirac. In a famous paper written over 50 years ago Dirac produced an argument to explain the quantization of electric charge, namely the fact that all observed particles appear to have an electric charge which is an integral multiple of that of an electron. Dirac postulated the existence of a magnetic monopole (which is allowed by Maxwell's equations) and proceeded to examine the quantum wave function of an electrically charged particle moving in the background field of the monopole. He showed that

$$\frac{2ge}{\hbar c} = \text{integer},$$

where $g$ is the magnetic charge, $e$ the electric charge ($\hbar$ and $c$ being Planck's constant and the velocity of light). Dirac's argument is a topological one and the topology (hence the integers) enters via the *circle* representing quantum mechanical phase.

## 3.5 Vortex atoms

So far my examples describe topology in a successful physical theory. Let me conclude with an interesting example from an unsuccessful theory. In the nineteenth century, before the days of modern atomic theory, the nature of atoms was a matter of much philosophical and theoretical speculation. One of the most ambitious and interesting ideas put forward was Kelvin's theory of Vortex Atoms. This theory was motivated by analogy with hydro-dynamics, and Kelvin's idea was that each atom was a knotted vortex tube. The general arguments in favour of this theory can be summarized as follows:

(i) *Stability*: The topological nature of knots could explain the stability of atoms.

(ii) *Variety*: The large variety of different knots provided ample room to accommodate all the different chemical elements.

(iii) *Spectrum*: The spectral lines of elements might in principle be related to vibration modes of vortex tubes.

In the light of subsequent knowledge we can even further strengthen point (i), because the break up and recombination of atoms at high energy could be envisaged as involving a cutting and rearrangement of a knot.

Kelvin's theory was taken fairly seriously. Kelvin himself, together with P. G. Tait, worked on it over a period of 20 years and Maxwell's view was that 'it satisfies more of the conditions than any atom hitherto imagined'. Moreover, it stimulated Tait to undertake a serious investigation into the classification of knots. This was in fact the beginning of knot theory and Tait made a number of conjectures which have only recently been established, using the new ideas from physics that I shall be explaining.

Of course, Vortex Atoms were in due course abandoned but the underlying motivation of Kelvin, using topology as a source of stability in physics, can be said to have surfaced again in a different guise in contemporary theory.

## 4 Gauge theories

As I mentioned at the beginning, in Einstein or Maxwell theory force is interpreted as curvature. Dirac's argument, described in Section 3, involves a global or integrated version of Maxwell theory relating curvature to topology. Closely related to Dirac's argument, and bringing topology even more into evidence, is the *Bohm–Aharanov* effect. This involves the following experiment. A solenoid carrying a magnetic flux is shielded so that there is no field outside it. A beam of electrons is then introduced and by examining interference patterns one finds that the solenoid produces a phase shift $\theta$ given by the formula

$$\theta = \frac{\varphi e}{\hbar c},$$

where $\varphi$ is the magnetic flux.

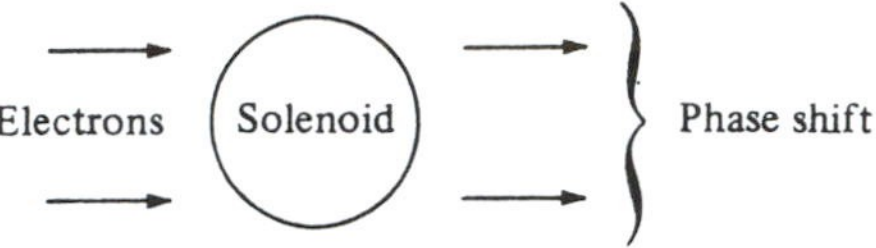


The remarkable thing about this Bohm–Aharanov effect is that the electrons travel in a force-free region and yet there is a physical effect which distinguishes electrons going round the back of the solenoid from those coming in front. Equivalently we could say that an electron that makes a complete circuit round the solenoid undergoes a phase shift $\theta$.

This phenomenon clearly ties the circle of quantum mechanical phase with the circle in space that winds round the solenoid. Both here and in Dirac's argument the presence of Planck's constant clearly shows that we are dealing with a quantum mechanical phenomenon. This demonstrates that topology is related to quantum rather than classical physics.

In electromagnetism phase is an angle and the corresponding 'gauge group' is the circle. Since this is an Abelian (or commutative) group Maxwell theory is an Abelian gauge theory. Current theories of elementary particle physics are based on *non-Abelian* gauge theories in which 'phase' is not an angle (rotation in the plane) but rotation in a higher dimension (internal) space. The simplest non-Abelian group is SO(3) or its double covering group SU(2) and we shall focus on this as a typical example.

In the Bohm–Aharanov experiment (in Maxwell theory), several solenoids would produce independent phase shifts $\theta$, and electrons weaving a path round such solenoids would undergo a total phase-shift $\theta$ given by the sum of the $\theta_i$. Since the circle group is Abelian the order or pattern of 'weaving' would be irrelevant.

Suppose now we envisage a gedanken Bohm–Aharanov experiment in a non-Abelian gauge theory. This time I shall assume my tube or solenoid carries some suitable non-Abelian flux and that it is actually *knotted* in 3-space (e.g. as a trefoil knot $K$). I now imagine taking a test particle on a closed path $\alpha$ that weaves in and out of the knot in some complicated way:

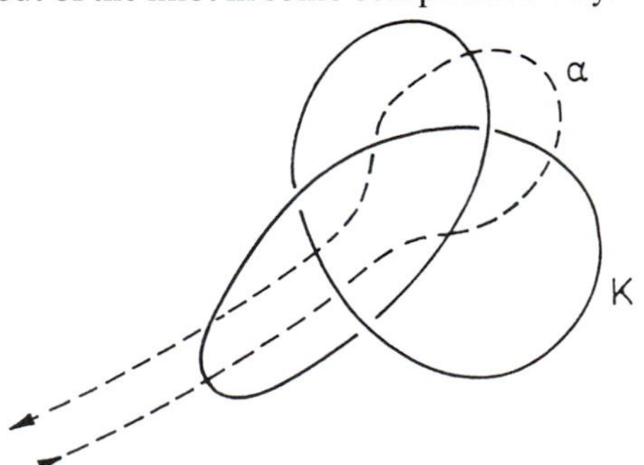

We would then expect our particle to have undergone some non-Abelian phase shift given by an element of SU(2). This will depend only on the topological nature of the path $\alpha$, i.e. it will be unchanged by a deformation that does not cross the knot or flux tube $K$. Let us denote this element of SU(2) by $g_\alpha$.

If $\beta$ is another such closed path it will produce a phase shift $g_\beta$. Suppose now we consider the composite path $\alpha\beta$ (i.e. $\beta$ followed by $\alpha$). The effect of this will be to produce the phase shift $g_\alpha g_\beta$ given by the (matrix) product in the group SU(2). Notice that since SU(2) is not commutative the order of multiplication is important: in general $g_\alpha g_\beta \neq g_\beta g_\alpha$.

If we want to express this situation more formally we introduce the *fundamental group* $\pi_1\ (R^3 - K)$ which consists of closed paths $\alpha$ outside $K$, with multiplication being composition of paths and inverses corresponding to reversal of direction. Here as before we identify paths which are topologically equivalent. The collection of phase shifts then describes a *representation* of $\pi_1(R^3 - K)$ by $2 \times 2$ unitary matrices.

We can therefore summarize our discussion by saying that the *different vacua of the classical gauge theory outside the knot* K *are described by representations of the fundamental group* $\pi_1(R^3 - K)$.

Before proceeding further with these ideas on non-Abelian gauge theories I have to digress to discuss some crucial ideas on critical point theory.

## 5 Critical point theory

The fundamental link between Geometry and Physics is provided by analysis in the form of differential equations associated with Hamiltonians. Topology in its turn ties up with analysis through critical point theory, i.e. the study of maxima, minima and saddle points. Let me review very rapidly the essential ideas involved.

Consider a real-valued function $f$ defined on the surface of a sphere or a torus. For example imagine our surface standing on the ground and let $f$ be the height function. On the sphere $f$ clearly has one maximum and one minimum, whereas on the torus it has in addition two saddle points:

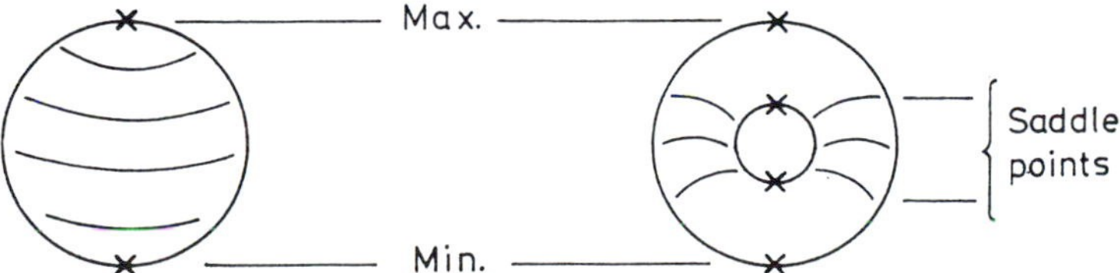

The topology of the torus differs from that of the sphere (it contains interesting closed curves) and this forces the existence of the saddle points.

In this example our function was very simple and so we had only a small number (two or four, respectively) of critical points. Suppose now we complicate things a bit, replacing the sphere by a dented surface as shown:

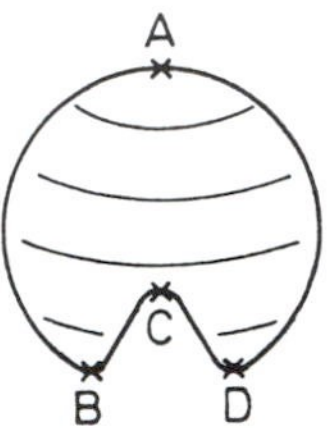


The height function now has four critical points, one maximum, two minima and one saddle point. Notice that the surface is still topologically a sphere. Thus we have increased the number of critical points from two to four. From a topological point of view we have some unnecessary or 'superfluous' critical points. Intuitively we can understand that these superfluous critical points can be cancelled off in pairs (e.g. C and D 'cancel').

This cancellation idea can be developed quite formally and it is standard theory that after such cancellation the number of critical points is always the same, depending only on the topology and not on the particular function. Thus for a sphere cancellation will always reduce the number to two while on the torus the number will be four.

This process of cancelling critical points was given a physical interpretation some years ago by Witten (1982) in a paper which has been very influential and whose basic idea I will now describe.

Consider first a classical particle moving on a surface (e.g. a torus) under a potential function of the form

$$V = |\operatorname{grad} f|^2. \tag{5.1}$$

The equilibrium or ground state will then be given by $V = 0$, i.e. by the critical points of $f$. As we have seen there will (on the torus) be at least four such points.

If we consider the corresponding quantum mechanical problem then the ground-state wave function $\psi(x, y)$ will be the solution of the differential equation

$$\left(\Delta + \frac{V}{\hbar^2}\right)\Psi = 0, \tag{5.2}$$

where $x$, $y$ are angular variables for the torus and

$$\Delta = -\left(\frac{\partial^2}{\partial x^2} + \frac{\partial^2}{\partial y^2}\right)$$

is the Laplace operator. Since Planck's constant $\hbar$ is small the potential term $V$ dominates this equation and this is the way quantum states are approximated by classical ones. However, as Witten has explained, this classical approximation works even better in a *supersymmetric* theory.

Supersymmetry is at present a popular idea in theoretical physics. Roughly speaking a theory is supersymmetric if the fundamental equations allow one to interchange the roles of bosons and fermions. In the simple quantum mechanical model for a particle on a surface one makes it supersymmetric by adjoining to the original scalar (bosonic) field, $\psi_0$, three new fields

$$\psi_1\,\mathrm{d}x, \quad \psi_2\,\mathrm{d}y, \quad \psi_3\mathrm{d}x\mathrm{d}y$$

the first two being fermionic and the last another bosonic field. There is then a natural generalization of the hamiltonian operator in eqn (5.2) and Witten observes that the *supersymmetric ground states* will always have the correct topological number, namely four for the torus. This is in fact just a reinterpretation of the basic theorem of Hodge relating harmonic differential forms to homology.

We can now ask how it is that, whatever the potential function $V$ given by eqn (5.1), with perhaps a large number of classical ground states corresponding to critical points of $f$, the quantum theory always succeeds in having the correct topological number of ground states? Witten's explanation is that the necessary cancellation process is now provided by *quantum tunnelling*.

I recall that a quantum ground state is approximated by a linear combination of classical ground states, the coefficients being probability amplitudes of 'tunnelling' from one classical ground state to another.

This principle of Witten's asserts therefore that, *in a supersymmetric quantum theory, quantum tunnelling acquires a topological significance*.

Now, Witten was really interested in supersymmetric quantum field theory, and the quantum mechanical version was merely intended as a motivating simple case. We can therefore try to apply Witten's principle to our favourite supersymmetric quantum field theory and see what topological information emerges from the case of quantum tunnelling.

Suppose we try to apply these ideas to SU(2) gauge theory outside a knot $K$ in 3-space. We saw in Section 4 that the classical vacua or ground states were described by SU(2)-representations of $\pi_1(R^3 - K)$. Actually some boundary conditions need to be imposed along the boundary of the flux tube which is the more realistic thickened version of the knot. I will not go into technical details except to illustrate the kind of boundary conditions one might impose in the analogous 2-D case when $K$ is just two points in a plane. In this case we might require our wave functions on two small circles surrounding the two points to coincide. Alternatively we could add a handle connecting the two circles and consider wave functions defined on the new surface:

Let me symbolically indicate these boundary conditions along the knot by writing K* instead of *K*. Then, applying Witten's ideas, A. Floer (1987) has been led to define new topological invariants of K* from the quantum ground states of SU(2) gauge theory. Although this quantum field theory may not exist in a rigorous mathematical sense the tunnelling approximations used by physicists do exist and Floer's theory is mathematically well defined. It provides subtle new topological invariants.

The approach to quantum field theory I have been describing is the Hamiltonian one, in which space and time are separated. However, in a Lorentz-invariant theory there is a more symmetrical Lagrangian formulation. The analogue of Floer's theory in a 4-D formulation (for closed manifolds) is due to S. Donaldson (1988) who has discovered deep and surprising results in 4-D geometry by using these ideas originating from physics.

Donaldson's work antecedes Floer's theory but the physical interpretation is not quite so clear in Donaldson's version. The two theories however connect in a natural and simple way. I have elsewhere (Atiyah 1988) given a fuller account of all these ideas.

I should perhaps say that the quantum field theory that lies behind the Floer/Donaldson work is not supersymmetric Yang–Mills theory as understood among physicists. In fact as a physical theory it suffers from the serious defect of appearing to contradict the important Spin-Statistics theorem according to which fermions have half-integral spin. However, Witten (1988) has discovered very recently how to formulate the Donaldson/Floer theory in an acceptable physical framework. This brings the geometry and the physics even closer together and will perhaps lead to deeper understanding on both sides.

To sum up so far: physical ideas dealing with the quantum versions of non-Abelian gauge theories have led to extremely important new results in Geometry and Topology which are still being actively explored. The significance for Physics is unclear but the significance for Geometry is not in doubt.

## 6 Knot polynomials

I will now turn to a different topic. It still involves the interaction of Geometry and Physics and it also deals with the topology of knots but it appears, at the moment, to be quite independent of the ideas in Section 5.

In the study of knots it is clearly helpful to have numerical invariants that can be used to help distinguish different knots. Of course when a knot is given by a plane diagram with over/under crossings there are many numbers that can be associated with it, e.g. the number of crossings. However, the same knot can be represented by many quite different diagrams, and a numerical *invariant* is a quantity which will be the same for all diagrams describing the same knot. Crude quantities like the number of crossings are not invariant.

Some 60 years ago J. W. Alexander introduced invariants which are conveniently regarded as coefficients of a Laurent polynomial in an indeterminate $t$ (and $t^{-1}$). This was subsequently refined by J. H. Conway and is now referred to as the Alexander–Conway polynomial. It has been one of the most basic and useful tools in the study of knots.

Then in 1985 Vaughan Jones astonished knot theorists by producing a second invariant polynomial. This differed from the Alexander–Conway polynomial in one important respect, because the Jones polynomial is capable of distinguishing (some) knots from their mirror images. In physical terms it is a *chiral* invariant.

Both polynomials can be characterized and computed by appropriate recurrence formulae. To explain these we should first extend knots to *links*. A link differs from a knot only in that it may consist of several different closed pieces of string. The simplest interesting link, having two pieces, is the following:

Moreover it is best to *orient* links. This simply means we consider each piece of string to be traversed in a particular direction, indicated in diagrams by an arrow. We are now ready to describe the recurrence formulae for the Alexander–Conway polynomial $A(t)$ and the Jones polynomial $V(t)$.

Given any oriented link diagram we focus attention on one crossing point. We then consider the three different links ($L_+$, $L_-$, $L_0$) which coincide away from this crossing and at this crossing are represented by one of the three local pictures:

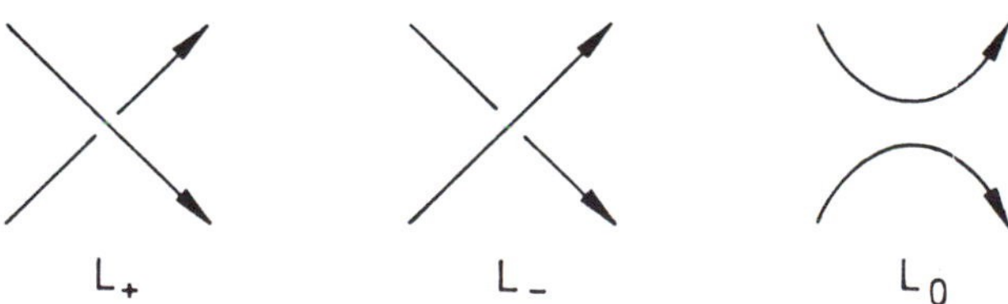

The Alexander polynomials for these three links will be denoted by $A_+$, $A_-$, $A_0$ with a similar notation for the Jones polynomials. The recurrence relations then take the form

$$A_+ - A_- = (t^{1/2} - t^{-1/2})A_0,$$
$$t^{-1}V_+ - tV_- = (t^{1/2} - t^{-1/2})V_0.$$

*Note*: For a general link $A$ and $V$ involve $t^{1/2}$, but the square roots disappear for a *knot*.

For the standard unknotted circle $V = A = 1$, while for the trefoil knot

$$A = t - 1 + t^{-1},$$
$$V = -t^4 + t^3 + t.$$

In fact there are two different trefoil knots, mirror images of each other. $A$ is the same for both but $V$ is altered by the substitution $t \to t^{-1}$.

The recurrence formulae are deceptively simple. The main difficulty is to show that they always lead to the same answer for a given link independently of the representing plane diagram.

The Jones polynomial arises from the use of *braids* and it is now becoming clear that the essential algebraic aspects of the theory were well known in Physics through the work of C. N. Yang (1968) and, independently, that of R. J. Baxter (1982). I will now try to explain the *Yang–Baxter equation* and its relevance to braids and links.

## 7 Braids and the Yang–Baxter equation

The fundamental braid relation is exhibited in the topological equivalence of the two following diagrams:

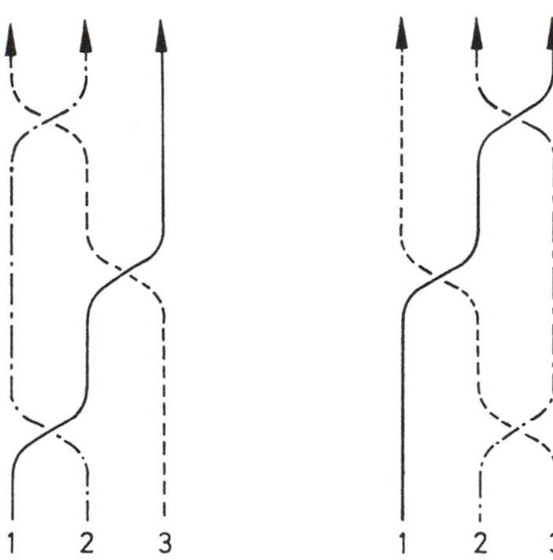

Symbolically this is written as:

$$\sigma_1\sigma_2\sigma_1 = \sigma_2\sigma_1\sigma_2, \tag{7.1}$$

where $\sigma_1$ or $\sigma_2$ is the basic two-strand braid

the index 1 or 2 indicating whether the two strands are (12) or (23).

Instead of braids we can now consider the graphs or world lines of particles moving on a line with constant velocities. For three particles we then get two diagrams depending on the order in which two-particle 'collisions' occur:

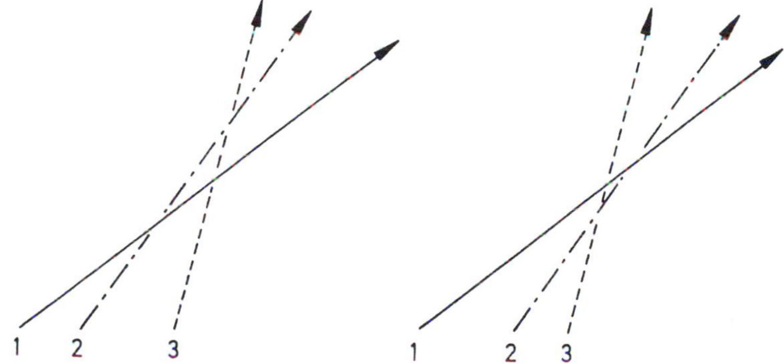


It is clear that these pictures coincide in their essential features with the braid pictures above.

Suppose now we consider a physical model in which each particle can be in one of $N$ internal states, and that two-particle collisions leave the trajectories unaffected but act quantum mechanically on the internal states by an $N^2 \times N^2$ matrix $R = (R_{ij}^{kl})$. Thus $R_{ij}^{kl}$ gives the amplitude of:

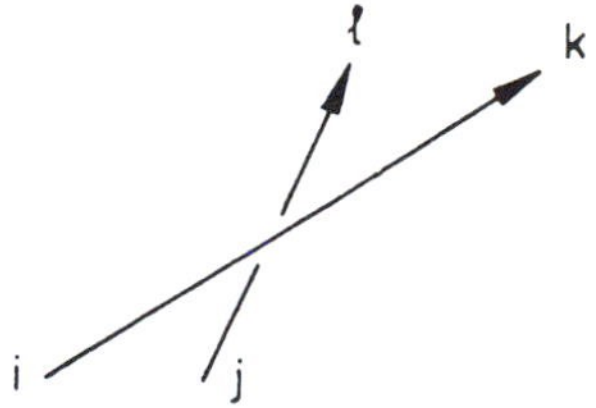


We assume for simplicity that $R$ does not depend on the momenta but only on the internal states and the order of the particles.

For three particles colliding (or interacting) in pairs we can then work out the total scattering matrix, which will be $N^3 \times N^3$. Let us assume that *this three-particle scattering is independent of the order in which collisions occur.* This means that the two pictures above must yield the same $N^3 \times N^3$ matrix. Algebraically this leads to the *Yang–Baxter equation*

$$R_1 R_2 R_1 = R_2 R_1 R_2, \tag{7.2}$$

where the index 1 or 2 indicates that $R$ acts on particles (12) or (23) respectively.

Clearly equation (7.2) is essentially the same equation as the braid relation equation (7.1). More formally, if we introduce the *braid group* $B_3$ on three strands, in which the product is given by composition of braids, and if $R$ is a solution of the Yang–Baxter equations then $\sigma_i \to R_i$ defines a matrix representation of $B_3$. This generalizes immediately to give representations of the braid group $B_n$ on $n$ strands.

From representations of the braid groups $B_n$ we can try to construct numerical invariant of links by taking the trace of the matrix associated with a braid. This is in principle how the Jones polynomial can be constructed, although one has to study the relation between $B_n$ and $B_{n+1}$. I shall not go into any details, but the essential point is that solutions $R(t)$ of the Yang–Baxter equation, depending on $t$ as parameter, eventually yield polynomial invariants of knots and links.

The model of colliding particles which I have described occurs in the work of Yang (1968). Baxter (1982) in his work on soluble models in 2-D statistical mechanics was led to essentially the same equation (7.2). Moreover Baxter's work has stimulated a great deal of activity leading to a general theory for constructing solutions of equation (7.2). These can then be used systematically to construct polynomial invariants of links, and the Jones polynomial is only the first of a whole series of such polynomials. At the present time the subject is being vigorously investigated and the final picture has still to emerge. What is already clear is that the Yang–Baxter equation, motivated by physical models, is highly relevant to the study of knots.

## 8 Conclusions

From the two topics I have briefly described it is clear that there are deep relations between 3-D topology and important non-linear models in quantum field theory and statistical mechanics. I have emphasized the mathematical benefits of this relationship, but the benefits to physics are certainly there, if more difficult to identify.

I have refrained from any mention of 'string theory' which is the latest theory of elementary particles. In this theory the topology of Riemann surfaces is vitally important and many other topological ideas are also relevant. To enter into this subject would, however, require another lecture and another speaker. I will however make a few brief comments on string theory.

Originally strings were investigated as a possible model for hadrons but this proved unsuccessful. Instead the mathematical model provided by strings has been given a quite different physical interpretation in terms of quarks. In this respect there may be an analogy with the use of knots in physics. Kelvin's vortex atoms were discarded but, as I have indicated, knots seem to be reappearing in quite different physical guises. They may yet turn out to play some fundamental role in physics.

## Acknowledgement

The Milne Lectures are sponsored by IBM UK Ltd.

## References

Atiyah, M. F., 1988. New invariants of 3 and 4-dimensional manifolds, *Amer. Math. Soc. Symposia in Pure Maths.*, **48,** 285–299.

Baxter, R. J., 1982. *Exactly solved models in statistical mechanics*, Academic Press, London.

Donaldson, S. K., 1988. The geometry of 4-manifolds, *Proc. Int. Congress of Mathematicians*, Berkeley (1986).

Floer, A., 1989. An instanton invariant for 3-manifolds. *Commun. Math. Phys.*, **118**, 215–240.

Witten, E., 1982. Supersymmetry and Morse theory, *J. Diff. Geom.*, **17,** 661–692.

Witten, E., 1988. Topological quantum field theory, *Commun. Math. Phys.*, **117**, 353–386.

Yang, C. N., 1968. S matrix for the one-dimensional N-body problem with repulsive or attractive delta function interaction, *Phys. Rev.*, **168,** 1920.

# Polarization—its message in astronomy

V. Radhakrishnan

> Light brings us the news of the Universe. Coming to us from the Sun and the stars it tells us of their existence, their positions, their movements, their constitutions and many other matters of interest.

So said William Bragg over half a century ago. The news of the universe today is brought to us by electromagnetic radiation spanning an enormous range from kilometre-long radio waves to gamma rays of TeV energies and more. And it is the *strength* of the radiation at different wavelengths from different directions in the sky that has told us most of what we know about what is out there. But electromagnetic waves are transverse in nature, and hence the *instantaneous* electric (or magnetic) field must have some particular orientation perpendicular to the direction of propagation of the wave. This preferred dirction and its behaviour as a function of time forms yet another characteristic of the radiation. If at one instant the field is at a particular angle, then at a later time it is likely to be different (Fig. 1). In natural radiation, the orientation of the vector displays a systematic variation over the period of the wave, and a random change to a new pattern over a longer time-scale, that of the inverse bandwidth. An understanding of this aspect of electromagnetic radiation began with an observation in 1808 by Malus who happened to be looking through a piece of Iceland spar at the setting Sun reflected in the windows of the Luxemburg Palace in Paris. The relative intensity of the two images varied as he rotated the crystal, and being a Frenchman I imagine he exclaimed 'Voilà polarisation'! In any case, that was the name he gave the phenomenon.

My first awareness of this property, as of many other things that later earned me a living, was gained from the *Radio amateurs' handbook*. It emphatically recommended that the receiving antenna be oriented in the same way as the transmitting one for the message to be received well. It followed from this and more in the Good Book that two different transmitters and receivers

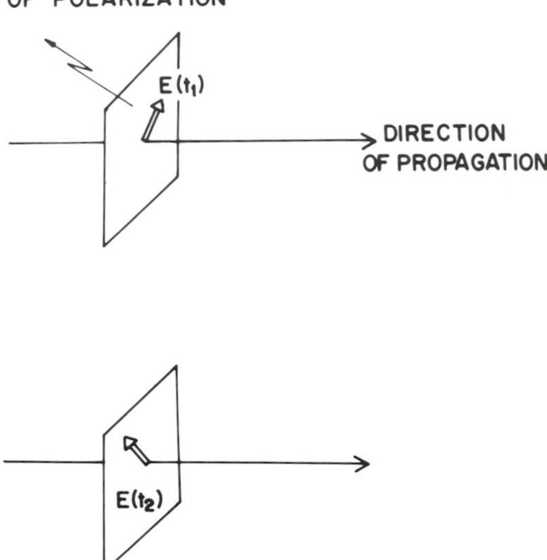


**Fig. 1.** The electric field as a function of time.

with their antennas appropriately oriented could use the same wavelength or channel without interfering with each other. I used to wonder why this wasn't standard practice to double the use of the densely crowded radio bands until I learnt about the imperfections of antennas, and particularly the strange things that can happen to the polarization of a wave as it is bounced around by objects in its path. Trying to obtain a null on a short-wave station by orienting a hand-held radio receiver inside a building can be an instructive experience.

My intention, in this lecture, is to give you a personal view of events as I saw them in the development of astronomical polarization studies over the years. Rather than attempting to be comprehensive I have chosen to concentrate on just a few particular examples to illustrate what we have learnt from such studies. And in this choice, I have leaned naturally towards those things with which I was familiar.

The most intense astronomical radiation that we receive on earth (about a kW per square metre) is the sunlight we are bathed in all day, at least in some countries. Barring a very feeble amount emanating from sunspot regions, and confined to very narrow frequency intervals, all of this visible radiation is randomly polarized. The meaning of this is that the field vectors have no preferred orientation, or sense of rotation, when averaged over long periods. Minnaert devotes less than 1% of his classic book *Light and colour in the open air* to discussing polarization phenomena associated with the light we

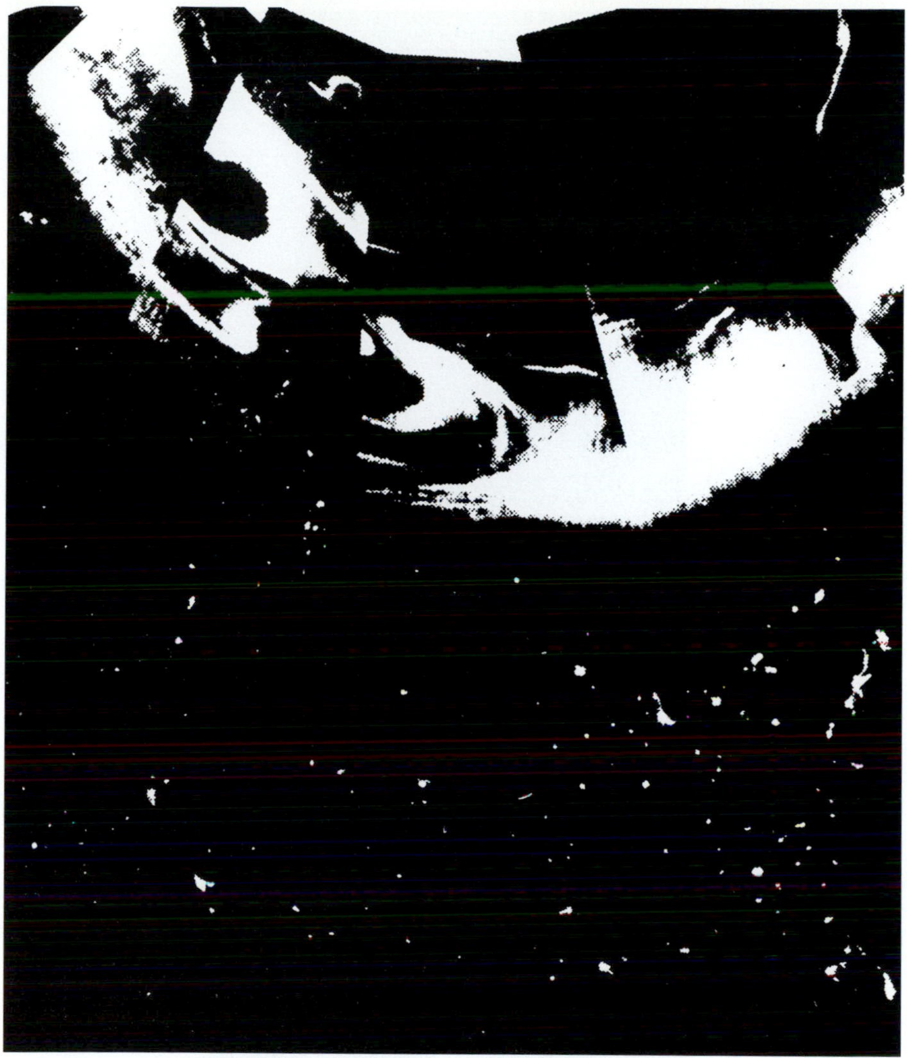

**Plate I** Aurora and city lights, United States and Canada, near midnight, 14 February 1972. Mosaic photograph from five satellite passes: the straight lines mark the dividing lines between passes. Reproduced by permission of the US National Geophysical and Solar-Terrestrial Data Center.

**Plate II** Edmond Halley (1656–1742), who in 1686 successfully deduced how atmospheric pressure decreases with height. From the portrait by Thomas Murray, reproduced by permission of the Royal Society.

**Plate III** Erasmus Darwin (1731–1802), who in 1788 explained how clouds form, and devised a realistic atmospheric model. From the portrait by Joseph Wright of Derby, reproduced by permission of the Master and Fellows of Darwin College, Cambridge.

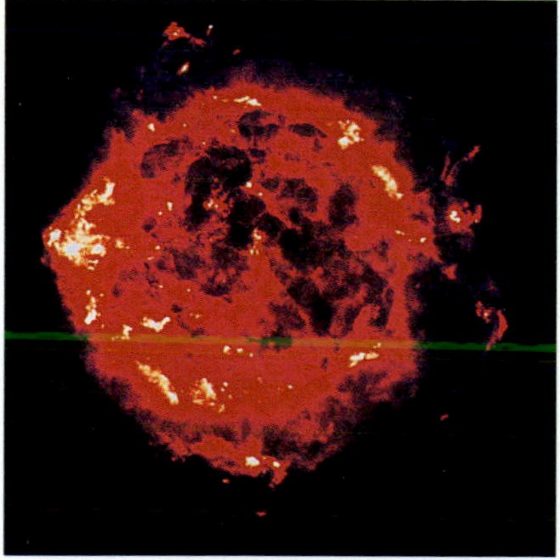

**Plate IV** Radiograph of the total intensity of Cassiopeia A. Photograph courtesy of NRAO/UI. Observers: P. E. Angerhofer, R. Braun, S. F. Gull, R. A. Perley and R. J. Tuffs.

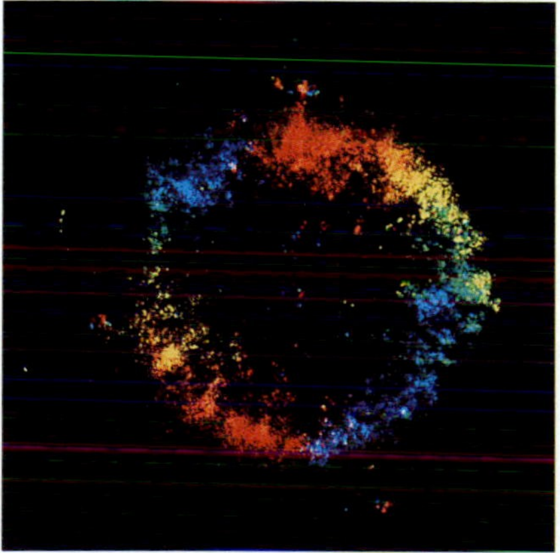

**Plate V** The intensity of linearly polarized emission of Cassiopeia A colour coded by the projected direction of the magnetic field. Photograph courtesy of NRAO/AUI. Observers: P. E. Angerhofer, R. Braun, S. F. Gull, R. A. Perley and R. J. Tuffs.

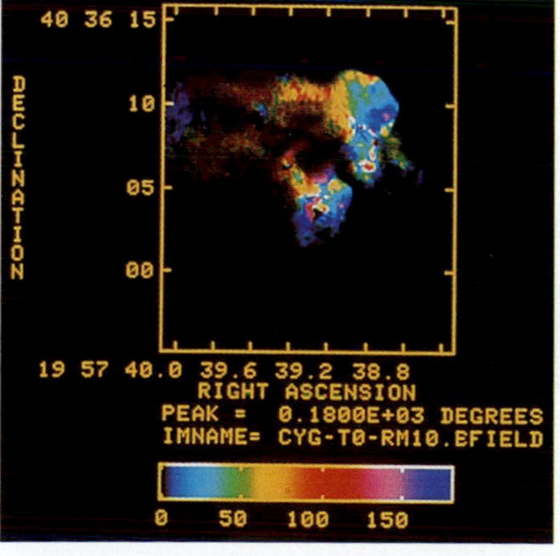


**Plate VI** The head of the NW lobe of Cygnus A. The intensity indicates the brightness of the radio emission, while the colour indicates the direction of the magnetic field. Photograph courtesy of NRAO/AUI. Observers: R. A. Perley, J. W. Drepher and J. H. Cowan.

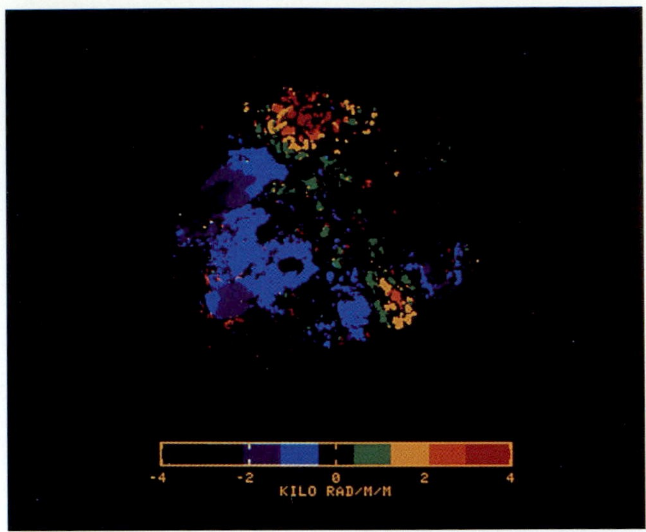


**Plate VII** The head of the SE lobe of Cygnus A. The intensity indicates the brightness of the radio emission, and the colour the value of the rotation measures. Photograph courtesy NRAO/AUI. Observer: R. A. Perley.

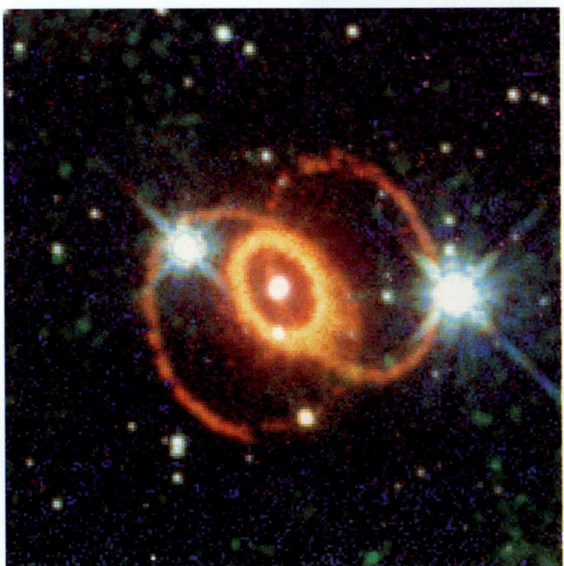

**Plate VIII** Image of supernova 1987A obtained with the Hubble Space Telescope. The central object is the debris of the exploded star, while the bright inner ring is formed of material lost from the pre-supernova star about 30 000 years before the explosion. The geometry of the inner ring yields a distance to the Large Magellanic Cloud which is independent of the Cepheid distance, but agrees with it.

live in. And all the cases of polarization that he does discuss, whether they be the appearance of scratches on the window panes of railway compartments, haloes, rainbows, or just skylight, refer to the modification of the symmetry of the incident radiation by the geometry of the scattering process involved. Scattered radiation will have a preferred orientation for the vector depending on the direction in which it is scattered. Such radiation is said to be *linearly* polarized (Fig. 2). Rainbows are polarized upto 95% but you would never know unless you were told or were as passionate and perceptive an investigator of what we see around us as Minnaert was.

The polarization of skylight is a more widely appreciated phenomenon, particularly since the invention of Polaroid and its incorporation into spectacles to ward off glare. Bees apparently have no problem in detecting this polarization, even without spectacles, and they put this ability to very good use. By performing a complicated dance they convey to other workers in the hive information as to which direction to fly, to find plentiful sources of nectar. The figure drawn by a colleague of mine purports to depict what might be taking place in a typical beehive (Fig. 3). For those not familiar with the jargon, $Q$ and $U$ are polarization symbols whose ratio gives the angle of the preferred orientation.

The only other practical application of the polarization of skylight that I

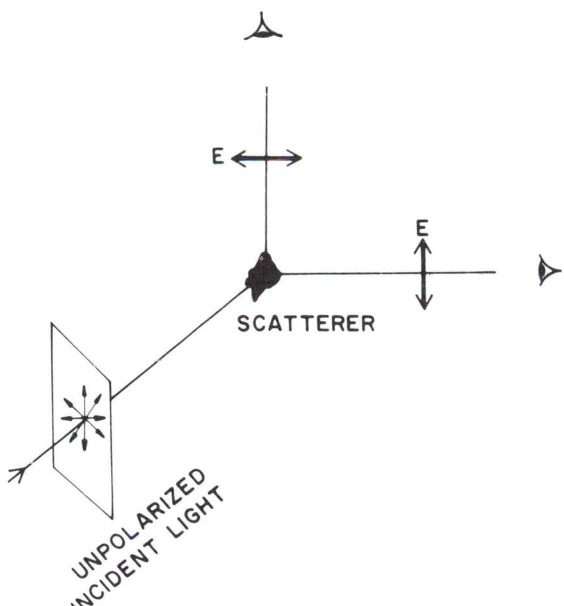


**Fig. 2.** The polarization of scattered radiation. When viewed at right angles to the incident radiation, the scattered fraction will be fully linearly polarized with the electric vector perpendicular to the two directions.

**Fig. 3.** The dance of the bees.

have heard of was by the ancient Viking mariners, who apparently made it a navigational aid. They used some form of natural birefringent crystal as a polarization analyser to determine the position of the Sun when it is not directly visible. And I have read a Swedish story of a young Viking who won the hand of his chieftain's daughter by challenging that he could do this without the aid of the traditional *sunstone* kept in the chief's custody, and proving that he could.

This has to do with the little-known but definite ability of the unaided human eye to detect linear polarization in light. All of us here can do this, but you may need a little practice staring through a sheet of Polaroid at a bright background and learning to recognise a fleeting yellow and blue form. It looks like a yellow bug with blue wings, it is called Haidinger's brush, and it is formed in the eye. With more practice you can discern this form in skylight and determine the direction of the Sun even if it is clouded over. But in the present day social climate, you are unlikely to win the kind of prize that the young Viking did. One wonders incidentally if the presently marginal human faculty also owes its origin to the *existence* of polarization in skylight, and did it perhaps serve a function in the hunting or food-gathering habits of an earlier stage in our evolution?

Returning to astronomy, almost all the radiation from celestial sources at wavelengths shorter than a centimetre or so is also of thermal origin. The electrons or atoms or molecules responsible describe chaotic motions and their radiation is randomly polarized. Starlight was found to be weakly polarized by Hiltner in the middle forties but this was subsequently understood to be also a propagation effect. It was identified as due to scattering by elongated particles of interstellar dust subjected to an alignment force, presumably a galactic magnetic field. If there were a radiation mechanism that reflected the strength and direction of this hypothetical magnetic field, such radiation should be *intrinsically* polarized. The extra information the wave would carry is the signature of the emission mechanism and the orientation on the sky of the magnetic field. Remarkably, such a candidate appeared on the scene just around this time, but from the new and emerging field of radio astronomy.

The cosmic radio radiation discovered by Jansky in 1933, first mapped by Reber in 1944, and studied intensively by many in the post-war years, clearly had a non-thermal component from its spectrum, but its origin was at first obscure. Alfvén and Herlofson suggested in 1950 that such radiation could be produced by relativistic electrons gyrating in a magnetic field—the synchrotron mechanism. Very soon thereafter, Kiepenheuer suggested that it was cosmic ray electrons trapped in the galactic magnetic field that produced the non-thermal halo component of the galactic radio emission. Because the acceleration of the particles would be perpendicular to the magnetic field, the radiation should be linearly polarized in the same direction. The detection of polarization in this radiation would provide the strongest support for the proposed synchrotron mechanism of cosmic radio emission, and Razin in the Soviet Union, where the theory had received its greatest development by Ginzburg and others, did make an attempt which indicated a positive result (Shklovsky 1960). The observational difficulties were considerable, however, and subsequent attempts by others to repeat these pioneering Soviet efforts unfortunately failed to confirm Razin's results.

Apart from the diffuse halo component, several so-called discrete sources of radio emission had also been discovered. John Bolton in 1949 made the first ever identification of such a discrete radio source, Taurus A, with an optical object. It was the Crab nebula, the visible remnant of the famous supernova of 1054 observed and recorded by the Chinese. The optical radiation from this nebula had already been noted to contain a diffuse component whose spectrum was featureless and whose origin was unknown. Shklovsky, the great Russian astrophysicist, advanced the bold hypothesis that both the radio and diffuse optical emission were due to the synchrotron mechanism. If this were so, then, as pointed out independently by both Gordon and Ginzburg in the Soviet Union in 1954, the optical radiation should be polarized. And in the same year this was detected by Dombrovsky and by Vashakidze, also from the Soviet Union. Subsequent work by Oort and

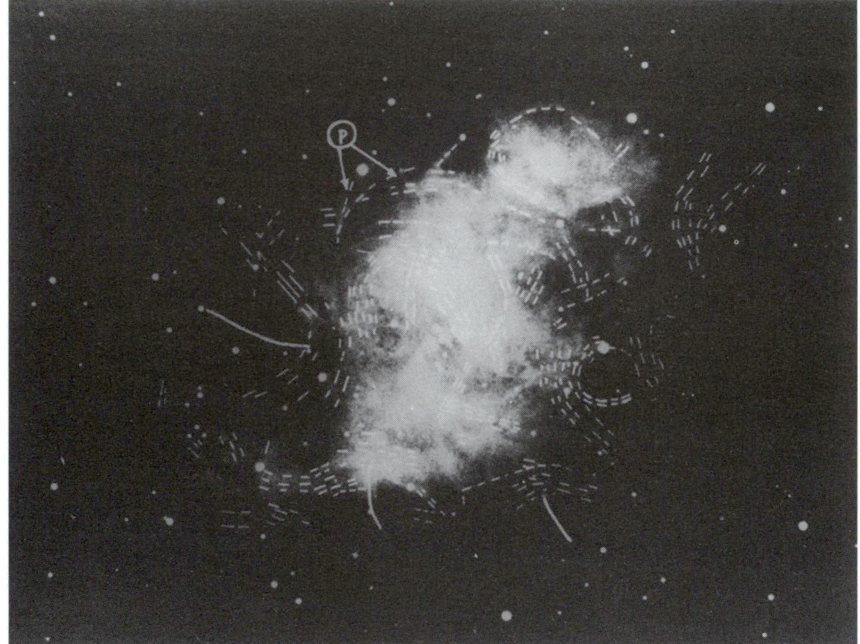


**Fig. 4.** Sketch of magnetic lines of force in the Crab nebula from Oort and Walraven (1956).

Walraven beautifully confirmed these early findings, and provided us with the first map of the magnetic field inside a distant astrophysical object (Fig. 4).

Very soon thereafter, two more radio sources were identified with extragalactic nebulae, or galaxies as we would call them now. One of these was Virgo A, identified with the peculiar galaxy NGC 4486, better known as M87, in whose central and brightest region Baade and Minkowski had found a striking jet-like feature at optical wavelengths (Fig. 5). Its spectrum showed no lines either, and once again Shklovsky suggested that, as in the case of the Crab nebula, the optical emission from the jet was due to synchrotron radiation from relativistic electrons. One should therefore expect the jet emission to be polarized, and this was indeed shown to be so by Oort and Walraven, and by Baade, both in 1956. Thus, the most crucial evidence in support of the existence of the synchrotron mechanism in both galactic and extragalactic radio sources was the detection of its *polarization*. But it was extraordinary that the only two known polarized emitters in the sky were in the optical, although they owed their detection to their associated radio properties. The implications of this took quite some time to be appreciated, and I will return to this point a little further on.

The outstanding pioneers in the search for polarization in radio sources

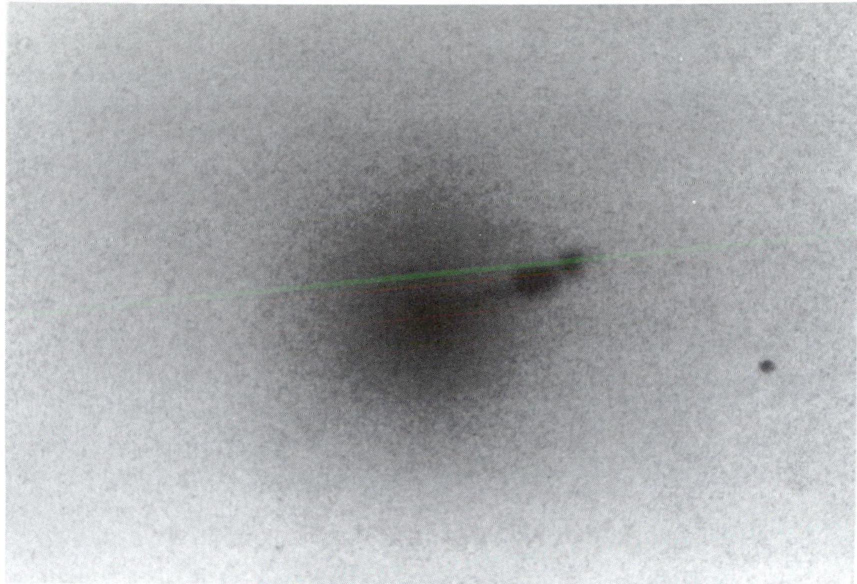

**Fig. 5.** The centre of NGC 4486 photographed with the 100 inch (2.5 m) telescope by Baade and Minkowski (1954).

were Cornell Mayer and his group at the Naval Research Laboratory (NRL) in Washington DC. They found the first such source in 1957, and it was none other than the Crab nebula, which has continued over the decades to hold its place as the most remarkable single object in the sky. Unlike the extragalactic non-thermal sources, its spectrum is relatively flat and the flux does not fall off as fast with increasing frequency. The NRL group had pioneered radio astronomical observations at very short wavelengths, and they showed that at $\lambda \sim 3$ cm, the radiation from the Crab nebula was substantially polarized (8%), at a position angle close to the average optical direction. But from then on, for the next five whole years, an incredible situation persisted. Hundreds of radio sources, both galactic and extra galactic, had been discovered, the spectra of most of which clearly indicated that they were non-thermal, and in all probability synchrotron radiators. But not one of them was detectably polarized, excluding, of course, the always extraordinary Crab nebula.

The longest wavelength at which polarization in this object had been measured was around 10 cm with a value of 3–4%. Some attempts at 20 cm wavelength had only placed a limit of around 1% but there seemed good reason to doubt that the instruments used could reach such a limit with confidence. Convinced that all synchrotron sources must show polarization, and that an improvement in measurement techniques should lead to success, my colleagues and I made the most strenuous efforts at the Caltech Observatory, where I happened to be working at the time. If I might digress a little I would like to describe some incidental aspects of this exercise. I for one suddenly discovered that my bible the *Radio amateur's handbook* could not see me safely through this lot, and it appeared unavoidable that I should learn about those strange things called Stokes parameters. The place to learn about them (in those days) was of course Chandrasekhar's classic treatise on Radiative Transfer, where the mathematics of how to deal with it all was laid out in detail. This was fine if you were one of those who loved equations. Also if you didn't mind that the simplest antenna—a single straight piece of wire—could respond to all four parameters, although they were supposed to represent different aspects of the radiation! Worse, there was no way in which you could orient the wire (or rotate your coordinate system) to get it to respond to *less* than three of the four Stokes parameters. I needed another way, preferably pictorial, to look at this difficult problem, and see the answer, so to speak. Fortunately, it turned out that there was one.

It is described in a series of papers on the generalized theory of interference which form part of the doctoral work of Pancharatnam, a former student of the Institute where I now work. The papers I refer to are in his collected works brought out by Oxford University Press some years after his premature death in 1969 while he was still a Research Fellow of St. Catherine's College. To deal with the complex polarization phenomena encountered in crystals involving double refraction, absorption, optical rotation, and partial coherence, he used and extended the geometric representation of polarization states on a sphere due to none other than Poincaré (Fig. 6). Elegance and simplicity characterized his approach, and his theorems and methods provide a powerful way of dealing not only with radiation in different states of polarization, but also with antennas (which are merely analysers) and interferometers, and their responses to arbitrary states of incident radiation. Speaking for myself, not only did these methods make it almost trivial to understand what would happen in even a complicated situation, but they revealed a certain beauty and aroused an enduring fascination for the subject of polarization in general.

As an example of the power of this method and the insight it provides I may mention one of Pancharatnam's surprising results. He showed that two beams of polarized radiation which are in phase with a third beam need *not* be in phase with each other! This led him to conclude that a phase change must result from describing a closed curve on the Poincaré sphere equal to half the solid angle subtended by the curve at the centre. This result, derived in the fifties, has now been identified with a phase factor derived a decade ago in a quantum mechanical context by Michael Berry of Bristol. (If I may be permitted an aside, the very large number of publications sparked off by

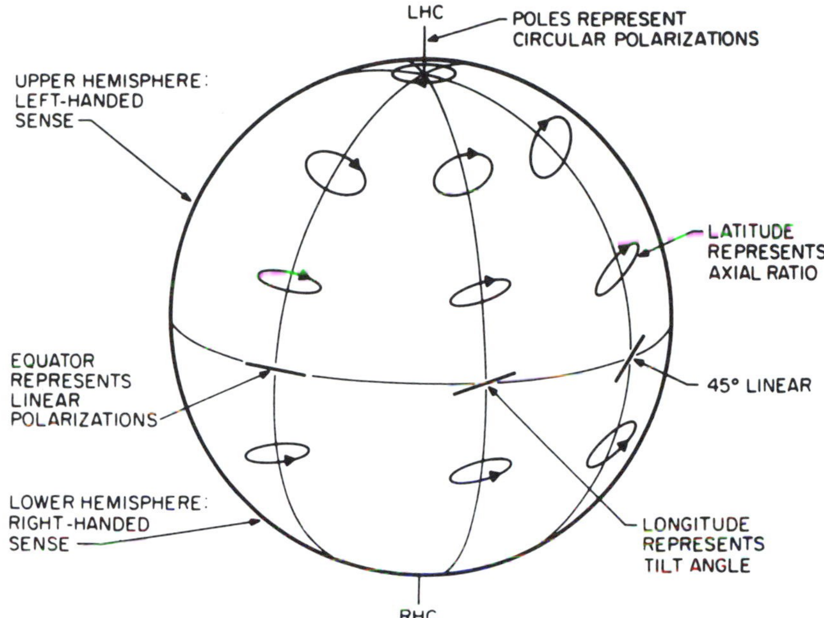


**Fig. 6.** Polarization states on the Poincaré sphere.

Berry's work prompted some wag to call it *the phase that launched a thousand scrips*.)

Returning to antennas, the immediate result of our efforts was the development of an interferometric null technique which provided an order of magnitude improvement, and we were able to measure with confidence the polarization of the Crab nebula at $\lambda \sim 20$ cm as $1.5 \pm 0.2\%$. But what was staggering was that the three other strongest non-thermal sources in the sky—Virgo A, Cas A, and Cygnus A—were polarized at less than a third, a fourth, and a *fifth* of a per cent respectively! (The theoretical maximum, under normal conditions, expected from synchrotron radiation is around 70%.) I cannot forget the major mystery these extraordinarily low limits posed at the time, particularly considering that the optical counterpart of Virgo A had been found to be polarized as described earlier. I remember trying to get five minutes' time at the 1961 IAU General Assembly in Berkeley to communicate our mostly negative, but nevertheless mystifying, observations just made. However, the radio astronomers there (as many to this day!) were more involved in heated arguments about number counts vs. flux density and such issues. Looking for radio polarization was clearly a waste of time, they felt, as had been established by earlier attempts! But in fact, things were just about to take off.

The breakthrough came next year in 1962 when, on the one hand, Mayer *et al.* found that two important extragalactic sources, Cygnus A and Centaurus A, were polarized several per cent at $\lambda \sim 3$ cm, and on the other hand Westerhout *et al.* in Holland detected polarization in the galactic background radiation at $\lambda \sim 75$ cm, which could not be attributed to spurious effects. The spell had been broken, and within weeks the bandwagon was rolling. A flood of reports poured in claiming practically every direction in the galaxy and every strong non-thermal source to be polarized, whether or not the particular telescope used was capable of such a measurement and the observer capable of telling one Stokes parameter from another. By the time of the next IAU General Assembly at Hamburg in 1964 the measurement of polarization in radio sources had become an industry. While many were busy adding to the list of sources in which some polarization could be detected, there were also a few trying to understand the implications of the observed low values. In any event, when the dust finally settled, the several reasons for the failure of our and other previous attempts appeared obvious in retrospect.

Firstly, in the design of radio telescopes and particularly feeds for them, little or no thought had been given to their ability to measure polarization, and this resulted in large instrumental effects. Secondly, the systematic or random changes in the direction of the magnetic field within a radiating region could drastically reduce the net polarization when the source was unresolved and contributions from its different parts were averaged together, as in our measurements of the integrated radiation from the three strong sources. Hiltner showed that polarization exists at all orientations in the Crab and varies in amount from zero to almost 60% over the face of the nebula. In the case of M87, Baade found up to 30% polarization in the jet, but also that it was broken up into knots polarized at considerable angles to each other. And lastly, there was the result of an effect discovered by Faraday and named after him, that radiation traversing a medium threaded by a magnetic field has its plane of polarization rotated by an amount proportional to the square of the wavelength (Fig. 7). The constant of proportionality, which depends on the longitudinal field strength and the medium traversed, is called the rotation measure. In the present context, the medium is the ionised gas in interstellar space or within the radio source itself, and the magnetic field is that present in these regions.

If the rotation takes place *outside* the source, different wavelengths appear to be polarized at different angles, but this can be measured and allowed for. On the other hand, if the rotation occurs within the source, *depolarization* occurs and cannot be corrected for. This was already being illustrated in the case of the Crab nebula where the mean polarization at optical wavelengths—where Faraday rotation is negligible—was as high as 9%.

But even in the case of *external* rotation, very large values can do harm by

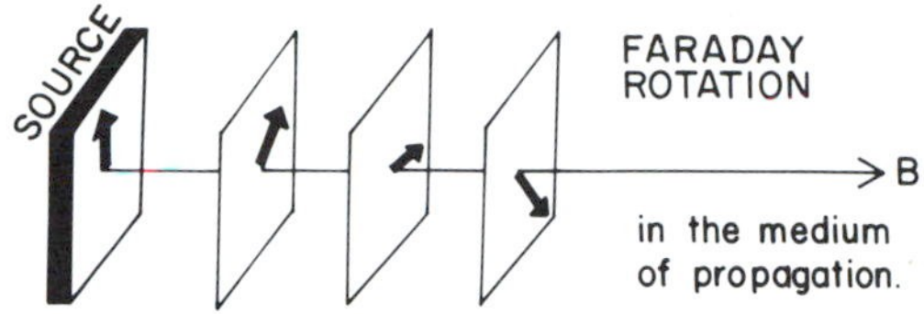


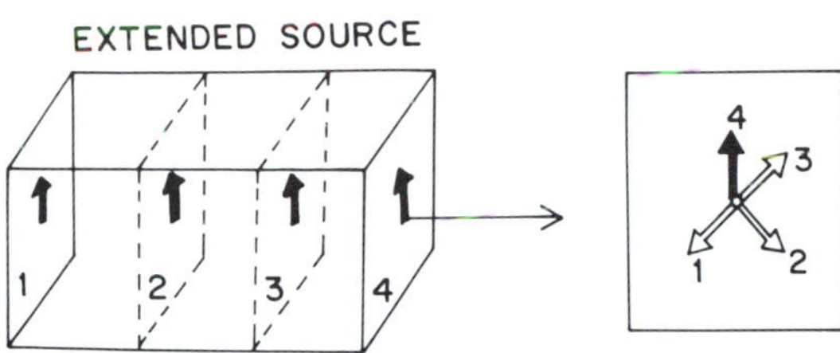


**Fig. 7.** Faraday rotation occurring in a medium outside the source turns the plane of polarization through angles proportional to $\lambda^2$ but does not affect the true degree of polarization. Rotation within the emitting region leads, however, to the incoherent addition of Stokes vectors in different orientations and thus to true depolarization.

introducing differential rotation within the band of wavelengths being observed. Similar effects can be caused by changes in the rotation measure within the resolution angle of the telescope. But unlike *internal* Faraday rotation, these last two effects can be overcome by achieving higher resolution both in wavelength and angle on the sky, as I shall discuss later.

As a result of the efforts of those early years, it became evident that the future lay in high frequency interferometric observations which could reveal the details of individual regions within sources with higher and different polarizations than the average. With modern aperture synthesis instruments this has now come to pass, and in the last part of my talk I shall describe some of these results.

In all of the discussion so far, the magnetic fields responsible for the polarization have been very weak, and of the order of micro- or milligauss, associated with currents in the plasma in which they are embedded. I shall turn now to a very different situation, that of the giant planet Jupiter, whose case was an important landmark in the study of radio polarization. In the early fifties it was believed that the planets, like the moon, would only emit thermal radiation characterized by their physical temperature and surface emissivity. Several planets had been observed at very short radio wavelengths and found to emit about the expected amount of thermal radiation. Jupiter, however, was known to put out bursts of radiation at decametre wavelengths, and it also showed an apparent excess over the expected thermal emission at centimetre wavelengths. This effect was found to become stronger at decimetre wavelengths, and led to the speculation that this excess could be cyclotron or synchrotron radiation from a magnetosphere containing trapped relativistic particles ejected by the Sun, similar to the belts around the earth that had been discovered by van Allen. In work that involved several people at the Caltech Observatory it was possible to demonstrate both the high linear polarization of the decimetric non-thermal radiation, and that the emitting region was several times larger than the visible planet. Follow-up observations with the same instrument showed a beautiful periodic rocking of the plane of polarization, and led to a determination of the angle between the magnetic and rotational axes of the planet; this also provided a direct measurement of the period of rotation of its *interior*, where the magnetic field was anchored. It may be recalled that optical features on the surface rotate at different speeds depending upon their latitudes.

Further detailed observations at the Parkes Observatory showed that a single 'belt' would not suffice to account for the polarization variation with rotation, and led to a more elaborate model. Finally, aperture synthesis measurements with large telescope arrays in the UK, Holland, and the USA, led to the equivalent of time-lapse *tomography*, and have provided us with an excellent detailed picture of the three-dimensional structure of Jupiter's magnetosphere (Fig. 8). *In situ* measurements made later by several spacecraft during their fly-by(s) have confirmed and complemented all that was learned from the ground-based measurements.

By astronomical standards Jupiter is a very tiny and 'local' object, but I have discussed it in some detail because its case was important in many ways. It was the second radio source found to be polarized, and it was the most highly polarized known for some years. The polarization decisively established the synchrotron nature of its high-frequency radio emission, but unlike other synchrotron sources, it had no internal Faraday rotation. Because of the physical rotation of the object the polarization varies periodically, and it was the first case of such a study revealing the geometry of a magnetic field which had a *separate* origin, and was therefore unaffected by the particles which traced it. But most remarkable in the present context is its similarity in many respects with the next class of objects I shall discuss, namely pulsars, which came along a decade later.

When the incredible objects we now call pulsars were discovered, there were, apart from dozens of minor questions, three major ones demanding urgent answers. What was the nature of the object? What caused the periodicity? And what was the radio emission mechanism? From the observed timescales and energies involved, the discoverers had already narrowed the

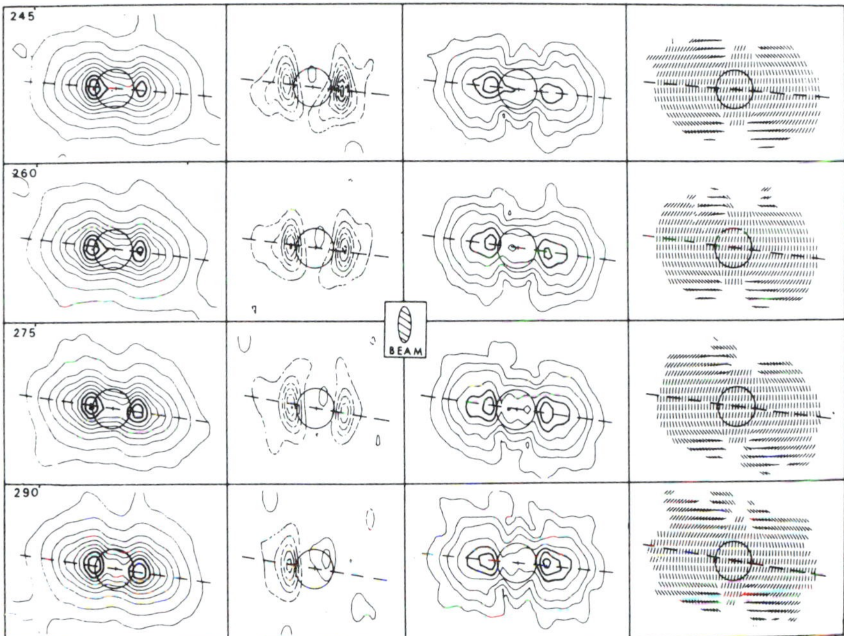


**Fig. 8.** One of a series of diagrams showing maps of Jupiter's radiation belts from all Rotational Aspects. From left to right are maps of the total intensity, circular and linear polarization fluxes, and a vector diagram of the magnetic field of Jupiter, from de Pater (1980).

choice for the object to one between white dwarfs seen in large numbers, and the theoretically predicted neutron stars which had never been observed. As one can see from the introductory articles in the collection of the first fifty or so papers on pulsars, white dwarfs oscillating in various ways were still the favourite candidate, although Gold had already proposed rotating neutron stars. The discoveries of the Vela and Crab pulsars with their very short periods (by the standards of the day), and the measurement soon after of the lengthening of these periods, as predicted by Gold, put the rotational neutron star hypothesis way ahead of the rest of the field. But the reason for the beaming of the radiation as from a lighthouse was as obscure as ever. The spectrum of the radiation had been shown quite early to be highly non-thermal, but the question of an emission mechanism was generally avoided. The theoretical papers of the time that attempted to address this issue now read like science fiction, all attesting to the strangeness of the objects whose nature we were struggling to fathom.

While the intensity of pulsar radiation had been noted to be highly erratic, there were two features which showed a very systematic dependence on time; one was the period and its lengthening already referred to, and the other was the polarization variation within the pulse. The radiation from the

Vela pulsar was found in studies at the Parkes Observatory and at Caltech to be highly linearly polarized, with the plane of polarization sweeping systematically through a large angle within the short pulse duration. It was further found that when Faraday rotation in the interstellar medium was corrected for, the instantaneous plane of polarization at any specific point within the pulse envelope was the *same* at all frequencies! When the erratic pulse to pulse variations of amplitude were averaged out, the integrated pattern always showed a highly linearly polarized but simple pulse shape of great stability. Taking Vela as the archetype, these observations led directly to the providing of *independent* answers to *all three* of the major questions mentioned earlier.

We have already seen in the case of Jupiter that there were two simple and natural consequences of the polarization depending on a dynamically dominant magnetic field associated with a rotating body. One is that the plane of polarization was a function of longitude as the object rotated; and the other was the stability of the period, since the field is anchored in the object (planet or star) and not related to surface or atmospheric phenomena like sunspots, aurorae, etc. Misalignment between the rotational and magnetic axes is essential to manifest the periodicity, and the shape of the position angle versus longitude curve would reflect the geometry of these axes relative to the direction of the observer. From the excellent agreement of the observed shape of this curve for Vela with that expected for a dipolar field, it seemed more than suggestive, almost compulsory, to identify pulsars with *rotating*—as opposed to vibrating—objects. To appreciate the strength of this argument against the vibrational models which were still rife, it should be noted that the polarization sweep does not reverse after the peak of the pulse, but continues in the *same* sense. Taking this as proof of rotation, white dwarfs could be ruled out, as the observed period of Vela of 89 ms was far too high a rotation rate for them. Pulsars *had* to be neutron stars.

The observed rate of position angle sweep was also very high compared to the rate of change of longitude, and, in this picture, gave a direct indication of the proximity of the radiating region to a magnetic pole. The symmetry of the polarization pattern also showed that the pulse straddled the longitude that contained the projected magnetic axis. Further, of all the field lines which emanate from each magnetic pole, only those which subtend less than a certain angle with the magnetic axis are *open* field lines, and would have relativistic particles streaming along them as predicted by Goldreich and Julian. This provided automatically a cone centered on each magnetic pole, and one had thus a simple explanation both for the location of a rotating beam of limited angular extent, and for the stability of its period (Fig. 9).

Given the high magnetic fields expected for neutron stars ($\sim 10^{12}$ G) the energetic particles discussed by Goldreich and Julian would radiate away any

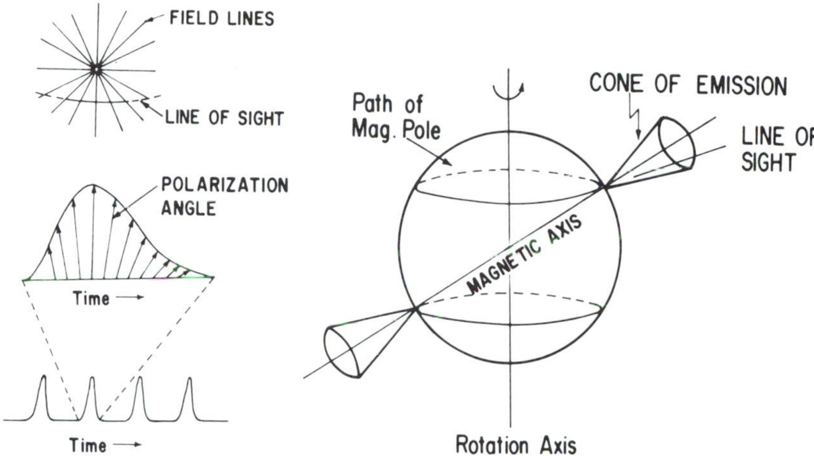


**Fig. 9.** The polarization of pulsar radiation and the magnetic pole model.

energy and momentum they had *transverse* to the magnetic field in a negligible time, and be forced to move strictly along the field lines in their lowest Landau levels. But as the field lines curve away from the magnetic axis, the acceleration associated with motion along them should produce radiation in the observed radio range. Given the high Lorentz factors estimated for the particles, the expected polarization would be almost totally linear. It was in fact so, as observed in Vela, and was much higher than the maximum of 70% expected in any synchrotron source. In this model the electric vector would be parallel to the projected field lines, and this incidentally enables one to fix the projected direction on the sky of the axis of rotation of the neutron star!

In putting forward this simple picture a number of important but awkward details were swept under the carpet, such as the coherence mechanism to explain the very high specific intensity of pulsar radiation, the intrinsic variability of this intensity over short time-scales, and most seriously in the present context, sudden and often drastic departures in the polarization behaviour from that expected of the simple model. These aspects all continue to be tackled in increasingly more elaborate models spanning the twenty-year period to the present time. My point here is only to illustrate how much could be gleaned about pulsars from just a study of their polarization.

In the last part of my talk, I would like to touch briefly on work in recent times which shows what can be revealed by high resolution observations at high frequencies with high sensitivity and dynamic range. As examples, I have chosen the same three strong sources where early attempts at longer wavelengths and with no resolution led to limits of a few tenths of a per cent

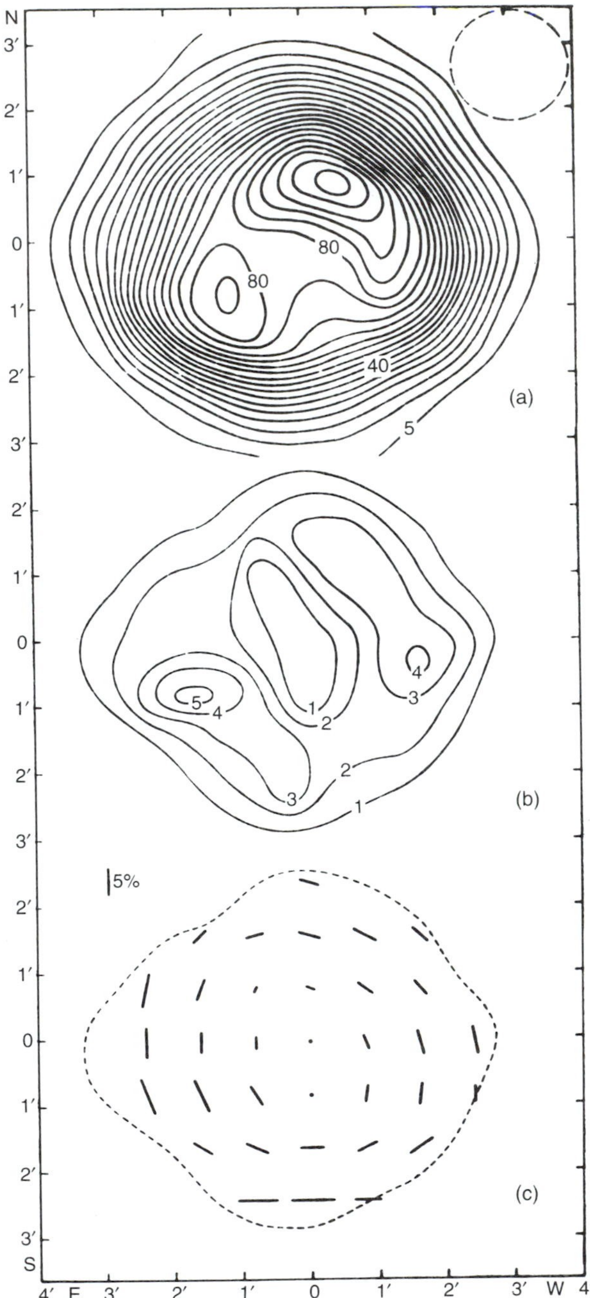


**Fig. 10.** Distribution of intensity and polarization over Cassiopia A. (a) Total intensity; (b) Linearly polarized intensity; (c) Degree of polarization and orientation of electric vector (Mayer and Hollinger 1968).

for the average polarization. Let me start with Cas A, which is the strongest source in the sky at low frequencies and is the remnant of a supernova outburst possibly witnessed by Flamsteed in 1680. The NRL group again were those who made the first breakthrough in 1968 and their beam size at 1.5 cm wavelength was just small enough to resolve the source. The source was found polarized in a beautifully symmetrical way with the electric vector circumferential and the polarization almost zero at the centre of the source (Fig. 10). The degree of polarization was about 4.5 per cent around the ring of polarized radiation and the remarkable symmetry of its distribution explained the failure to observe any polarization using lower resolution, as in the longer wavelength measurements I just mentioned. It is a tribute to the perception of these early investigators that they recognised that the unexpected result of a radial component of magnetic field in the source

> suggests that the magnetic field has been carried out with the expansion of the supernova envelope, and we observe polarized radiation associated with a component which has been stretched out in the radial direction during the expansion of the shell.

The new observations that I shall discuss were all made with the Very Large Array* in New Mexico, the largest and most powerful synthesis radio telescope in the world today (1988). Recent pictures obtained by Rick Perley and colleagues with this instrument had a quarter of a million picture points within *each* of the beam areas of the original NRL measurements and have been put together to make a map of the total intensity from this supernova remnant representing the most detailed radio image yet constructed (Plate IV). Vindicating the astrophysical interpretation of Mayer and Hollinger based on a radio map with a mere 16 beam areas within the source, the VLA image reveals among numerous other details conical extensions of its circular outline. These have been interpreted as more slowly expanding material from deeper within the star breaking through the shell which has been decelerated sufficiently by sweeping up surrounding interstellar material. A similar picture of the intensity of *linearly polarized* emission, colour-coded by the projected direction of the magnetic field, is shown in Plate V. The overall radial field orientation referred to earlier is shown by the systematic change in colour from blue through green, yellow, red and back to blue on the opposite side. Local departures from this pattern clearly define the regions of tangential and turbulent fields at the apex of the conical extensions. Such pictures show changes in their details over time-spans as short as a year and make it possible to recognize processes that shape such supernova

---

*This is part of the National Radio Astronomy Observatory of the USA, operated by Associated Universities, Inc., under contract with the National Science Foundation.

remnants; they lend strong support to current theories that the fields are built up through turbulent action near the expanding boundaries of the shell as shown by Braun *et al.* (1987).

I turn next to Cygnus A, the second strongest source in the sky at low frequencies, whose distance determination by Baade and Minkowsky in the early fifties was the first decisive indication of the potential of radio astronomy to probe large volumes of the observable universe. Cygnus A is the prototype of the double-lobed radio source with an optical galaxy in the middle feeding the lobes by a jet. The NRL measurements of 20 years ago had revealed that the lobes were polarized at different angles and that the Faraday rotations in their directions (up to 1400 rad $m^{-2}$!) were too large to be caused by the ISM in our own Galaxy and thus had to be associated with the source itself. Perley and his colleagues have made many observations over several years with the VLA and combined them to produce a number of extraordinary images with sub-arc-second resolution. These images reveal incredibly fine details including filamentary structure in the two lobes.

Maps of the polarization show that the direction of the magnetic field varies considerably within the source and that in several places it changes direction to align with the filamentary structure seen in the radio brightness distribution (Plate VI). In such maps the field direction shown by the colour indicates the flows involved, as for example along the boundaries of the lobes. These lobes are giant versions of those that occur in miniature in Cas A. The field there also becomes tangential at the *apices* of the conical extensions where matter was pushing through the shell and flowing back along the sides.

As the mapping of the field direction requires the determination of the intrinsic polarization angles, this involves correctly allowing for the rotation measures which, as I said before, were known to be large and the cause of confusion and ambiguity in interpreting earlier measurements over the years. Dreher *et al.* (1987) have finally succeeded in sorting out this problem and have produced maps displaying enormous rotation measures which vary from −4000 to +3000 rad $m^{-2}$ over the face of the source (Plate VII). Their achievement in this exercise can be appreciated by noting that the gradients in the rotation measures in both lobes commonly exceed 300 rad $m^{-2}$ per arc second and that this number is typical of the *maximum* RM encountered due to passage of radiation even through the plane of our Galaxy.

The overall pattern in rotation measure suggests that the magnetic field is ordered on scales of 20–30 kpc. But most remarkably there is no evidence for internal depolarization despite these high RM values. In fact, there are regions where the linear polarization actually reaches 70%, the theoretical maximum for synchrotron radiation. The operative Faraday screen must therefore be outside the radiating region, either just surrounding the lobes or in the intracluster gas from which X-radiation is observed. If it is in this gas, the minimum fields required are of the order of several microgauss comparable to what is found in interstellar regions within our galaxy! If, on the other hand, the Faraday screen is in a dense sheath around the lobes, Dreher *et al.* have calculated that the field required is very much higher, approximately 100 microgauss, and not too different from what is estimated for the field in the lobes themselves.

There is one aspect of the distribution of rotation measures in the two lobes, namely that it was twice as great in one lobe than in the other, the significance of which was only appreciated later. Laing (1988) and Garrington *et al.* (1988) have shown very recently that in a sizeable sample of double-lobed sources with a single jet, the depolarization at longer wavelengths is systematically and considerably greater in the lobe not associated with the jet. If it is due to an externally depolarizing Faraday screen, then it affects the jetless lobe more than the other. This has been interpreted by Laing as the result of a halo of hot gas around the associated galaxy or quasar, and supports the Doppler beaming hypothesis according to which only that jet is seen which is pointing towards us. What is interesting and important here is that observations of the polarization are supplying a third dimension to the previously two-dimensional distributions of total intensity.

I turn next and last to M87, the extraordinary galaxy in the Virgo cluster in which Baade had found an optical jet I referred to earlier, and whose radio counterpart is beautifully seen in VLA observations by Frazer Owen and colleagues. At the time of the Cygnus A observations it was believed that this source had a relatively low rotation measure. This was puzzling as M87, like Cygnus A, is also immersed in a dense halo of X-ray emitting gas. Very recent observations by Frazer Owen at the VLA have solved this puzzle. Observations around 6 cm wavelength have revealed Faraday rotation measures typically 1000–2000 rad $m^{-2}$ in small regions. And as in the case of Cygnus A the higher rotation measures are in the lobe away from the jet. But most of the source is significantly polarized, showing again that the Faraday screen is in front of the radio-emitting region rather than mixed with it.

An estimate of the field required to explain the observations is of the order of 40 $\mu$G, a value approaching an energy density not too far below the thermal gas energy density. It should be mentioned that in the case of Coma and several other such rich clusters, all estimates for the field strength gave very low values, generally much less than a microgauss, suggesting that Cygnus A may be unique in some way. These new observations of Virgo A, however, support the idea that very high rotation measures may be typical of all radio sources in clusters with dense X-ray emitting gas but relatively fewer Galaxies in them. As Cygnus A and Virgo A are the first extragalactic sources that have been studied in such detail, there are sure to be others with similar characteristics, and it is clear that these observations have revealed something very important whose full significance is yet to be appreciated. In particular it is not obvious how much of the magnetic fields found in these observations is primordial, and whether it played any role in the evolution of the structure of the universe as we see it now.

The spectacular observations I have just described demonstrate the state-of-the-art in obtaining radio pictures of remote objects in the universe with the biggest and most complicated camera in the world. They represent a magnificent technical feat, showing what can be achieved by stretching both hardware and software to the limit of present ingenuity and in that sense they need no further justification. But in the context of today's discussion, one may ask for the *message*, i.e. what has one learnt about the workings of the objects we are looking at in increasing detail. My personal assessment is that with this kind of detail the observations have overtaken the theories. The plasma magneto-hydrodynamicists face a formidable challenge if they are hoping to explain every detail that is seen in these images. But there is no question that such pictures of the magnetic field are indispensable in improving and refining current models. And as we have seen, it is the study of the polarization which can and is providing us with them.

As I indicated at the beginning, I have necessarily had to omit discussion of a number of topics connected with polarization all of which form part of the whole story. But in every case that I have discussed, and many which I have not, it is the polarization that is laying bare the magnetic field which Eugene Parker, its highest priest, has called 'the radical element responsible for the continuing thread of cosmic unrest'. Not bad, you will agree, for what started as the strange behaviour of two images in a window pane of the setting Sun, viewed through a crystal of calcite.

## Acknowledgements

I thank the Milne Society for inviting me to deliver this lecture, and the Mathematical Institute, Wadham College and IBM UK Ltd. for their hospitality during my visit to Oxford. In the preparation of the lecture I received much help in many different ways, for which I would like to thank R. Nityananda, R. D. Ekers, F. N. Owen, G. Srinivasan, C. V. Vishveshwara, S. Sridhar, W. M. Goss, and D. F. Malin.

## Addendum (1995)

A number of topics related to the theme of the lecture had necessarily to be omitted at the time it was given. Now, seven years later, the editors of the present volume have graciously provided the possibility of adding a little something, to make the story less incomplete or dated than it would otherwise be.

The message of polarization in revealing the geometry of the symmetry breaking agent—the magnetic field—was carried by the radiation, which in all the cases I discussed was continuum, or broadband, in nature. Consequently, it was only the orientation of the component of the field in the plane of the sky which could be determined, without any clue to its strength. The change of this orientation with the observation of frequency due to Faraday rotation depended, on the other hand, on the line-of-sight component of the field, whose strength could be estimated based on various assumptions (see, for example, Carilli 1995). In the case of pulsars, for example, the independent determination of the column density of electrons, the dispersion measure, enables the calculation of the mean value of the line-of-sight field strength, weighted by the electron density (see, for example, Rand and Kulkarni 1989).

In sharp contrast are the *narrow* band emission and absorption processes characteristic of atoms and molecules, aggregates found in plenty in interstellar space. As pointed out by Bolton and Wild as early as 1957, the phenomenon of Zeeman splitting of the 21 cm line radiation from neutral hydrogen provides a direct means of measuring the *strength* of interstellar magnetic fields. In this case again, it is the longitudinal or line-of-sight component of the field whose magnitude would now be reflected in the separation of the two oppositely *circularly* polarized features of the line radiation. The technical difficulty of measuring this separation is inversely proportional to the sharpness of the spectral line observed. In early attempts this consideration was an important one in the selection of absorption rather than emission lines, as they tend to be narrower than the latter, and led to the determination of the field strength in certain gas clouds as a few tens of microgauss (Verschuur 1968, Brooks *et al.* 1971).

In later attempts, numerous emission lines have also been observed for Zeeman splitting with single telescopes equipped with circularly polarized feeds (Heiles 1989). But the results obtained by different observers with different telescopes on the same regions of sky are so discrepant as to cast doubt on any serious astrophysical interpretation of the observations (Verschuur 1995 and references therein). I find the situation reminiscent of the early attempts to measure linear polarization, as described in the lecture, and this leads me to feel that interferometric methods are the ones that will finally succeed here, as in the previous case, for the reliable detection of *minute* amounts of circular polarization associated with weak fields.

Zeeman splitting from relatively stronger fields in star-forming regions has been successfully measured with aperture synthesis telescopes such as the

VLA, e.g. see Troland *et al.* (1989). In recent observations, Roberts *et al.* (1995) have looked at several such regions and measured line-of-sight fields of order 50 to 500 μG. Even stronger fields have been found in circumstellar regions of late type giant and supergiant stars, as measured through observations of the circular polarization of the molecular masers found in such regions. Values range from ~5 mG for OH masers found at $\sim 10^{16}$cm from the star to ~50 G for the SiO masers that reside above the photosphere (Reid 1989). For the sake of completeness it should be mentioned that fields of order $10^5$ to $10^7$ gauss have also been measured in white dwarfs through a study of their optical circular polarization (Angel 1978).

## References

Alfvén, H. and Herlofson, N. (1950). Cosmic radiation and radio stars, *Phys. Rev.*, **78**, 616.

Angel, J. R. P., (1978). Magnetic white dwarfs, *Ann. Rev. Astron. Astrophy.*, **16**, 487–519.

Baade, W., (1956). Polarization in the jet of Messier 87, *Ap. J.*, **123**, 550–551.

Baade, W. and Minkowski, R., (1954). On the identification of radio sources, *Ap. J.*, **119**, 215.

Berry, M. V., (1984). *Proc. R. Soc.*, London, **A392**, 45–57.

Bolton, J. G. and Stanley, G. J., (1949). The position and probable identification of the source of galactic radio-frequency radiation Taurus A, *Austral. J. Sci. Res.*, **2A**, 139–148.

Bolton, J. G. and Wild, J. P., (1957). *Ap. J.*, **125**, 296.

Bolton, J. G., Stanley, G. J., and Slee, O. B., (1949). Positions of three discrete sources of galactic radio-frequency radiation, *Nature*, **164**, 101–102.

Bragg, W., (1959). *The universe of light*, Dover Publications, New York.

Braun, R., Gull, S. F., and Perley, R. A., (1987). Physical process which shapes Cassiopeia A, *Nature*, **327**, 395.

Brooks, J. W., Murray, J. D., and Radhakrishnan, V., (1971). Zeeman splitting of the 21-cm absorption line in Orion A, *Astrophys. Lett*, **8**, 121–123.

Carilli, C., (1995). Rotation measure variations in radio sources, *IAU Symp.*, **175**.

Chandrasekhar, S., (1950). *Radiative transfer*, Oxford University Press.

de Pater, I., (1980). *Astron. Astrophys.*, **88**, 175–183.

Dombrovsky, V. A., (1954). On the nature of the radiation from the Crab Nebula (in Russian), *Dokl. Akad. Nauk SSSR.*, **94**, 1021–1024.

Dreher, J. W., Carilli, C. L., and Perley, R. A., (1987). The Faraday rotation of Cygnus A: magnetic fields in cluster gas, *Ap. J.*, **316**, 611.

Gardner, F. F. and Whiteoak, J. B., (1966). The polarization of cosmic radio waves, *Ann. Rev. Astr. Astrophys.*, **4**, 245.

Garrington, S. T., Leahy, J. P., Conway, R. G., and Laing, R. A., (1988). A systematic asymmetry in the polarization properties of double radio sources with one jet, *Nature*, **331**, 147–149.

Gehrels, T. (ed.), (1976). *Jupiter (Studies of the interior, atmosphere, magnetosphere and satellites)*, University of Arizona Press, Tucson, Arizona.

Gold, T., (1968). Rotating neutron stars as the origin of the pulsating radio sources, *Nature*, **218**, 731.

Goldreich, P. and Julian, W. H., (1969). *Ap. J.*, **157**, 869.

Heiles, C., (1989). A survey of recent magnetic field measurements in HI features of the galaxy, *IAU Symp.*, **140**, 35–40.

Hewish, A., Bell, S. J., Pilkington, J. D. H., Scott, P. F., and Collins, R. A., (1968). Observation of a rapidly pulsating radio source, *Nature*, **217**, 709–713.

Hiltner, W. A., (1949). On the presence of polarization in the continuous radiation of stars II, *Ap. J.*, **109**, 471–478.

Jansky, K. G., (1933). Electrical disturbances apparently of extraterrestrial origin, *Proc. I.R.E.*, **21**, 1387–1398.

Kiepenheuer, K. O., (1950). Cosmic rays as the source of general galactic radio emission, *Phys. Rev.*, **79**, 738–739.

Laing, R. A., (1988). The sidedness of jets and depolarization in powerful extragalactic radio sources, *Nature*, **331**, 149–151.

Maran, S. P. and Cameron, A. G. W. (eds), (1964). *Physics of nonthermal sources*, Goddard Institute for Space Studies, NASA, **53**.

Mayer, C. H., and Hollinger, J. P., (1968). Polarized brightness distribution over Cassiopeia A, the Crab Nebula and Cygnus A at 1.55 cm wavelength, *Ap. J.*, **151**, 53.

Mayer, C. H., McCullough, T. P., and Sloanaker, R. M., (1957). Evidence for polarized radio radiation from the Crab Nebula, *Ap. J.*, **126**, 468–470.

Mayer, C. H., McCullough, T. P., and Sloanaker, R. M., (1962). Evidence for polarized 3.15-cm radiation from the radio galaxy Cygnus A, *Ap J.*, **135**, 656–658.

Minnaert, M., (1940). *Light and colour in the open air*, G. Bell and Sons Ltd., London.

Morris, D. and Radhakrishnan, V., (1963). Tests for linear polarisation in the 1390 MC/S radiation from six intense radio sources, *Ap. J.*, **137**, 147.

Oort, J. H. and Walraven, T., (1956). Polarization and composition of the Crab Nebula, B.A.N. **12**, 285–308 (No. 462).

Owen, F. N., Hardee, P. E., and Cornwell, T. J., (1989). High resolution, high dynamic range VLA images of the M87 jet at 2 cm, *Ap. J.*, **340**, 698–707.

Owen, F. N., Eilek, J. A., and Keel, W. C., (1990). Detection of large Faraday rotation in the inner 2 kiloparsecs of M87, *Ap. J.*, **362**, 449.

Pancharatnam, S., (1975). *Collected works*, Oxford University Press.

Parker, E. N., (1979). *Cosmical magnetic fields—their origin and their activity*, Clarendon Press, Oxford.

Perley, R. A., Dreher, J. W., and Cowan, J. J., (1984). The jet and filaments in Cygnus A, *Ap. J.*, **285**, L35.

*Pulsating stars* (1968). Nature reprint, Macmillan Press, London.

Radhakrishnan, V., (1969). Fifteen months of pulsar astronomy, *Proc. Astron. Soc. Austr.*, **1**, 254.

*The radio amateurs' handbook*, published by the Headquarters Staff of the American Radio Relay League, West Hartford, Connecticut, USA.

Rand, R. J. and Kulkarni, S. R., (1989). The local galactic magnetic field, *Ap. J.*, **343**, 760.

Reber, G., (1944). Cosmic static, *Ap. J.*, **100**, 279–287.

Reid, M. J., (1989). Masers and stellar magnetic fields, *IAU Symp.*, **140**.

Roberts, D. A., Crutcher, R. M., and Troland, T. H., (1995). The distribution of molecular and neutral gas and magnetic fields near the bipolar HII region S106, *Ap. J.*, **442**, 208–227.

Shklovsky, I. S., (1953). On the nature of the radiation from the Crab Nebula (in Russian), *Dokl. Akad. Nauk SSSR.*, **90**, 983–986.

Shklovsky, I. S., *Cosmic radio waves*, Harvard University Press, Cambridge, Massachusetts.

Troland, T. H., Crutcher, R. M., Goss, W. M., and Heiles, C., (1989). Structure of the magnetic field in the W3 core, *Ap. J. Lett.*, **347**, L89–92.

Vashakidze, M. A., (1954). On the degree of polarization of the light near extragalactic nebulae (in Russian), *Astr. Tsirk.*, **147**, 11–13.

Verschuur, G. L., (1968). Positive determination of an interstellar magnetic field by measurement of the Zeeman splitting of the 21-cm hydrogen line, *Phys. Rev. Lett.*, **21**, No. 11, 775–778.

Verschuur, G. L., (1995). Zeeman effect observations of HI emission profiles. II. Results of an attempt to confirm previous claims of field detections, *Ap. J.*, **451**, 645–659.

Westerhout, G., Seeger, C. L., Brouw, W. N., and Tinbergen, J., (1962). *Bull. Astron. Inst. Neth.*, **16**, 187.

# Carbon—the element of life: what is its origin on Earth?

Thomas Gold

The Earth is a body with a most complex history. None of its sister planets, not even its neighbours Venus and Mars, display signs of the processes that appear to have been the major ones to shape this planet of ours. On the other solid planetary bodies, Mercury, Venus, the Moon, Mars, and the satellites of the major planets, the effects of impact cratering can be seen very clearly. Craters spanning a range of sizes from a few kilometers to several thousand can be recognized. Impacts of solid bodies upon other planets must have been a very frequent occurrence during the formation stage of our planetary system.

Strangely, just on our Earth, similar cratering events can only be seen in a very subdued form. We see arcs of circles appearing in the midst of a topography shaped by other effects. It seems reasonable to interpret such circular features as the remains of impacts, but then one has to suppose that other events occurred here later, to obscure much of the evidence of this bombardment.

Nevertheless it appears that the Earth also formed by a solid accretion process, but partial melting then took place, causing melts of lower density to make their way to the surface while, presumably, melts of higher density sank down towards the center. The heat for this melting was the result of radioactivity contained in the material, as well as just the heat resulting from compression. Once partial melting occurred, two other sources of heat came into play. Firstly, there is the gravitational energy that is released as materials can move and sort themselves out according to density. Secondly, there is the chemical energy that results from chemical reactions that can now take place, either between different liquids, or between liquids and solids. The original diverse materials, accreted as cold objects, would certainly not have been in their lowest chemical energy configuration; and they would have been in an uneven distribution as left by the impact events. After gaining mobility by melting, chemical reactions would take place that make an approach to chemical equilibration.

Both these last two sources of energy have the interesting property that they make the heating unstable: where more heating has occurred, and more melt produced, more of these actions can take place, and therefore still more heat will be produced there. One may speculate that the very uneven distribution of internal heat sources which we recognize at the surface derives from such an instability. The circumpacific 'belt of fire' is the most striking example.

The low density melts produced the rocky layer which we call the crust. Since this covers almost all the surface, it created the impression that the Earth had frozen from an initial melt; every basement rock could be seen to have once been a liquid magma. This picture of a once liquid Earth became adopted, and it shaped much of the discussion in the early days of geology. Even though by now it has become quite clear that only a partial melting was involved, and that the major volume had never been molten, there has not been that thorough re-evaluation of geologic theory that such a change should have caused. Nowhere is this more in evidence than in the discussion of the origin of the volatile substances on the surface: the water of the oceans, the nitrogen of the atmosphere, and the carbon-bearing fluids that appear to have been responsible for a great enrichment of the surface with carbon.

On an initially molten Earth, the volatiles would have come to the surface in the first phase. Later, when such a body had cooled, there would be little expectation of a renewed supply of volatiles from below.

Quite the opposite would be the expectation on a cold body that is heating up, on which successive layers would reach temperatures at which volatiles would be driven off. Outgassing processes would be expected to continue as long as internal temperatures were increasing in any part of the body. Furthermore there would be quite different expectations of the chemical nature of the various volatile substances. On a hot Earth, most fluids would be brought to the lowest chemical energy configuration, and would not later provide any source of energy. On a cold body that is heating up, the fluids that are produced would often be out of chemical equilibrium with their surroundings and thus be a source of chemical energy.

The manner in which these differences of outlook translated into research work in the Earth sciences can still be seen. Very little scientific literature is devoted to the discussion of the mechanisms of outgassing, that is the transport of gases and liquids through the rocks towards the surface. There appears to be a great hesitation to attribute natural phenomena to an outgassing process, even when that would seem to be the most direct explanation. For example when a volcanic eruption results in a gigantic gas explosion, most discussions in the literature attempt to attribute this to sudden massive releases of gas from solution in the magma, rather than simply the escape of gases from deep, through the channels held open by the magma. Similarly when an earthquake is preceded by the emission of gas from the region, it is the stress-pattern that is invoked as responsible for generating gas from the rocks, rather than the invasion of gas from below, and the consequential fracturing of the rock.

Also the discussion of the chemical nature of liquids and gases that make their way up from depth can now be evaluated differently. Firstly there is the question of the source materials that were incorporated in the Earth, that give rise to outgassing, and secondly there is the question of the chemical modifications that the fluids will suffer at depth and on their way up. For the element carbon and its various compounds, this change of outlook is particularly significant.

Carbon is the fourth most abundant element in the universe, and also in our solar system (after hydrogen, helium, and oxygen). In the planetary system it is found mostly in compounds of carbon and hydrogen, which, at different temperatures and pressures, may be gaseous, liquid, or solid. Thus the gaseous envelopes of the giant planets contain massive amounts of methane and other hydrocarbon gases; also gaseous, liquid, and solid hydrocarbons are identified on many of the satellites, as well as on Pluto, and on asteroids and comets. The meteorites, which represent debris left over from the formation of the solid bodies, or samples subsequently exploded off their surfaces by impacts, also show carbon predominantly in solid compounds, largely unoxidized, and with hydrocarbons a major component. Compounds of oxidized carbon, such as carbonate rocks, exist in meteorites, but represent a small fraction only. From this information one would conclude that the Earth at its formation acquired its carbon predominantly in unoxidized form also, and probably largely as solid hydrocarbons. Any major addition to the forming Earth of gaseous substances can be ruled out, simply because any mix of solar system gases would have contained some gases that could not have been selectively removed, but which we have on the Earth only in trace amounts; neon is the most striking example.

The geologic record on the Earth has some remarkable things to tell us about carbon. The basement rocks, whose erosion was responsible for most of the sediments, contain very little carbon. The sediments, however, contain a high proportion, mostly in the form of the carbonate rocks (calcium and magnesium carbonates), but also some in unoxidized carbon compounds: coal, oil, gas, but predominantly the widely dispersed coal-like substance called kerogen. A large supply of carbon must have been available, at least a hundred times greater than the eroded rocks could have provided.

Since most of the carbonate rocks obtained their carbon from atmospheric $CO_2$, one may compare the $CO_2$ at present available in atmosphere and oceans with the total amount of carbon in these rocks, laid down in all geologic epochs. The surprising result is that the rocks contain about 2000 times more than the present surface $CO_2$ could provide. Did the Earth have an enormous carbon dioxide atmosphere, which has been diminishing and has almost been depleted now? Certainly not, since the geologic record would clearly show that. There have been fluctuations in atmospheric $CO_2$ levels, but only by factors of a few; certainly not by anything approaching thousands. So it can be concluded that carbon must have been deliverered into the atmosphere in a form that would end up as $CO_2$, on a continuous basis throughout geologic time (and carbon isotope studies strongly confirm this and argue against any major carbon supply being due to recycled sediments rather than a fresh supply). Infall of external material, long after the main accretion of the Earth, cannot be invoked on this large scale, as it too would have left its mark in the record. The answer must lie in fluids coming up from below.

Like the water of the oceans and the nitrogen of the atmosphere, the surface carbon must be largely a product of outgassing. The small amount of scientific literature devoted to this problem was still very much influenced by the concept of an initially hot Earth, and it therefore discusses mainly carbon fluids in chemical equilibrium with the crustal rock, and those would be $CO_2$ almost exclusively. The source material in the Earth for this $CO_2$ was not known, but hydrocarbons could not come into consideration, since they would not survive in molten rock. Within this outlook one was forced to ascribe any hydrocarbons we now find on the Earth to processes that reduced $CO_2$ and made hydrocarbons out of it. Since a large amount of energy is needed for this process, only photosynthesis in plants, using sunlight, could be held responsible. There seemed to be no alternative to the view that hydrocarbons on the Earth were all a product of surface life.

Several other factors helped to strengthen this viewpoint. Hydrocarbons were considered to be biological products, unlikely to be generated without biology; the prevalence of them in the solar system was not known. Molecules of quite clearly biological origin ('molecular fossils') were found in petroleum and in coal, and seemed to provide further confirmation. But many recent observations argue strongly against such a viewpoint.

Hydrocarbons have been found in many locations where no sediments can be suspected. The great cracks in the ocean floor are known to give out hydrocarbon fluids. Deep boreholes in the Soviet Union show hydrocarbons at depths of more than seven kilometers in metamorphic rock. Similarly the pilot borehole of the German deep drilling program found methane at depth and in metamorphic rock.

The deep borehole drilled in a meteorite crater in Sweden, the 'Siljan Ring', was drilled entirely in granitic and igneous rock and yet saw hydrocarbon gases clearly increasing with depth, in step with the inert gas helium, and a pumping operation delivered 12 tons of crude oil from a depth range between 5.5 and 6.5 km. In addition a sludge was pumped up that was identified as similar to a known bacterial product often seen in petroleum-bearing areas. Very clear identifications of surface outgassing of hydrocarbons have been obtained both in the Swedish meteorite crater, and also in a similar one in Eastern Canada, and these effects are attributed to the destruction of the rock by the meteorite impact, opening up pathways from deeper levels.

Now the outlook should be quite different. The original carbon-bearing material accreted by the Earth would be judged to be solid hydrocarbons, very likely the material represented by the carbon-bearing meteorites, the carbonaceous chondites. Embedded in solids and heated under great pressure, these materials would recombine in a great variety of ways, and generate a great range of hydrocarbon molecules, according to the calculations of thermodynamics. (The widely held view that the high temperatures would destroy these molecules, and turn this material into graphite and hydrogen, did not take account of the corresponding pressures, at which the denser compounds are favoured.) The lightest and least viscous components will be subject to the forces of buoyancy in the denser rock. If a sufficient quantity of these fluids accumulates in any one location, they will fracture the rock and force their way upwards. As lower temperatures and pressures are reached, various compounds become permanently stable, and it is the mix of these that we finally find as natural petroleum. Methane, the molecule that is the most stable at low pressures, will be a major partner in such a stream, and carry the heavier molecules with it in solution. At very low pressures the heavier molecules will be shed from the stream, leading to the concentrated deposits of oil that we find. The modern observation that almost invariably these deposits have substantial seepage of methane on the surface above them tends to confirm this picture.

What is the stability of these molecules against oxidation in the deep hot ground? Many rocks are very highly oxidized, and in chemical equilibrium the hydrocarbons would be burnt. However, where the pathways are cracks in solid rock, the available oxygen is limited to that available on the surfaces of these cracks, and this is quickly exhausted. Where, however, the fluids make their way up through liquid rock, fresh oxygen becomes available in every bubble, and oxidation can proceed. The carbon gases coming out of volcanoes are consequently largely $CO_2$ (although a significant proportion of methane is often present also). Close to volcanic regions, where the temperature of the ground is high at shallow depths, methane tends to predominate, because of the lesser temperature stability of the other hydrocarbons at low pressure.

What, then, is the origin of the biological molecules that are generally to be found in petroleum, if they are not derived from biological debris? The answer to that has become available through the discoveries of the profuse life on the ocean floor, life that is fed by the supply of chemical energy from streams of upwelling fluids. This shows that biology will penetrate into any region where there is a supply of chemical energy, and where the temperature is not too high (120°C may be the upper limit). In these ocean vents it is the bacteria that first utilize the available chemical energy, and they in turn provide the food for the larger animals and plants that live there. But why should such bacterial life be limited to cracks in the ocean floor? Where hydrocarbons stream up in cracks in the rocks on land, bacteria can thrive just as well, only there is no space for the larger life forms by which such life was first recognized. No place on Earth has been kept sterile from bacteria for geologic times, if it is capable of providing living conditions for them. It is to be expected that every petroleum-bearing area will have profuse bacterial colonies, whose 'molecular fossils' can later be found in the petroleum.

This picture of the origin of hydrocarbons brings the discussion of the Earth into line with that of the other planets. No longer do we have to think that the Earth has its hydrocarbons for a unique and different reason from the other planets, namely the existence of biology here. The origin of the great 'surface carbon excess' can be traced back to a common planetary material. Many other features of the occurrence of hydrocarbons have an explanation, such as the general association of petroleum with the noble gas helium. We can now see this as the consequence of the petroleum having travelled up from great depth, and having swept up, therefore, the helium on these pathways, either from primordial or from radioactive production. If biological debris in the sediments were the origin of the petroleum, no reason for the observed close association with helium could be seen.

In the search for petroleum and especially for natural gas, this picture has important implications. Many areas that were ignored because of the absence of massive deposits can now come under consideration. Major fault-lines in the crust are favorable locations, as is the proximity to lines defined by volcanoes and earthquakes. Porosity at accessible depths will remain a major criterion. Surface prospecting for methane and other hydrocarbons, as well as for various trace elements that come up with these, now assumes a major importance, when one realizes that deposits are not permanently bottled up, but represent merely concentrations in an ongoing flow. Vertical stacking of hydrocarbon deposits has long been identified: hydrocarbon-rich areas tend to have some hydrocarbons at every level; therefore it is worth looking underneath known deposits. The deposition of coal seams appears to have been favored by petroleum outgassing, either by the direct deposition of carbon, or by the effects on the local biological environment. At any rate, coal often overlies, in totally different strata, deposits of oil and gas. Also, the emission of gas from coalbeds often exceeds by large factors the amount of gas that the coal itself could have generated. Coal-bearing regions are therefore candidates for investigation at deeper levels.

It is time now for a new wave of exploration, using the modern knowledge, rather than a continuation of the dependence on the oil and gas fields discovered before the new information became available.

## Addendum (November 1995)

Since this Milne Lecture was delivered, several new items of information have become known that clearly relate to the subject. I will mention here two of them that seem particularly significant.

The biological molecules that are known to exist in all petroleum have been regarded as the strongest evidence, and indeed the last remaining evidence, for a biological origin of the substance itself. My view had been that they represented a biological contamination, rather than a leftover component from a source material. An ongoing massive biological activity would be required for this explanation to hold.

I assembled all the evidence for such biological activity at depth, and published this in the *Proceedings of the National Academy of Sciences* (in the June 1992 issue), under the title: 'The deep, hot biosphere'. I gave there the reasons why this activity would account for all biological molecules in petroleum that we knew, and that therefore the argument for a biological origin of petroleum was no longer secure.

In our deep boreholes in Sweden some bacteria had already been identified in a totally non-biological setting deep in granitic rock. Since then, many further items have been published demonstrating widespread microbial life at depth, and in particular in petroleum-bearing regions.

Recently there have been several reports that oil and gas fields appear to be recharging themselves when production from them would have exhausted their original content. Evidence from the Mid-Eastern oil fields, from deep Oklahoma gas fields, and from fields off the Gulf of Mexico coast have been named. Also several investigators have calculated the rate of natural seepage that would allow fields to leak out their content, and they have concluded that no gas field could have been maintained over the long periods of geologic time that had initially been assumed. Times of only a few tens of thousands of years seemed possible now, instead of the 50 million years or so that had frequently been estimated for the ages. These observations imply that enormously larger quantities of hydrocarbons than had previously been considered were the sources of supply of hydrocarbon fields, and if previous evidence had pointed to an apparent shortage of biological 'source material', then this new evidence would aggravate this by factors of at least 100.

# Cosmology and particle physics: a new synthesis

Dennis Sciama

## 1 Introduction

I feel very honoured to have been invited to give the Milne Lecture. After a brilliant career in astrophysics Milne turned his attention to cosmology. I particularly admire his bold attempts to create a synthesis between fundamental physics and the structure of the Universe. Today scientists are using the latest results and speculations from elementary particle physics to try and create a new synthesis with cosmology. These attempts are mainly concentrated either on the early Universe ('our best high energy physics laboratory') or on the role of neutrinos in cosmology and in astrophysics. In this lecture I will concentrate on the neutrinos. These particles were pair-created in the hot big bang, had a major influence on the synthesis of the light elements in 'the first three minutes' and, if they have a non-zero rest-mass, could have survived to make the major contribution to the dark matter in galaxies and in the intergalactic medium. In addition neutrinos have been detected in the laboratory coming from the Sun and from supernova 1987A in the Large Magellanic Cloud. Perhaps one day soon they will be detected coming from quasars.

One of the triumphs of the standard model of the hot big bang is that one can limit the number of different types of neutrinos $N_\nu$ to between 2 and 3, since otherwise one would obtain the wrong primordial abundance of $He^4$. This result was originally obtained by cosmologists when the laboratory limit on $N_\nu$ ran into the thousands. Now very precise experiments at SLAC and at CERN have shown that indeed $N_\nu = 3$. Three neutrino types were already known individually: the electron type $\nu_e$, the muon type $\nu_\mu$ and the $\tau$ type $\nu_\tau$. Therefore no more types remain to be discovered. This clear vindication of the cosmological prediction has led to physicists generally taking big bang cosmology very seriously.

This story of the three neutrino types has often been told, and I do not want to repeat it here in detail. I prefer to discuss the consequences of assuming that neutrinos have a non-zero rest-mass and that they make the main contribution to the dark matter in the Universe.

## 2 Do neutrinos have a non-zero rest-mass?

This is one of the key questions in both elementary particle physics and in cosmology. We do not yet know the answer. All we have at the moment are the following upper limits (Holzschuch 1992):

$$m_{\nu_e} < 9\ \text{eV},$$
$$m_{\nu_\mu} < 250\ \text{keV},$$
$$m_{\nu_\tau} < 31\ \text{MeV},$$

in units where the velocity of light $c = 1$.

There are three arguments which favour $m_{\nu_i} \neq 0$, where $\nu_i$ means any of the three.

(i) There is no known symmetry principle in elementary particle physics which would enforce $m_{\nu_i} = 0$. By contrast, gauge invariance would require that the photon mass $m_\gamma$ is exactly zero. There now exist many (speculative) models which naturally lead to $m_{\nu_i} \neq 0$.

(ii) The most attractive explanation of the solar neutrino anomaly (the deficit of the observed solar neutrino flux compared to the predictions of the standard solar model) involves $\nu_e$ changing its type as it propagates from the centre of the Sun, either via matter-enhanced oscillations (Bahcall 1989) or from spin precession in a magnetic field into right-handed neutrinos (or into neutrinos of a different type (Akhmedov, Lanza and Petcov 1993)). These mechanisms would work only if $m_{\nu_i} \neq 0$.

(iii) Neutrinos would be a very attractive candidate for most of the dark matter, but again only if $m_{\nu_i} \neq 0$.

Let us pursue this last point further. What would $m_{\nu_i}$ need to be if $\nu_i$ were a significant component of the dark matter? Consider first the dark matter in galaxies like the Milky Way, which possess flat rotation curves extending far beyond their visible matter. There is a nice argument due to Tremaine and Gunn (1979) which gives a lower limit on $m_{\nu_i}$ if $\nu_i$ is assumed to dominate the mass of the Galaxy. This argument involves an application of Liouville's theorem, and is based on comparing the phase space density of the neutrinos in the Galaxy and in the early universe. For a simple model of the neutrino distribution in the Galaxy one finds that the most massive of the neutrino types has

$$m_{\nu_i}^4 \geqslant \frac{1}{6\pi}\left(\frac{3}{2\pi}\right)^{3/2}\frac{h^3}{Gv_0 a^2},$$

where $h$ is Planck's constant, $G$ is Newton's constant, $v_0$ is the velocity dispersion of the neutrinos and $a$ is the core radius of their distribution. This relation would have fascinated Milne. It involves a strange mixture of microscopic constants of nature ($h$, $m_{\nu_i}$), a macroscopic constant ($G$) and astronomical properties of the galaxy ($v_0$, $a$). Models of the Milky Way, for example, suggest that we should take $v_0 \sim 220\ \mathrm{km\,s^{-1}}$ and $a \sim 8$ kpc. One then obtains

$$m_{\nu_i} \geqslant 27\ \mathrm{eV}.$$

Since $m_{\nu_i} \propto v_0^{-1/4} a^{-1/2}$, this constraint is fairly insensitive to the choice of the parameters $v_0$ and $a$.

Let us now consider what $m_{\nu_i}$ would need to be if $\nu_i$ is the main contributor to the total density $\rho$ of the Universe. We do not yet know the actual value of $\rho$, but it is likely to exceed the value of the baryon density $\rho_b$ which can be derived from considerations of the primordial synthesis of the light elements. A particularly attractive possibility is that $\rho$ has the critical value $\rho_{crit}$ for which the Universe would just expand forever with no excess kinetic energy. This possibility is supported by a recent analysis of the observed angular diameter-redshift relation for compact radio sources (Kellermann 1993). General relativity tells us that

$$\frac{8\pi}{3} G\rho_{crit} = H^2,$$

where $H$ is the Hubble constant, which determines the expansion rate of the Universe. $H$ is usually parametrized by

$$H = 100\,h\ \mathrm{km\,s^{-1}\,Mpc{-}1},$$

where the uncertainty in the observed value of $H$ is usually expressed as $\tfrac{1}{2} \leqslant h \leqslant 1$.

We now ask the question, what would $m_{\nu_i}$ need to be if

$$\rho_{\nu_i} \sim \rho_{crit}?$$

To answer this question we need to know the present cosmological number density $n_{\nu_i}$ of neutrinos which have survived from their pair-creation in the hot big bang. In the early universe the neutrinos were in thermal equilibrium with the hot radiation which has now cooled down to form the famous 3K cosmical microwave background. On general grounds we would then expect a rough equality of particle and photon densities $n_{\nu_i} \sim n_\gamma$ at all cosmic epochs, since each of these quantities decreases more or less at the same rate in the expanding Universe. There is an exception to this relation commencing at the epoch when electron pairs, which were also originally in thermal equilibrium with the heat bath, annihilated permanently (when the temperature $T$ dropped to below $\tfrac{1}{2}m_e$). Allowing for this effect one finds that today

$$n_{\nu_i} = \tfrac{3}{11} n_\gamma.$$

The quantity $n_\gamma \sim T_0^3$, where $T_0$ is the present temperature of the microwave background. A recent *COBE* measurement (quoted in Silk 1993) is

$$T_0 = 2.726 \pm 0.01 \text{ K},$$

and so $$n_\gamma = 414 \pm 4 \text{ cm}^{-3}.$$

We thus obtain $$n_{\nu_i} = 113 \pm 1 \text{ cm}^{-3}.$$

Notice that this value of the present neutrino density is independent of both $m_{\nu_i}$ and the Hubble constant.

We can now derive $\rho_{\nu_i}$, since this quantity is just $m_{\nu_i} n_{\nu_i}$. The result is usually expressed in the form

$$m_{\nu_i} = 93\Omega_{\nu_i} h^2 \text{ eV},$$

where $$\Omega_{\nu_i} = \rho_{\nu_i}/\rho_{\text{crit}}.$$

Hence $\Omega_{\nu_i} \sim 1$ if $$m_{\nu_i} \sim 93\, h^2 \text{ eV}.$$

We have seen that $h$ is uncertain to a factor 2, and so the derived value of $m_{\nu_i}$ is uncertain to a factor 4. However, if we demand on observational grounds that the age of the Universe $t_u$ should exceed 12 billion years (this question of the age of the Universe was discussed in Fowler's Milne Lecture (Fowler 1987)) then

$$h < \sim 0.56$$

if $\Omega_{\nu_i} \sim 1$, since in that case $t_u = \tfrac{2}{3}H^{-1}$. Since $h \geqslant \tfrac{1}{2}$ one has, for $\Omega_{\nu_i} \sim 1$,

$$23 \lesssim m_{\nu_i} \lesssim 28 \text{ eV}.$$

This result leads to the first of several numerical coincidences which we will meet. If we compare our upper limit on $m_{\nu_i}$ with the lower limit following from the phase space constraint for our Galaxy, we find that the two constraints are only just compatible and would lead to

$$m_{\nu_i} \sim 28 \text{ eV}.$$

This then is the condition that:

(i) the dark matter in the Galaxy consists mainly of neutrinos;

(ii) $\Omega_{\nu_i} \sim 1$.

## 3 The radiative decay of massive neutrinos

Particle physicists tell us that if $m_{\nu_i} \neq 0$, one has the decay

$$\nu_i \to \gamma + \nu_2.$$

Conservation of energy and momentum in the decay then determines the energy $E_\gamma$ of the photon in the rest frame of $\nu_1$:

$$E_\gamma = \tfrac{1}{2}m_1\left(1 - \frac{m_2^2}{m_1^2}\right).$$

The decay photon is thus monochromatic.

If $m_1 \sim 30$ eV and $m_2 \sim 10^{-3}$–$10^{-2}$ eV (as is suggested by analysis of the solar neutrino problem) one would have $m_2^2/m_1^2 \ll 1$ and so

$$E_\gamma = \tfrac{1}{2}m_1.$$

Thus if $\Omega_{\nu_i} \sim 1$ we would have

$$E_\gamma \sim 14 \text{ eV}.$$

Here we meet our second numerical coincidence. The decay photons are just able to ionize hydrogen (whose ionization potential is 13.6 eV). Whether this coincidence is of interest in astronomy or cosmology depends of course on the decay lifetime $\tau$. Cowsik (1977) pointed out that cosmological lines of sight are so long that even a lifetime much greater than the age of the Universe could lead to an appreciable flux of decay photons. In fact a variety of arguments, from both observed ultraviolet backgrounds and from ionization considerations (de Rujula and Glashow 1980; Stecker 1980; Kimble, Bowyer and Jakobsen 1981; Melott and Sciama 1981) impose the constraint

$$\tau > 10^{23} \text{ s}.$$

This lower limit is about a million times longer than the age of the Universe. A reliable theoretical value of $\tau$ is not yet available, since there does not exist a definitive particle theory of massive neutrinos.

We now recall that there exist a number of ionization problems in both astronomy and cosmology. In all these problems more ionization of hydrogen is observed than would be expected—in our Galaxy, in other spiral galaxies such as NGC891, in Lyman $\alpha$ clouds and in the intergalactic medium.

I have suggested in a number of papers, beginning with Sciama (1990*a*), that all these problems might be solved at one stroke by using decay photons. A comprehensive discussion of this theory is given in my book *Modern cosmology and the dark matter problem* (Sciama 1993*c*).

What is the ionization problem in our Galaxy? One knows from H$\alpha$ and pulsar dispersion data that there is a widespread distribution of free electrons in the Galaxy, with $n_e \sim 0.03$ cm$^{-3}$. One usually supposes that this ionization is produced by ultraviolet radiation emitted by O stars and supernovae. This conventional explanation faces four serious problems:

(i) The atomic hydrogen in the interstellar medium leads to a high opacity for ionizing photons (mean free path ~ 1–10 pc). How do these photons travel hundreds of parsecs from their sources to a typical point in the interstellar medium?

(ii) The power required to maintain the ionization probably rules out supernovae as the source, leaving only O stars (Reynolds 1990*a*).

(iii) The scale height of the free electrons ~ 670 pc (Nordgren, Cordes and Terzian 1992) whereas the scale height of the O stars ~ 100 pc.

(iv) The electron density $n_e$ is remarkably constant in warm regions within 1–2 kpc of the Sun (Reynolds 1990*b*; Sciama 1990*c*; Spitzer and Fitzpatrick 1993; Sciama 1993*b*). If a tortuous geometry of the HI solves the opacity problem, it would be unlikely to lead to a constant $n_e$.

These problems have been compounded further by the recent discovery by Spitzer and Fitzpatrick (1993) that the free electrons and photons are probably mixed up with the HI rather than being in separate regions (which are known as the warm ionized medium or WIM and the warm neutral medium or WNM). Ionizing photons from distant O stars would not penetrate far into these regions even if they were able to reach them.

All these problems would be solved immediately if the ionization were due to decay photons.

(i) There are neutrinos everywhere in the galaxy, and so in particular there will be neutrinos within one mean free path of every point. One can thus account for $n_e \sim 0.03\ \text{cm}^{-3}$ with a reasonable choice for the decay lifetime $\tau$ (see below).

(ii) There is no power problem if this value of $\tau$ is assumed.

(iii) The scale height of the neutrinos would be expected to exceed 670 pc.

(iv) The constancy of $n_e$ in warm opaque regions near the Sun follows from the equation for ionization equilibrium

$$\frac{n_\nu}{\tau} = \alpha n_e^2,$$

where $\alpha$ is the appropriate recombination coefficient. With $n_\nu \sim$ constant in limited regions and $\alpha(T) \sim T^{-0.75}$, one sees that $n_e$ is nearly constant if $T$ remains within a factor 2 of $10^4$ K in warm regions of the interstellar medium, as is observed to be the case.

We now derive the required value of $\tau$ for this theory. Observations of the flat rotation curve of the Galaxy determine the column mass density of neutrinos at right angles to the galactic plane at the Sun's position. Dividing this by a neutrino mass ~ 30 eV we obtain the column particle density of neutrinos, which is $\sim 10^{30}\ \text{cm}^{-2}$. The H$\alpha$ observations tell us that there are $4 \times 10^6$ ionizations $\text{cm}^{-2}\,\text{s}^{-1}$ at right angles to the galactic plane at the Sun's position (Reynolds 1984). Hence

$$\tau \sim 2.5 \times 10^{23}\ \text{s},$$

which is compatible with the lower limit of $10^{23}$s. Finally, to account for $n_e \sim 0.03\ \text{cm}^{-3}$ locally we need to know $n_\nu$ near the Sun. If $T \sim 10^4$ K we have $\alpha \sim 2 \times 10^{-13}\ \text{cm}^3\,\text{s}^{-1}$ and so $n_\nu \sim 5 \times 10^7\ \text{cm}^{-3}$. Hence the scale height of the neutrinos at the Sun's position $\sim$ 3 kpc. Since the core radius of the neutrinos in the galactic plane $\sim$ 8 kpc, this indicates that the dark matter halo of the Galaxy is rather flattened. Such flattening has already been considered for other reasons and is an interesting prediction of the neutrino decay theory, which we will not go into here.

## 4 The intergalactic flux of ionizing particles

We now consider the implications of this theory for the intergalactic ionizing flux $F$ due to the cosmological distribution of neutrinos (Sciama 1990*b*). A naive formula would be

$$F \sim \frac{n_\nu(z=0)}{\tau}\frac{c}{H},$$

where $c/H$ represents the radius of the Universe. With $n_\nu \sim 100\ \text{cm}^{-3}$ and $h \sim 0.5$ we would obtain

$$F \sim 10^7\ \text{cm}^{-2}\,\text{s}^{-1}.$$

This result is apparently a disaster for the theory because it would violate an observational upper limit on $F$ based on H$\alpha$ observations of an intergalactic H1 cloud in Leo (Reynolds *et al.* 1986). This upper limit gives

$$F \lesssim 6 \times 10^5\ \text{cm}^{-2}\,\text{s}^{-1}.$$

Fortunately we can save the theory if it happens that $E_\gamma$ is close to 13.6 eV, because the red shift in the expanding Universe will reduce the photon energy to below the ionization potential of hydrogen in a distance much less than $c/H$. Writing

$$E_\gamma = (13.6 + \varepsilon)\ \text{eV},$$

we would have

$$F = \frac{n_\nu}{\tau}\frac{c}{H}\frac{\varepsilon}{13.6}.$$

Hence we require that

$$\varepsilon \lesssim 1$$

$$E_\gamma \lesssim 14.6\ \text{eV}.$$

This represents another numerical coincidence because then

$$m_{\nu_i} \lesssim 29.2\ \text{eV},$$

which is almost identical to our previous constraint $m_{\nu_i} < 28$ eV from the assumptions that $\Omega_{\nu_i} \sim 1$ and $h < \sim 0.56$.

After these ideas were developed observational evidence from *COBE* was published (Wright *et al.* 1991) indicating that nitrogen is also ionized throughout the interstellar medium of our Galaxy. Since the ionization potential of nitrogen exceeds that of hydrogen the same opacity and scale height problems arise. We conclude that the decay photons must also be able to ionize nitrogen (Sciama 1992). Hence

$$E_\gamma > 14.53 \text{ eV}.$$

This leads to another numerical coincidence, since this constraint is only just compatible with our previous constraint $E_\gamma < \sim 14.6$ eV. Of course if the theory is correct no coincidence is involved, but at this stage of our discussion we are still testing its validity. We see that up to this point the theory has a small but definitely non-zero domain of validity.

We would like to mention here one final numerical coincidence (Sciama 1993*a*). This would arise from Tarafdar's (1991) suggestion that decay photons could solve the so-called $C^0$/CO ratio problem. This problem arises because the abundance ratio of atomic carbon $C^0$ to carbon monoxide CO in the deep interiors of dense molecular clouds in the galaxy is observed to be $10^5$ times greater than expected on the basis of simple equilibrium models. Tarafdar calculated that the decay photons from neutrinos in the clouds would suffice to dissociate the CO to the required extent. However, this could be true only if these photons are unable to dissociate $H_2$, which is by far the dominant constituent of the dense molecular clouds. Otherwise the opacity of the clouds would be far too great. For this mechanism to work we therefore require that $E_\gamma$ be less than the threshold for the continuous photodissociation of $H_2$. This means that

$$E_\gamma < 14.68 \text{ eV}.$$

So for the third time we obtain an upper limit on $E_\gamma$ of the same order. Since the $H_2$ and N thresholds are accurately measured, the limits on $E_\gamma$ coming from them are numerically firm (of course only if the underlying assumptions are correct). With this supposition we can then write

$$14.53 < E_\gamma < 14.68 \text{ eV}$$

or

$$E_\gamma = (14.605 \pm 0.075) \text{ eV},$$

with an essentially exact range of uncertainty. If we continue to assume that $m_2/m_1 \ll 1$ we would also have

$$m_{\nu_i} = 29.21 \pm 0.15 \text{ eV}.$$

Thus the heaviest neutrino type (probably but not certainly the $\tau$ neutrino) would have its mass determined astronomically with a precision of ½ per cent.

We can also use this result to obtain an accurate value for Hubble's constant $H$. The main uncertainty in the total density $\rho$ of the Universe would now come from the uncertainty in the baryon density $\rho_b$ even though $\rho_b$ is only a few per cent of $\rho$, according to the standard argument from primordial nucleosynthesis (Walker *et al.* 1991). We find that $\Omega h^2 = 0.3$ with a precision of $\sim 2$ per cent. We would now like to determine $\Omega$ and $h$ separately. Our original argument (Sciama 1990*b*) used information concerning the age of the Universe. If, for example, we assume that $\Omega \sim 0.2$, the resulting value of $h$ leads to an age for the Universe of only $8 \times 10^9$ yr. This is far too low. As $\Omega$ increases and so $h$ decreases we find that the derived age increases. The lowest permitted age, from arguments concerning globular cluster stars, is $12 \times 10^9$ yr. This would require that $\Omega \sim 1$, in agreement with Kellermann's (1993) recent observational result that indeed $\Omega \sim 1$.

It is also tempting to assume at this point that $\Omega \sim 1$ exactly, since this is the only value of $\Omega$ that would not vary with the cosmic epoch.

If $\Omega = 1$ exactly we would then have

$$H = 56.3 \pm 0.5 \text{ km s}^{-1}\text{ Mpc}^{-1},$$

so that $H$ would be constrained with a precision of 1 per cent. It will be interesting to see whether this prediction agrees with future reliable observational determinations of $H$.

A further consequence of our analysis is that the intergalactic flux $F$ of ionizing photons must be close to the observational upper limit, so that

$$F \sim 6 \times 10^5 \text{ cm}^{-2}\text{ s}^{-1}.$$

This would be the value of the ionizing intergalactic flux at the present epoch. One also has observational information about $F$ at $z \sim 2$–5 from both Lyman $\alpha$ clouds and the intergalactic medium. In the neutrino decay theory one has $n_\nu(z) = (1 + z)^3\, n_\nu\,(0)$ and, if $\Omega = 1$, $H(z) \sim (1 + z)^{3/2}$. Hence

$$F(z) = (1 + z)^{3/2} F(0).$$

This simple relation must be modified at the highest values of $z$ to allow for absorption of the ionizing photons by intervening Lyman limit systems (Madau 1992).

In this way one arrives at a theoretical value for $F(z)$ without adjustable parameters which turns out to be in agreement with observational requirements. By contrast, the integrated ionizing flux from quasars falls short by a factor of $\sim 10$. I therefore conclude that the large-scale high degree of ionization of the Universe could be mainly due to decay photons.

This conclusion invites further questions, having to do with the ionization state of the early Universe. At small cosmic epochs the temperature was so high that the pregalactic gas would have been almost completely ionized. At a redshift of 1000 when the temperature $\sim$ 3000 K the gas would have become mainly neutral—one speaks of the recombination era, although strictly speaking the gas had never been neutral at earlier times. At that stage decay photons would not have been important. What happens later has been discussed by Scott, Rees and Sciama (1991). The Universe starts to become partially reionized, and switches to essentially complete ionization at a redshift $\sim 25$.

The partially ionized medium at redshifts exceeding 25 would be of great interest, because it would act as a thick scattering screen tending to smooth out angular fluctuations in the temperature of the microwave background arising from the primordial irregularities in the pregalactic gas. The resulting smoothing would be substantial on small angular scales but would be cut off, for causal reasons, at the angle subtended by the horizon at the last scattering surface, where the optical depth for Thomson scattering becomes unity. For a decay lifetime $\sim 2 \times 10^{23}$s the redshift of the last scattering surface is 450, and the critical angle is 3°. This is less than the angle ($\sim 10°$) at which *COBE* has observed temperature fluctuations, at a level $\sim 10^{-5}$. We therefore predict that these fluctations should decrease appreciably as the angular separation is reduced below 3°. This prediction will probably be tested in the near future.

## 5 Observational searches for the predicted decay line

Various attempts have been made by space experiments to detect a decay line from dark-matter neutrinos in clusters of galaxies. All these attempts have been unsuccessful. The most sensitive test was made by Davidsen *et al.* (1991) using the Hopkins Ultraviolet Telescope on the *Astro I* mission. They observed the rich cluster of galaxies A665 and placed a lower limit on the decay lifetime $\tau \sim 3 \times 10^{24}$s on the assumption that the dark matter in this cluster consists mainly of decaying neutrinos.

This assumption has recently been questioned (Sciama, Persic and Salucci 1992, 1993) as a result of the publication of new X-ray data for this cluster (Hughes and Tanaka 1992). These data indicate that the dark matter in A665 is more centrally concentrated than the visible matter in galaxies and hot X-ray-emitting gas. A similar conclusion follows for other rich clusters from both X-ray and gravitational lens data. This central concentration of the dark matter would not be expected for non-dissipative neutrinos, and it suggests that the dark matter in these clusters is mainly baryonic. By contrast, the dark matter in galaxies like the Milky Way is more extended than the visible matter, and so could still be mainly neutrinos whose decay photons could be ionizing the interstellar medium.

A decisive space experiment to test this theory would consist of searching for a decay line from neutrinos within one mean free path of the Sun. The Sun is believed to be immersed in a partially ionized cloud whose HI density $\sim 0.1\ \mathrm{cm}^{-3}$. Unit mean free path would then correspond to $\sim 1$ pc. The predicted flux in the decay line would then be $\sim 600\ \mathrm{photons\,cm}^{-2}\,\mathrm{s}^{-1}$.

A proposal to search for this line has been made to ESA by Stalio *et al.* (1992). We hope to have space on board *EURECA 2* to enable us to perform this experiment. Our detectors have been designed to have sufficient sensitivity, and the experiment should be decisive since absorption effects have been fully allowed for. A successful outcome would validate the new synthesis between cosmology and particle physics which I have outlined in this lecture.

## Acknowledgements

I am grateful to Miranda Weston-Smith, the Milne Society and Wadham College for their hospitality and for making the necessary arrangements for my lecture. My work is supported by the Italian Ministry for Universities and Scientific and Technological Research. The Milne Lectures are sponsored by IBM UK Ltd.

## References

Akhmedov, E. K., Lanza, A. and Petcov, S., 1993. *Phys. Lett.*, **B303**, 85.
Bahcall, J. N., 1989. *Neutrino astrophysics*, Cambridge University Press.
Cowsik, R., 1977. *Phys. Rev. Lett.*, **39**, 784.
Davidsen, A. F. *et al.*, 1991. *Nature*, **351**, 128.
Fowler, W. A., 1987. *Q. J. R. astr. Soc.*, **28**, 87.
Holzschuch, E., 1992. *Rep. Prog. Phys.*, **55**, 1035.
Hughes, J. P. and Tanaka, K., 1992. *Astrophys. J.*, **398**, 62.
Kellermann, K. I., 1993. *Nature*, **361**, 134.
Kimble, R., Bowyer, S. and Jakobsen, P., 1981. *Phys. Rev. Lett.*, **46**, 80.
Madau, P., 1992. *Astrophys. J.*, **389**, L1.
Melott, A. L. and Sciama, D. W., 1981. *Phys. Rev. Lett.*, **46**, 1369.
Nordgren, T., Cordes, J. and Terzian, Y., 1992. *Astron. J.*, **104**, 1465.
Reynolds, R. J., 1984. *Astrophys. J.*, **282**, 191.
Reynolds, R. J., 1990*a*. *Astrophys. J.*, **349**, L17.
Reynolds, R. J., 1990*b*. *Astrophys. J.*, **348**, 153.
Reynolds, R. J., Magee, K., Roesler, F. L., Scherb, F. and Harlander, J., 1986. *Astrophys. J.*, **309**, L9.
Rujula, A. de and Glashow, S. L., 1980. *Phys. Rev. Lett.*, **45**, 942.

Sciama, D. W., 1990*a*. *Astrophys. J.*, **364**, 549.
Sciama, D. W., 1990*b*. *Phys. Rev. Lett.*, **65**, 2839.
Sciama, D. W., 1990*c*. *Nature*, **346**, 40.
Sciama, D. W., 1992. *Int. J. Mod. Phys.*, D., **1**, 161.
Sciama, D. W., 1994*a*. *Astrophys. J. Lett.*, **415**, L 31.
Sciama, D. W., 1994*b*. *Astrophys. J. Lett.*, **409**, L 25.
Sciama, D. W., 1993*c*. *Modern cosmology and the dark matter problem*, Cambridge University Press.
Sciama, D. W., Persic, M. and Salucci, P., 1992. *Nature*, **358**, 718.
Sciama, D. W., Persic, M. and Salucci, P., 1993. *Publs Astr. Soc. Pacific*, **105**, 102.
Scott, D., Rees, M. J. and Sciama, D. W., 1991. *Astr. Astrophys.*, **250**, 295.
Silk, J., 1993. *Nature*, **361**, 111.
Spitzer, L. and Fitzpatrick, E. L., 1993. *Astrophys. J.*, **409**, 299.
Stalio, R., Bowyer, S., Sciama, D. W. and Gimenez, A., 1992. *Adv. Space Research*, **13**, No. 12.
Stecker, F. W., 1980. *Phys. Rev. Lett.*, **45**, 1460.
Tarafdar, S. P., 1991. *Mon. Not. R. astr. Soc.*, **252**, 55*P*.
Tremaine, S. and Gunn, J. E., 1979. *Phys. Rev. Lett.*, **42**, 407.
Walker, T. P., Steigman, G., Schramm, D. N., Olive, K. A. and Kang, H., 1991. *Astrophys. J.*, **376**, 51.
Wright, E. L. *et al.*, 1991. *Astrophys. J.*, **381**, 200.

*Note added in proof:*
ESA subsequently cancelled the flight of EURECA 2, but a detector called EURD was successfully launched in the first Spanish minisatellite on 21 April 1997. At the time of writing (16 June 1997), EURD is just about to start obtaining data. If all goes well a clear result should be available in a few months.

# Hunting for comets and planets

Freeman J. Dyson

## 1 Two theories of comets

We are here tonight to honour the memory of Arthur Milne. I will not say much about Milne, because I did not know him personally. Almost everything I know about him comes from the Milne Lecture given here in Oxford in 1979 by Chandrasekhar (Chandrasekhar 1980). This lecture was published in the *Quarterly Journal of the Royal Astronomical Society* and those of you who did not hear it can read it. It is a sensitive account of Milne's life and work, told by a man who reveres Milne deeply as a friend and colleague.

I decided to talk about comets. I apologize to those of you who think that anything smaller than a galaxy is unworthy of our serious attention. I happen to believe that nature is equally interesting on all scales, that scientific importance and beauty are to be found in objects of all sizes. So I ask you now to shrink your imaginations down to the size of our solar system and to things we might do with small and cheap instruments.

I begin with some historical background about comets. Independently and almost simultaneously, 40 years ago, two great astronomers propounded two different theories about the place of comets in the solar system. The first was Jan Oort (Oort 1950) and the second was Gerard Kuiper (Kuiper 1951). They started from different points of view and arrived at different conclusions. Oort began from the observational evidence that long-period comets arrive at the inner solar system with orbits that are hardly distinguishable from parabolas and from directions that are distributed rather uniformly over the sky. They arrive at an average rate of several per year and appear to come from great distances. These are called long-period comets to distinguish them from the other kind which have short periods and are seen repeatedly as they return to the neighbourhood of the Sun. The name 'long-period' is misleading, since most of the so-called long-period comets are only seen once and may have no regular period at all. Oort remarked that a long-period comet passing through the inner solar system is likely to suffer major perturbations of its motion by the gravitational effects of Jupiter and Saturn. With approximately equal likelihood, it may lose a small amount of energy and become a new short-period comet trapped in an orbit within the planetary system, or it may gain a small amount of energy and escape from the Sun, moving out for ever along a hyperbolic orbit. In either case the comet is lost from the population of long-period comets. From these facts Oort deduced that there must exist a very large reservoir from which the observed long-period comets have been supplied at the rate of several per year during the $4 \times 10^9$ years of the life of the solar system. He described the reservoir that has become known to the astronomical community as the 'Oort Cloud', a very extended and roughly spherical swarm of small objects loosely attached to the Sun. He estimated the number of objects of cometary size to be at least $2 \times 10^{10}$, with distances extending out from $10^3$ to $4 \times 10^4$ astronomical units from the Sun.

Meanwhile, Gerard Kuiper was thinking about a different question. Why does the series of big planets, Jupiter, Saturn, Uranus and Neptune, stop so abruptly at Neptune? Kuiper assumed the generally accepted picture of the origin of the planetary system, beginning from a cool disk of gas and dust orbiting around the Sun. In this picture, the cool dust in the outer part of the disk, mostly consisting of ice, condenses first into a huge number of planetesimals, objects of cometary size orbiting within the disk. Then, over a longer time, the planetesimals collide with one another, stick together, and gradually accumulate into planets. According to this picture, no big planet formed in the disk beyond Neptune because the planetesimals in that region were too far apart and too slow to accumulate. But it is unlikely that the disk itself was cut off sharply beyond Neptune. Presumably the disk and the population of planetesimals continued smoothly outward. Therefore, Kuiper argued, the population of planetesimals beyond Neptune is probably still there. Kuiper thus invented a second reservoir of comets, much closer to us than the Oort Cloud and concentrated in the plane of the planets. The second reservoir is known as the 'Kuiper Belt'. He imagined it to lie in a wide ring outside Neptune's orbit, extending roughly from 40 to 100 astronomical units from the Sun.

The logic of Kuiper's argument also led him to an estimate of the population of his belt. We know from the observed densities of Uranus and Neptune that they contain massive cores composed of elements heavier than hydrogen and helium. The cores are presumably made of rock and ice, the same material that originally constituted the planetesimals in that region of the disk. Each core contains roughly ten Earth masses. Kuiper estimated that the planetesimals still remaining in the region outside Neptune should likewise contain about 10 Earth masses. This is a much larger total mass than Oort required for his cloud. If we divide 10 Earth masses into objects with a diameter of 1 km, the size of a typical comet, we find the number of objects to be about $10^{13}$, a thousand times the estimated population of the Oort Cloud.

The Kuiper Belt and the Oort Cloud have different scientific rationales. The Cloud is deduced directly from observations, while the Belt is deduced from a speculative theory. Because of this difference, the Cloud has generally received more attention than the Belt. Especially in the popular press and in science fiction stories, the Cloud plays a prominent role while the Belt is hardly mentioned. Even among professional astronomers the same bias is noticeable. Until a few years ago, the general belief among astronomers was that the Kuiper Belt, if it existed, was not directly observable as a source of visible comets. The prevailing view envisioned the history of a visible comet as proceeding through four successive stages as follows:

*Stage 1.* A comet originating in the Kuiper Belt gradually diffuses outward away from the Sun, mainly as a result of slow perturbations of its orbit by Uranus and Neptune, until it reaches the distance of the Oort Cloud.

*Stage 2.* Random perturbations by passing stars change the plane of the comet's orbit until it is no longer aligned with the plane of the planets. The comet has then become a full member of the Oort Cloud.

*Stage 3.* A random perturbation of the comet's orbit may occasionally, over eons of time, bring its angular momentum about the Sun close to zero, so that it falls into the inner part of the solar system and is seen as a typical long-period comet with an apparently parabolic orbit.

*Stage 4.* Rapid perturbation of the long-period comet by Jupiter and Saturn will either throw it out of the solar system or convert it into a short-period comet. As a short-period comet it will remain active and visible for only a few thousand years before its volatile constituents are boiled away by the heat of the Sun.

According to this picture, the evolution of all comets follows a single track, from the Kuiper Belt to the Oort Cloud to the inner solar system, and finally from long-period to short-period to final extinction. If this view were correct, we would have no direct evidence that the Kuiper Belt still exists. So far as the visible evidence is concerned, the Kuiper Belt might have existed long ago in the primeval solar system and gradually disappeared as the comets moved out into the Oort Cloud. The Kuiper Belt would be needed only as a hypothetical ancestor of the Cloud.

The single-track theory of comets was generally believed until a few years ago. The alternative double-track theory, believed by almost nobody, had the long-period and short-period comets originating independently from separate reservoirs. According to the double-track theory, the long-period comets come from the Oort Cloud but the short-period comets do not. The short-period comets have their own reservoir in the Kuiper Belt and diffuse inwardly from the Belt into the inner solar system without making the big detour through the Oort Cloud. If the double-track theory is right, the short-period comets give us direct evidence that the Kuiper Belt is still alive and well.

## 2 Scott Tremaine and the second reservoir

Three years ago, Scott Tremaine and his colleagues Thomas Quinn and Martin Duncan at the Canadian Institute for Theoretical Astrophysics in Toronto proved that the single-track picture of the evolution of comets is wrong and the double-track theory is right (Duncan, Quinn and Tremaine 1988). They proved this by an analysis of the orbits of short-period comets. They compared the observed statistics of short-period comets with the two theories of their origin. The theories were tested by computer simulations of the behaviour of hypothetical comets entering the inner solar system either from the Cloud or from the Belt. The results of the simulations decisively favour direct entry from the Belt. The most striking feature of the observed distribution of short-period comets is the strong concentration of their orbits toward the plane of the planets. Not only do they stay close to the ecliptic plane, but the majority of them move around in the same direction as the planets. Halley's comet belongs to the minority which go round the wrong way, but Halley's is in many ways not a typical member of the short-period club.

When Tremaine simulated the long-period comets arriving from random directions in the sky and encountering Jupiter or Saturn, he found that those captured into short-period orbits showed almost no tendency to concentrate toward the ecliptic plane or to move around the Sun in the same direction as Jupiter and Saturn. People who believed the one-track theory had supposed that Jupiter and Saturn would somehow drag the captured comets into orbits similar to their own. The computer simulations show that nothing of the sort happens. Comets arriving from random directions in the sky are captured into randomly oriented orbits. This result of the computer simulation ought not to have come as a surprise. There is an old theorem of Joseph Liouville which says that the density of an ensemble of objects in phase-space can never be changed by any non-dissipative force. Since the attraction of Jupiter is non-dissipative, Liouville's theorem would make it very difficult for Jupiter to squeeze a randomly oriented collection of initial long-period orbits into a collimated stream of short-period orbits. Tremaine's simulation of the single-track theory was in essence only verifying that Liouville was right.

The results of Tremaine's simulation of the double-track theory were equally convincing. He started with comets in the Kuiper Belt and integrated their motions over long periods of time until the perturbations produced by the major planets caused them to move inward into short-period orbits. After repeating this procedure many times with various initial conditions, he obtained a distribution of final orbits agreeing well with the observed short-period orbits. He found in particular that the final orbits are concentrated in the plane of the planets to exactly the same extent as the real short-period comets. This concentration of orbits is allowed by Liouville's theorem because the original distribution of orbits in the Kuiper Belt was governed by dissipative forces in a condensing disk of gas and dust. Once again, Tremaine confirmed by laborious calculation what Liouville had understood intuitively. Only one detail of Tremaine's calculation was unrealistic. Since he did not have time or resources sufficient to calculate cometary orbits for $10^9$ yrs, he arbitrarily multiplied the masses of Jupiter, Saturn, Uranus and Neptune by a factor of 40. These big masses speeded up the diffusion of cometary orbits from the Kuiper Belt inward, so that the calculations could be finished in a reasonable time. Tremaine freely admits that his calculation is a swindle. His simulated planets are not the planets that we see in our sky. However, it seems unlikely that the distribution of his cometary orbits would have been qualitatively different if he had been able to use the correct planetary masses and integrate the equations of motion a thousand times longer. Since the art of integrating orbits over long stretches of time is advancing rapidly, it should be possible within a few years to repeat Tremaine's simulations with the correct masses.

As a result of this work of Tremaine, the time has come to take a new look at an old question. Is it possible to see comets in their natural habitat, in the reservoir where they live when they are not making their final and suicidal dive toward the Sun? So long as we believed that the main reservoir was the Oort Cloud, the answer to this question was clearly negative. The Oort Cloud is too far away and the comets are too small. But if, as Tremaine's analysis indicates, there is a second large reservoir in the Kuiper Belt, the prospects for direct observation of comets are vastly improved. There are three reasons why the Kuiper Belt is easier to observe. First, it is a hundred times less distant. Second, it is concentrated in the ecliptic plane instead of being scattered all over the sky. Third, if Kuiper's estimate of the population is correct, the Belt contains a larger number of comets than the Cloud.

## 3 Occultation astronomy

Several astronomers have studied seriously the possibility of seeing comets in the Kuiper Belt by detecting their reflected sunlight with a large telescope. Unfortunately the brightness of reflected sunlight decreases with the inverse fourth power of the distance from the Sun. It is easy to estimate the brightness of a comet in the Belt by comparing it with Neptune. Neptune, with a diameter of $4 \times 10^4$ km and a high optical albedo, has visual magnitude 7.6. If we assume optimistically that a comet is situated at Neptune's distance at the inner edge of the Belt, with an albedo as high as Neptune and a diameter of 10 km, as large as Halley's comet and much larger than the majority of comets, then its visual magnitude turns out to be 26. A magnitude 26 object is barely detectable in a large telescope under conditions of excellent seeing. It would be difficult to justify the use of prime time on a large telescope for comet-hunting. The most we can hope for in this direction is that Kuiper Belt comets may occasionally be found on exceptionally deep images of patches of sky made for other purposes. It is possible that the Belt may contain a substantial population of larger objects with diameters greater than 100 km and visual magnitudes around 20. Such objects, if they exist, could be found and identified as a by-product of various sky-survey projects that are now going forward.

So far as the great majority of ordinary comets is concerned, even if they are as near to us as Neptune, optical detection using reflected sunlight is hopeless. This brings me to the main subject of my talk, occultation astronomy. Occultation astronomy means looking for dark objects in the sky by observing occultations of bright objects that are further away. So far as I know, the first suggestion that one could look for comets by observing occultations of stars was published in *Nature* 15 years ago by Mark Bailey, at that time a student in Edinburgh (Bailey 1976). Bailey stated clearly the essential facts, that occultations by comets in the Kuiper Belt could be observed but that occultations by comets in the Oort Cloud could not. Everything I shall say in this talk is based on Bailey's two-page letter. My proposals are substantially the same as his. It is unfortunate that Bailey's ideas received little attention at the time they were published. Bailey suffered the usual fate of the premature discoverer.

A simple way to calculate the occultation frequency is to look at the situation from the point of view of a distant star. Suppose that the star lies close to the ecliptic plane. Then it sees the Kuiper Belt as a thin edge-on disk with the Earth moving inside it. The dimensions of the disk are about 200 astronomical units or $3 \times 10^{10}$ km from East to West and about 10 astronomical units or $15 \times 10^{8}$ km from North to South. The area of the disk as seen from the star is $5 \times 10^{19}$ km$^2$. If we consider a comet with diameter 1 km occulting the star as seen from the Earth, the shadow of the comet sweeps out a track 1 km wide on the Earth. The Earth is moving in its orbit at 30 km s$^{-1}$, much faster than the comets in the Belt. The rate at which the shadow sweeps out area on the Earth is 30 km$^2$ s$^{-1}$. Dividing the total area of the disk by the sweep-rate, we deduce that an average comet occults the star, as seen from any single point on the Earth, approximately once every $2 \times 10^{18}$ s. If we adopt Kuiper's optimistic estimate of $10^{13}$ for the population of the disk, we find that the star is occulted once every $2 \times 10^{5}$ s, or once every 2 days. If we set up a small telescope with automatic detecting equipment to watch 100 stars continuously, we might expect to see occultations at an average rate of two per hour. That would be frequent enough to make the search for occultations interesting. It might even be frequent enough to justify spending a modest amount of money on small telescopes dedicated to the job of searching.

How small does a star have to be to be eclipsed by a comet? The angular size of a 1 km comet in the Kuiper Belt is about $10^{-10}$ radians, roughly the same apparent size as the famous golf ball that one of our astronauts took with him to the Moon. Fortunately it turns out that there are plenty of stars with angular size as small as this. Nearby stars will not do. A star like Sirius or Alpha Centauri at a distance of a few parsecs is about a hundred times too fat, and a red giant star like Betelgeuse is even worse. To be eclipsed, a star similar to the Sun needs to be a few hundred parsecs away. This means that any star with the same colour as the Sun or bluer, and of visual magnitude 13 or fainter, will have angular size small enough for occultations to be seen. If we consider stars of magnitude 13, bright enough to be easily observed in a small telescope, there are several million of them in the sky. They are about ten times more abundant in the part of sky near to the plane of the Galaxy than they are at the galactic poles. Near the galactic plane there are about a hundred of them per square degree of sky. It would be easy to find places where a hundred suitable stars of magnitude 13 are available within the one-degree field of a small telescope. The best places to observe are the two places in the sky where the plane of the ecliptic (with the highest concentration of comets) intersects the plane of the Galaxy (with the highest concentration of stars). The planes intersect in the constellations Taurus and Sagittarius. In the northern hemisphere, Taurus is the place to observe in winter and Sagittarius is the place to observe in summer. These also happen to be places in the sky where there are many other interesting things to observe besides occultations.

Since the shadow of a comet travels over the Earth at about 30 $\mathrm{km\,s^{-1}}$, a 1 km comet will occult a star for only 30 ms. The star will blink out for a thirtieth of a second. To detect such a quick blink, the telescope must be provided with a multichannel photometric detector, with one channel for each star, counting the photons as they come in and reading out the numbers of photons every hundredth of a second. When a star is occulted, its photon count should be low in at least three consecutive read-outs. We can estimate roughly the size of telescope required to give enough photons to detect an occultation reliably. A star of magnitude 13 supplies about $3 \times 10^5$ photons $\mathrm{m^{-2}\,s^{-1}}$. If we use a 30 cm or 12″ telescope, and detect the photons with 50% efficiency, this will give a hundred detected photons per read-out for each star, enough to make an accidental coincidence of three consecutive low counts very unlikely. But the statistical fluctation of photon-counts is not the most serious cause of spurious events. The more serious problem is the twinkling of the star caused by turbulence of the atmosphere. The frequency of false occultation signals caused by atmospheric turbulence must be measured empirically before an effective observation-system can be designed.

No matter whether the effects of atmospheric seeing turn out to be mild or severe, we shall need to deploy several telescopes along the track of a comet's shadow and correlate their outputs in order to be sure that we have a genuine occultation. We know that the shadow-track should run from East to West at about 30 km $s^{-1}$. If we put three or four telescopes a few km apart in an East–West line, the occurrence of three or four blinks separated by a few tenths of a second in sequence will give a secure identification of a comet. There will also be sequences of blinks produced by shadows travelling at the wrong speed for a comet. These will give evidence for occultations of stars by other objects such as nearby asteroids, fragments of debris in orbit around the Earth, high-flying aeroplanes, meteorological balloons, birds and bats. One person's noise is another person's signal. As a by-product of a system built to observe comets, we may obtain important information about the movements of migrating birds and bats.

The telescopes and photometers required for the monitoring of occultations are simple and fairly cheap. The telescopes need only be programmed to stare at the same group of stars, night after night, without human intervention, as the stars drift across the sky. High optical precision is not needed. The difficult part of the operation is the distribution and coordination of huge quantities of data. Each telescope will be reading out photon-counts from a hundred stars a hundred times a second. All these numbers must be fed by datalinks with accurate timing to a central computer. The computer must then correlate them, pick out the events which show correct time-delays between responses of several telescopes, and send the records of all interesting events to a databank for permanent storage. The most expensive part of the project will probably be the installation and maintenance of the communication and data-processing equipment.

We shall need to try out the system first on as small a scale as possible, with two or three telescopes looking at a small number of stars. If all goes well and some real occultations are seen, the system will grow larger, with the changes and improvements dictated by practical experience. I hope that we might in this way arrive at a large array of small telescopes, perhaps a hundred telescopes arranged in 25 rows of four, the four in each row defining an East–West track, and the 25 rows giving the array a North–South coverage of 50 km. The optimistic estimate for the frequency of occultations observed by the array would then be about one per minute. We could expect to see many of the rarer types of object in the Kuiper Belt, not merely the common-or-garden comets.

## 4 Diffraction and lensing

Another happy coincidence makes the observation of occultations even more interesting. The distance of the Kuiper Belt is just great enough for the diffraction of light around the edges of a comet to produce a noticeable smearing of the shadow. For a comet at a distance of 50 astronomical units, the scale of the smearing on the Earth is about 2 km, which happens coincidentally to be the same size as a typical comet. The smearing has two consequences, one good and one bad. The bad news is that the occultations will be harder to see because the shadows are not sharp. The good news is that the effects of diffraction will be different for light of different colours, and will depend on the distance of the diffracting object. If we measure the brightness of a star separately in two or three colours, we can measure the differential effects of diffraction during an occultation, and this will enable us to determine both the size and the distance of the comet. In this way, if all goes well, an array of telescopes observing occultations in three colours will give us a true three-dimensional map of the Kuiper Belt.

Until now I have been talking only about comets and not about planets. Occultation astronomy gives us a way to search for dark objects of any kind, including planets as well as comets. The obvious place to look for planets is in orbit around stars. But it is also possible that the Galaxy is populated by huge numbers of loose planets unattached to stars. I put forward, as a hypothesis to be tested, the idea that planets may form before stars when an interstellar dust-cloud condenses. Then we might have a large population of loose planets, of which only a few grow big enough to accrete major quantities of gas and turn into stars. For example, the mass of elements other than hydrogen and helium now in the Sun might have made 5000 planets as big as the Earth, if the Sun had not happened to gobble it up first.

One good place to look for loose planets is the Hyades, a region where substantial numbers of stars have formed recently not too far from the Sun. It happens that the Hyades lie in the constellation Taurus, in the area where comet occultations are likely to be most frequent. If we are lucky, we may find an occasional loose planet as a by-product of the search for comets.

Just as optical diffraction helps us to identify comets and measure their distances, gravitational lensing helps us to identify planets. Gravitational lensing is the focusing of light from a background object by the gravitational field of the planet. Because of the lensing effect, the signature of a planet passing in front of a distant star is not a darkening but a brightening of the light from the star. The lensing dominates the shadowing effect of the planet. The apparent diameter of the gravitational lens, the region within which substantial brightening is seen, turns out to be about ten times the diameter of the planet. For a planet as massive as the Earth at the distance of the Hyades, the diameter of the lens is $10^{-10}$ radians, equal to the apparent diameter of a comet in the Kuiper Belt. We can use the same stars to look for planetary lensing and for comets. But the lensing by a planet is a slow process. Since the diameter of the lens is of the order of $10^5$ km, the lensing lasts for about an hour instead of a fraction of a second. In looking for planets, we could examine many more faint stars in a more leisurely fashion, reading out light-intensities every few minutes instead of a hundred times a second. There would be plenty of time to collect enough light from a magnitude-18 star to identify a lensing event unambiguously. The measurement is slow enough that it is not seriously affected by atmospheric turbulence.

As planetary lensing is slow, the expected event-rate is low. Even with the optimistic assumption that our Galaxy contains a thousand loose planets for every star, we expect each background star to be lensed about once in 100 000 years. This sounds discouraging. However, the situation is not hopeless. The slow read-out of data allows us to use in the search for planets the cheap and convenient Charge Coupled Device or CCD detector instead of a multichannel photometer. With a modern CCD detector, a small telescope could monitor a thousand magnitude-18 stars in a one-degree field. Unlike the search for comets, which requires every telescope in an array to stare at the same stars, the search for planets would have each telescope looking at a different set of stars. An array of a hundred small telescopes could monitor $10^5$ stars for lensing events, and could expect to see one event per year.

The search for loose planets is a highly speculative undertaking. The Galaxy is probably not swarming with planets. It would make no sense to embark on the building of an array of telescopes dedicated to the search for planetary lensing events alone. The probability of total failure is too high. But it would make sense to add a capability for planet-search to an array dedicated to the search for comets. The comets in the Kuiper Belt are known to exist and can certainly be detected. An array dedicated to comet-search will have enough success to justify its existence—and it happens that a single array can efficiently search for comets and for planets by creative use of time-sharing. Since planetary lensing events are slow, the array does not need to look at stars continuously to catch them. It is enough to look at stars for a few seconds in each minute. I envisage a combined search for comets and planets, with the detection equipment in each telescope switching on and off, using 54 s in each minute for comets and 6 s for planets. We would then have complete coverage of planetary lensing events using only 10 per cent of the time, while losing only 10 per cent of the comets. The possible discovery of an occasional planet comes as an almost free bonus, while the exploration of the Kuiper Belt continues. The fraction of time and effort devoted to planets would be adjusted after the array is operating. If some real lensing events were seen we would probably decide to look at them for more than 10 per cent of the time.

Several ongoing ventures in occultation astronomy are already in progress, aimed at the discovery of dark objects of various kinds. Charles Alcock at the Lawrence Livermore Laboratory is running a comet-hunt using tiny one-inch telescopes with wide-field CCD cameras (Alcock, Axelrod and Park 1990). He will monitor bright stars and will only see occultations by objects bigger than ordinary comets in the Kuiper Belt. Another project, suggested 5 years ago by Bohdan Paczynski, is going ahead in France (Paczynski 1986*a*). Paczynski's idea is to look for putative black holes or other dark objects in our Galaxy by observing lensing of background stars in areas of the sky where we can find $10^5$ stars within the field of a single large CCD detector. The best places to look are in the Magellanic Clouds or in the window where we can see the central bulge of our Galaxy close to the galactic plane. The stars to be monitored are very faint and the project is definitely not for amateurs. It requires a large telescope with modern detection equipment. A third project, also following Paczynski's suggestion and using large telescopes, is an international collaboration rejoicing in the name MACHO (Massive Compact Halo Objects). The MACHO group includes Charles Alcock and eleven other astronomers in California and Australia. They are now ready to begin the search for lensing events in the central bulge of our Galaxy (Griest *et al.* 1991).

I have not attempted a complete survey of projects in occultation astronomy. So far as I know, none of them has yet produced any positive results, and none of them exploits the sociological opportunities that have been opened by recent developments in the technology of small telescopes.

## 5 Multiple telescope robotic observatories

The international community of amateur astronomers has changed its character over the last 10 years. Until 10 years ago, the typical serious amateur astronomer was somebody who loved to spend long hours grinding and polishing mirrors. The chief object of the game was to build a home-made telescope that could take photographs of celestial objects beautiful enough to publish in magazines such as *Sky and Telescope*. But the devoted lens-grinder is a vanishing species. Today the typical serious amateur is a computer hacker, somebody whose chief pleasure in life is messing around with electronics and software. As a consequence of this sociological revolution, there is now a network of amateur astronomers capable of doing good quantitative measurements, with a precision and sophistication previously available only to professionals. These people are also capable of processing large amounts of data and shipping it around the world at electronic speeds. They have already paid out of their own pockets the costs of adequate computers and software. The main reason why I am enthusiastic about occultation astronomy is that it could engage the resources of the modern amateur community in an enterprise of major scientific importance. It is an activity well suited to the fruitful collaboration of amateurs and professionals. Amateurs can provide small telescopes and local data-processing equipment, with operation and maintenance at no cost to the public. Professionals can provide overall guidance, central computing facilities, and funds for items such as multichannel photometers which amateurs may not be able to make for themselves.

One of the leaders in the new wave of amateur astronomy is Russ Genet, who operates the Fairborn Observatory in Arizona. Genet began as an amateur but has now turned professional, running his observatory as a commercial enterprise. He calls it MTRO, Multiple Telescope Robotic Observatory. He has published a book with the title, *Robotic observatories* (Genet and Hayes 1990). To give you the flavour of it, I quote a passage from the first chapter, 'Astronomy While You Sleep'.

> Observational astronomy is entering a new era which we call Remote-Access Personal-Computer Astronomy. Soon research astronomers, faculty, students and amateurs will have direct access to unattended fully-automatic telescopes at prime sites around the world. Access to these telescopes will require only a personal computer, a modem, and a phone connection. Using only a few minutes of telephone time, they will send a request for observations to be made for all or part of a night, and they will be able to sleep while the request is executed. Their personal computer will automatically retrieve the results for them the following morning, again in a couple of minutes over the phone. When desired, they will be able, with their personal computers and ordinary phone lines, to take direct, real-time control of these same telescopes. The cost of this will be so low that astronomers will be able to conduct certain research programs that would heretofore have been totally impractical. Undergraduate college students will be able to get sufficient observing time on telescopes located at excellent sites to be able to do first-class science projects. Even high-school students and amateurs will be able to use these facilities to explore the world of astronomy.

Since Russ Genet is running the MTRO as a profit-making venture, we must excuse him for talking like a salesman. I do not believe that his claims are exaggerated. He put the first MTRO into commercial operation in 1985 with a simple photometry service which he called Rent-a-star. The fee was \$2 per observation. The service was used extensively by variable-star observers who had grown tired of sitting up all night to obtain a single light-curve. The next level of service is a system called ATIS, Automatic Telescope Instruction Set. With the ATIS software package, the customer can telephone instructions to the observatory for a wider variety of photometric observations. The ATIS service is now in operation. The third level of service is coming soon and will include CCD imaging and spectroscopy. The main virtue of this system is that it is user-friendly. The user does not need to put on winter clothes and struggle with frozen fingers. Even Russ Genet himself does not need to struggle. The telescopes on the mountain work by themselves. Except for maintenance visits, and occasional emergencies when squirrels eat the telephone-lines, nobody is there.

The world-wide network of MTRO observatories of which Russ Genet dreams does not yet exist. I do not know how long it will take to grow. If one can judge by the speed with which electronic-mail networks have grown during the last few years, the MTRO network should not take long to become world-wide. The advantages of a world-wide network are obvious and immediate. A network of this sort would allow amateurs and students from all over the world to collaborate in the search for comets and planets. They already do collaborate in many of their observations. The coming of MTRO will make their collaboration closer and more effective. The MTRO technology will make arrays of small telescopes all over the world scientifically productive. It gives students and amateurs a chance to do something important. Occultation astronomy is only one of the important things that they can do.

## 6 Gravitational lens interference

For the last section of this talk I return from sociology to science. Jeffrey Peterson and Toby Falk, two young professionals at Princeton University, suggested recently that the search for gravitational lenses might include a search for effects of wave interference (Peterson and Falk 1991). When light or radio-waves arrive at the Earth from a distant source along two optical paths through a gravitational lens, the two waves may interfere so as to produce an observable pattern of alternating high and low intensities. Just as the diffraction pattern of light occulted by a comet can reveal details of the comet's size and distance, so can the interference pattern of waves focused by a gravitational lens reveal details of the lensing object. The analysis of this effect by Peterson and Falk shows that the interference patterns will be observable only when the radiation has wavelength comparable with the Schwarzschild radius of the lens. The Schwarzschild radius of any object is its mass converted into a length according to the rate of exchange: 8 mm per Earth mass. Thus the Schwarzschild radius of the Earth is 8 mm while that of Jupiter is 2 m. If we want to see interference patterns in gravitational lenses produced by planets, visible light is useless. This is a job for professional radio-astronomers, observing with m waves if they are looking for Jupiters, with cm waves if they are looking for duplicate Earths. Amateurs with small optical telescopes will not be able to help.

There are many reasons why a search for planetary lenses using radio-waves may be impractical. You have first to find a background radio-source with angular size as small as $10^{-10}$ radians. The only known sources that may be suitable are quasars and masers, and these have complicated structures not yet understood. They are not, like the stars that one uses for optical occultation, little round uniform spheres. Radio-waves, unlike light waves, may be refracted by inhomogeneities of the interstellar plasma to such an extent as to smear out lens interference effects completely. The only hopeful aspect of the situation is the fact that radio-telescopes routinely achieve far higher signal-to-noise ratios than optical telescopes. Because of the superior sensitivity of radio detectors, Peterson and Falk estimate that a distant quasar source will show detectable lens interference effects with high probability, if as much as one part per million of the total mass of the universe happens to be in the form of loose planets, small black holes or other dark objects of planetary mass. So far as I know, no radio-astronomers are rushing to find out whether our universe is infested with such a high abundance of interstellar vermin.

A typical comet has a mass about $10^{-12}$ times the mass of the Earth. Its Schwarzschild radius is about $10^{-14}$ m, 10000 times smaller than an atom. Therefore, we all said until recently, gravitational lensing by comets is utterly negligible. The lensing effect would be lost in the comet's shadow. This conclusion is still correct so far as occultation of visible starlight is concerned. But Andrew Gould, another young colleague of mine at Princeton, pointed out a few days ago that gravitational lens interference effects by objects of cometary mass might be detectable in the gamma-ray bursts that are now being observed at a rate of about one per day by the Gamma-Ray Observatory in orbit (Gould 1991). The Gamma-Ray Observatory is now called the Compton Observatory in honour of Arthur Compton, who discovered the process of electron–photon scattering that the observatory uses to detect gamma-rays. The Schwarzschild radius of a large comet, with a physical diameter of 5 or 10 km, lies conveniently in the range of wavelengths observed by the Compton Observatory. If we imagine both the gamma-ray source and the comet to be at cosmological distances, the size of the lens is of the order of $10^4$ km, well outside the area obstructed by the body of the comet. Gould then finds, in agreement with the calculation of Peterson and Falk, that the probability of observing lens interference in each gamma-ray burst is roughly equal to the fraction of the mass of the universe contained in the lensing objects. The interference effect would be manifested as a series of equally-spaced maxima and minima in the observed spectrum of the gamma-rays.

The origin of the gamma-ray bursts is one of the great unsolved problems of astronomy. Five years ago, Bohdan Paczynski argued on the basis of fragmentary evidence that they must be at cosmological distances (Paczynski

1986*b*). Now that the Compton Observatory is in orbit, we have at last some reliable information about the distribution of the sources in the sky and about their spectra and luminosities. The evidence, still preliminary, seems to indicate that Paczynski was right. If they are in fact at cosmological distances, they must be events of extreme violence, surpassing supernovae and quasars in the instantaneous intensity of their radiation. They are perhaps giving us a glimpse of a new universe, as revolutionary as the new universe revealed by the primitive radio-telescopes of Martin Ryle and Bernard Lovell 40 years ago. As usually happens, when we put up an instrument that opens a new window on the universe, we see a new and unexpected universe. We do not know whether the gamma-ray bursts are connected with any of the familiar objects that are seen in visible light or in radio-waves. All that we know for sure is that Nature's imagination is richer than ours.

How splendid it is, to think that these vast and mysterious outbursts of extreme violence may also carry information in their spectra about objects as parochial as comets. And how splendid it is, that the search for comets may bring together in fruitful collaboration the professional space-astronomers with their massive orbiting observatories and the world-wide fraternity of students and amateurs with their cheap little telescopes and personal computers. In astronomy, just as in music and drama and all the other fine arts, it is the amateurs who ultimately sustain the culture within which the professionals can flourish.

## Addendum (December 1995)

In the four years since this lecture was given, almost every paragraph has been superseded by new discoveries. The most dramatic of these discoveries is the identification of at least fifty objects belonging to the Kuiper Belt, putting an end to any doubts about the real existence of this reservoir of comets. The first two objects were found by J. X. Luu and D. Hewitt, using a 2 m telescope in Hawaii. At the last count, 28 objects have been found by ground-based observations (see Weissman (1995)). About 25 more objects were found by Cochran *et al.* (1995), using the Hubble Space Telescope. Since these objects are seen in abundance by reflected sunlight, the idea of trying to detect them by occultation of distant stars becomes much less attractive.

A second revolutionary advance during the last four years has been the success of searches for gravitational microlensing of stars, both in the Large Magellanic Cloud and in the central bulge of our own galaxy. The groups of astronomers mentioned in Section 4 of the lecture, together with others, have identified more than fifty microlensing events. [For details of twelve events, see Udalski *et al.* (1995). For a review of all the observations, see Paczynski

(1996)]. Since the duration of each event is typically a few weeks, the foreground objects that create the lensing effect are probably ordinary stars. The frequency of observation of background stars is not high enough to detect lensing by objects of planetary mass. Plans are afoot to use the lensing effect to search for planets, but this requires much more frequent measurement of the brightness of each star. For the detection of planets by gravitational lensing to become a routine operation, the existing programs of observation would need to be greatly expanded.

The third revolutionary advance since 1991 has been the discovery of planets around alien stars. Alexander Wolszczan discovered a family of planets orbiting around a neutron star, using the Arecibo radio-telescope (1994). Michel Mayor and Didier Queloz of the Geneva Observatory in Switzerland discovered a planet orbiting the main-sequence star 51 Pegasi, using a 2 m optical telescope in France (Mayor and Queloz 1996). These discoveries will stimulate further searches using similar techniques, and will undoubtedly lead to the identification of more planetary systems.

Since the professional astronomers have been leaping ahead with spectacular discoveries, the gap between professionals and amateurs is in some respects wider today than it was in 1991. Both for hunting comets in the Kuiper Belt and for hunting planets by gravitational lensing, big professional telescopes are the weapon of choice. Nevertheless the amateurs are making rapid progress too. The main conclusion of the 1991 lecture, that collaboration between amateurs and professionals will become increasingly fruitful in the future, is still valid.

## References

Alcock, C., Axelrod, T. and Park, H.-S., 1990. *Sky and Telescope*, **80**, 12–13.

Bailey, M. E., 1976. *Nature*, **259**, 290–291.

Chandrasekhar, S., 1980. *Q.J.R. astr. Soc.*, **21**, 93–107.

Cochran, Levison, Stern and Duncan, 1995. *Astrophys. J.*, **455**, 342–346.

Duncan, M., Quinn, T. and Tremaine, S., 1988. *Astrophys. J. Letters*, **328**, L69–L73.

Genet, R. M. and Hayes, D. S., 1989. *Robotic observatories: remote-access personal-computer astronomy*. Autoscope Corporation, Mesa, Arizona.

Gould, A., 1991. *Femto-microlensing of gamma-ray bursters*. Institute for Advanced Study Astrophysics Preprint, IASSNS-AST 91/47, Princeton, New Jersey.

Griest, K. *et al.* (Twelve authors) 1991. *Astrophys. J. Letters*, **372**, L79–L82.

Kuiper, G., 1951. On the origin of the solar system. In *Astrophysics, a Topical Symposium*, pp. 357–424, ed. Hynek, J. A., McGraw-Hill, New York.

Mayor, M. and Queloz, D. 1996. News item in *Sky and Telescope*, January issue, 38–40.

Oort, J., 1950. *Bull. Astron. Inst. Netherlands*, **11**, 91–110.

Paczynski, B., 1986*a*. *Astrophys. J.*, **304**, 1–5.

Paczynski, B., 1986*b*. *Astrophys. J.*, **308**, L43–L46.

Paczynski, B., 1996. Gravitational microlensing in the local group. *Ann. Rev. Astronomy and Astrophysics*, **34**, 419–459.

Peterson, J. and Falk, T., 1991. *Astrophys. J. Letters*, **374**, L5–L8.

Udalski, A. *et al.*, 1995. *Acta Astronomica*, **45**, 237–257.

Weissman, P. R., 1995. The Kuiper Belt. *Ann. Rev. Astronomy and Astrophysics*, **33**, 327–357.

Wolszczan, A., 1994. *Science*, **264**, 538–542.

# Modern cosmology—a critical assessment*

Malcolm S. Longair

## 1 Preliminaries

When quality newspapers such as *The Independent* devote the whole of a front-page spread to the detection of fluctuations in the Cosmic Microwave Background Radiation, we can be sure that cosmology has come of age. The story of the last 30 years has been one of quite remarkable progress in cosmology and I am one of that very lucky generation who began to carry out research in this area just as the flood was about to break. Progress has been spectacular and has opened up completely new ways of addressing problems of cosmology and fundamental physics. In their enthusiasm, however, I believe that some cosmologists have been carried away and a number of exaggerated claims have been made about how much we really understand about topics related to the origin of our Universe. My objective here is to attempt to redress the balance and to address four questions simultaneously:

1. How much of what is 'known' can we really trust?
2. Which will be the lasting contributions?
3. What can we hope to achieve?
4. What should everyone know?

My approach will be not to exaggerate but to take a long hard look at those bits of the story which we can really believe. In my view, the story is quite remarkable enough without speculation beyond what has already been established with a good deal of confidence. To pose the question another way, which bits of the story are so convincing that we would be prepared to die in the ditch defending them?

*This is an abbreviated version of the lecture which was published in the *Quarterly Journal of the Royal Astronomical Society*, **34**, 157–199, in 1993.

Before embarking upon this analysis, let me note what I consider to be the four developments which have made the advances possible.

1. In my view, by far the most important development has been the opening up of the whole of the electromagnetic spectrum for astronomical observations. Since 1945, the disciplines of radio, millimetre, infrared, ultraviolet, X-ray and $\gamma$-ray astronomy have all become major astronomical disciplines and each has had its own unique contribution to make to filling out the cosmological picture. Some of these astronomies can only be carried out from above the Earth's atmosphere—far-infrared, ultraviolet, X-ray and $\gamma$-ray astronomy—and so space and ground-based observations are complementary in the information they provide about our Universe (for more details, see Longair (1989)).
2. Going hand-in-hand with the new astronomies has been the development of new technology. The computing and microelectronic revolution, as well as the astounding developments in instrument and detector technology, mean that the telescopes for essentially all the electromagnetic wavebands are approaching their ideal efficiencies.
3. The astrophysicists and cosmologists have rapidly absorbed all the great discoveries of modern physics into the battery of tools which they use to study the cosmos, to great advantage.
4. Finally, and by no means least, the astronomical discoveries have led to completely new astrophysical disciplines which have provided new tools for studying key astrophysical problems of cosmological importance.

To my regret, I will make little further allusion to instruments and telescopes but they are the prime source of the great discoveries and new understandings we will discuss. However ephemeral the theories, the observational and instrumental achievements are outstanding lasting contributions.

## 2 Observational cosmology in 1963

Why are observational and astrophysical cosmology feasible at all? When we look at the Universe, it is of quite daunting complexity. Within our own Galaxy, we observe the complexities of the birth, life and death cycle of stars—on the scale of galaxies, we observe configurations of stars and gas which range from the completely regular to the totally pathological—on the scales of whole collections of galaxies we observe regular clusters of galaxies as well as 'stringy' structures and huge 'holes' which seem to be devoid of galaxies. Despite the complexity, it turns out that the Universe as a whole has some very simple large-scale features and it is these which make the subject of astrophysical cosmology possible as a science.

When I began research in radio astronomy as a student in 1963, my supervisor Dr Peter Scheuer gave me a copy of Sir Hermann Bondi's classic text *Cosmology* to absorb and warned me that 'There are only 2½ facts in cosmology'. The point is a very important one, in that most of the mass of observations which can be made of gas, stars and galaxies tell us nothing of real cosmological significance. We have to select from this plethora of data those pieces which establish real facts about the Universe as a whole. We now know many more real facts about the Universe but they are still a small finite number. My personal view is that the uncertainties are greater than many professionals would wish to believe. I will therefore start with these 2½ facts and see how much more we have learned over the last 30 years.

In 1963, these facts were as follows:

**Fact 1 The sky is dark at night.**

This is the well-known observation which leads to what is known as *Olbers' paradox*, although the paradox was well known to earlier cosmologists. Sir Hermann in his text *Cosmology* gives a thought-provoking discussion of the meaning of the paradox (Bondi 1952). The fact that the sky is not as bright as the surface of the Sun provides us with some very general information about the Universe. Probably the most general way of expressing the significance of this observation is that the Universe must, in some sense, be far from equilibrium although in what way it is in disequilibrium cannot be deduced from this very simple observation.

The second fact was as follows:

**Fact 2 The galaxies are receding from each other as expected in a uniform expansion.**

This was Hubble's great discovery of 1929 and I will say much more about it in a moment. The 2½th fact was as follows:

**Fact 2½ The contents of the Universe have probably changed as the Universe has grown older.**

The reason for the ambiguous status of this fact was that the evidence for the evolution of extragalactic radio sources as the Universe grows older was then a matter of considerable controversy, particularly with the proponents of Steady-State cosmology. I was plunged straight into this debate as soon as I began my research programme with Martin Ryle and Peter Scheuer. As we will see, this is no longer a controversial issue—there is no question at all but that many classes of object exhibit evolutionary changes as the Universe grows older. Thus, in 1963, the number of real facts which characterised the Universe as a whole was very small and relatively modest progress had been made since the 1930s and the time of Milne.

To the chagrin of the observers, the standard world models which we use in our everyday work were discovered when only the first of these facts was known. Einstein completed his General Theory of Relativity in 1915 and quickly realised that he had a tool which could be used to construct meaningful models of the Universe as a whole, unlike the Newtonian theory of gravity which does not take account of the fact that the speed of light is finite, and for which satisfactory boundary conditions at infinity cannot be found. Einstein's static model of the Universe was published in 1917 but the real breakthrough came with the work of the Russian meteorologist and theoretical physicist Aleksander Aleksandreyevich Friedman who, in the period 1922 to 1924, solved Einstein's field equations of General Relativity for the complete set of homogeneous, uniformly expanding models of the Universe. At that time, this was a theoretical exercise and the deep significance of his results was only appreciated when Friedman's papers were later publicised by Georges Lemaître. Friedman died of typhoid in 1925 during the years of the civil war in Leningrad and so did not live to see the observational validation of his solutions of the field equations.

In 1929, Hubble announced his discovery of the relation between the velocity of recession of the galaxies and their distances. It had been known for some years that, when the velocities of galaxies are measured, they are all observed to be receding from our own Galaxy. Hubble's great discovery was that the further away a galaxy is from our own Galaxy, the greater its velocity of recession. Figure 1 shows a modern version of what is now known as the Hubble diagram and is due to Sandage (1968), who used the brightest galaxies in clusters to define the relation. This relation is known as Hubble's Law and is written $v = H_0 r$ where the constant $H_0$ is known as Hubble's constant. All classes of extragalactic objects obey this law.

Milne realised that the observation of a velocity–distance relation meant that the objects observed must be participating in a uniform expansion. This is illustrated in Fig. 2 which shows a uniform distribution of galaxies participating in a uniform expansion. A uniform expansion means that the distance between neighbouring galaxies increases by the same factor in a given time interval. The result is that, if we fix our attention upon any one galaxy in the distribution, the further it is away from the chosen point, the greater the distance it has to travel in the same time in order to preserve the uniform expansion. Thus, the observation of the velocity–distance relation for all extragalactic systems simply means that the distribution of galaxies is participating in a uniform expansion in which the distances between neighbouring galaxies continually increase with time.

In the 1930s, Milne and McCrea (1934) explained the physical content of Friedman's solutions of Einstein's field equations in terms of a simple Newtonian picture. They realised that, in completely homogeneous, isotropic universes, global physics must be the same as local physics since every point

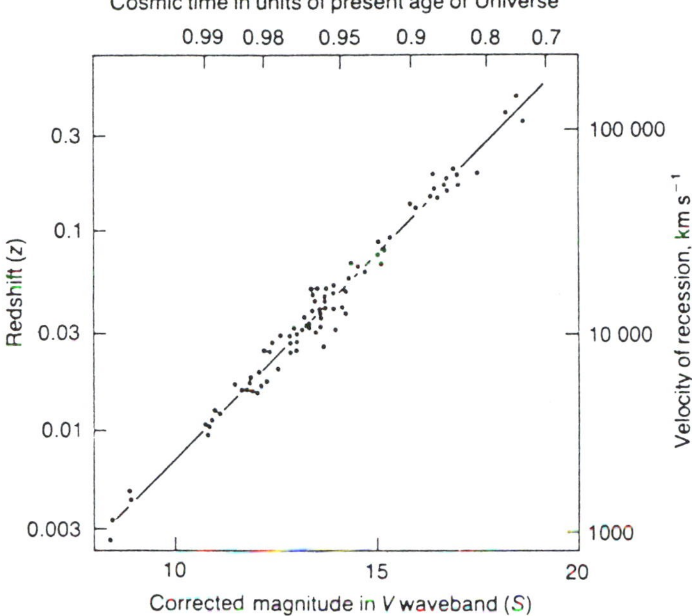


**Fig. 1** A modern version of the velocity–distance relation (or Hubble diagram) for the brightest galaxies in clusters (after Sandage 1968). In this logarithmic plot, the corrected apparent magnitudes (that is $-2.5$ times the logarithm of the flux density $S$) of the brightest galaxies in clusters are plotted against their redshifts $z$. The straight line shows what would be expected if $S \propto z^{-2}$. This correlation indicates that the brightest galaxies in clusters have remarkably standard properties and that the distances of the galaxies are proportional to their redshifts which, for small redshifts, implies that velocity is proportional to distance.

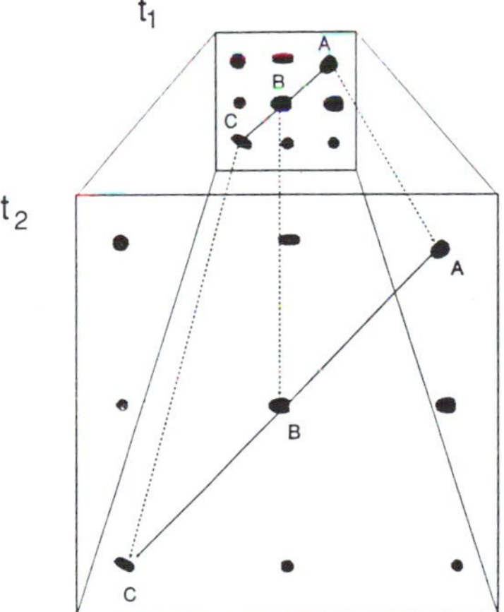


**Fig. 2** Illustrating the origin of the velocity–distance relation for an isotropically expanding distribution of galaxies. The distribution of galaxies expands uniformly between the epochs $t_1$ and $t_2$. If, for example, we consider the motions of the galaxies relative to Galaxy A, it can be seen that galaxy C travels twice as far as galaxy B between epochs $t_1$ and $t_2$ and so has twice the recession velocity of galaxy B relative to A. Since C is always twice the distance of B from A, it can be seen that the velocity–distance relation is a general property of isotropically expanding Universe.

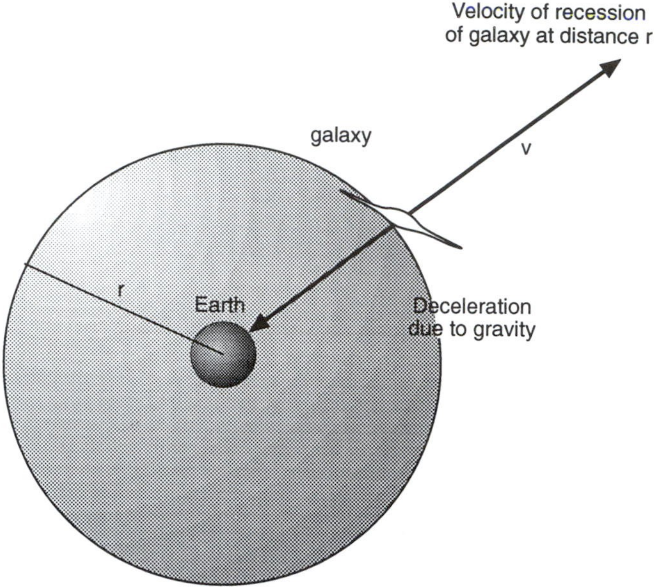


**Fig. 3** Illustrating the Newtonian model for the dynamics of the classical Friedman models of General Relativity. The velocity of recession of a galaxy at distance *r* can be considered to be decelerated by the gravitational attraction of the matter within distance *r* of our own Galaxy. Because of the assumption of isotropy, an observer on any galaxy participating in the expansion of the Universe would carry out exactly the same calculation.

in the Universe has to be equivalent to every other point at a given time and so the same physics applicable locally must be applicable at all points in the Universe. They showed how to derive the essential content of the Friedman models in terms of the simple Newtonian picture illustrated in Fig. 3. We observe the same velocity–distance relation in whichever direction we look and so we ask what would be the deceleration of a galaxy at distance $r$ from us due to all the mass within the sphere of radius $r$. Because of the spherical symmetry of the problem, Gauss's theorem tells us that we obtain the correct expression for the deceleration of the galaxy by placing all the mass within radius $r$ at the location of our Galaxy. Thus, the dynamics of the galaxy depend upon how much mass there is within radius $r$ and hence upon the average density of matter within that sphere.

The classical Friedman models describe the expansion of the Universe in terms of a scale-factor, which describes how the separation between points partaking in the universal expansion changes with time. There are three types of solution. If the Universe is of high density, the force of gravitational attraction is sufficiently great to halt the expansion and the Universe eventually collapses back to a high-density, high-temperature state, a state often

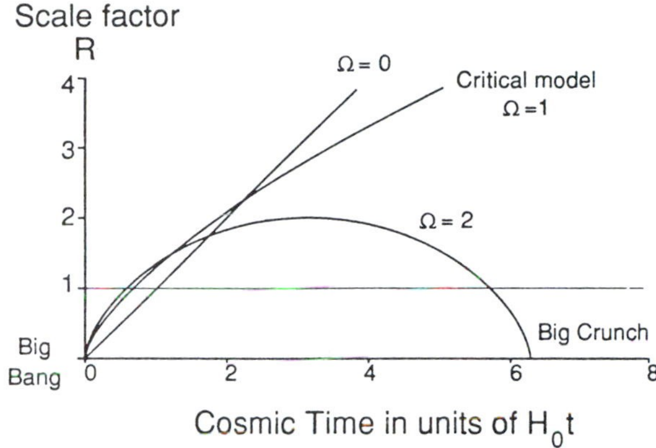


**Fig. 4** The dynamics of the classical Friedman models parameterised by the density parameter $\Omega = \rho/\rho_{crit}$. If $\Omega > 1$, the Universe collapses to $R = 0$ as shown; if $\Omega < 1$, the Universe expands to infinity and has a finite velocity of expansion as $R$ tends to infinity. In the case $\Omega = 1$, $R = (t/t_0)^{2/3}$ where $t_0 = (2/3)H_0^{-1}$. The time axis is given in terms of the dimensionless time $H_0t$. At the present epoch $R = 1$ and in this presentation, the three curves have the same slope of 1 at $R = 1$, corresponding to a fixed value of Hubble's constant. If $t_0$ is the present age of the Universe corresponding to $R = 1$, then for $\Omega = 0$, $H_0t_0 = 1$; for $\Omega = 1$, $H_0t_0 = 2/3$; and for $\Omega = 2$, $H_0t_0 = 0.57$.

referred to as the 'Big Crunch'. If the Universe is of low density, the force of gravitational attraction is not sufficient to halt the universal expansion and, in the limit of infinite time, the expansion velocity remains finite. Separating these models, there is a unique model known as the critical or Einstein–de Sitter model which has zero velocity of expansion as the time tends to infinity. In other words, the Universe just possesses its own escape velocity. There is a corresponding critical density $\rho_0$ associated with this model which depends only upon the value of Hubble's constant, $\rho_{crit} = 3H_0^2/8\pi G$. It is often convenient to compare the mass densities of world models (or any constituent of the models) with the critical density $\rho_{crit}$ so that a density parameter $\Omega = \rho/\rho_{crit} = 8\pi G\rho/3H_0^2$ can be used to parameterise the world models. These solutions are illustrated in Fig. 4.

What cannot be incorporated into the Newtonian arguments developed by Milne and McCrea is the dependence of the global geometry of space upon the density except in some special cases. According to General Relativity, the geometry of space–time is determined by the mass–energy distribution throughout the Universe. In the homogeneous Friedman models, the density distribution is the same at all points in space at a given time and so the spatial curvature is the same everywhere at that epoch. Formally, the curvature of the geometry is given by $\kappa = 1/\mathscr{R}^2 = (\Omega - 1) / (c / H_0)^2$, where $\mathscr{R}$ is the radius of curvature of the spatial sections. In the case of the high density models, the geometry is closed and spherical while in the low density models, it is open and hyperbolic. The case of the completely empty Universe was analysed by

Milne and he showed that its spatial geometry is hyperbolic (see Longair (1992) for an elementary derivation of this result). Appropriately, the empty model having $\rho = 0$ is known as the Milne model. It turns out that only in the case of the critical model is the geometry flat Euclidean space. As we will see, the unique features of the critical model have a certain theoretical appeal. In principle, the geometry of space is a measurable quantity—one simply needs to measure accurately the sum of the angles of a triangle over significant cosmological distances and find out whether the sum is equal to 180° or not. In practice, this test is not feasible.

## 3 The Friedman models in 1992

It will be noted that the standard Friedman models use only two pieces of observational evidence, the velocity–distance relation for galaxies and the isotropy of the Universe as a whole on the large scale. The models also assume that, on the large scale, the dynamics of the Universe can be described by General Relativity. There is also an implicit assumption made which is known as the Cosmological Principle, according to which the large-scale features of the Universe which we observe would also be observed by any other suitably chosen observer who looks at the Universe at the same cosmic epoch. In other words, an astronomer on a distant galaxy would also observe the same Hubble's law and an isotropic Universe if the observations were made at the present epoch. As we have argued above, the interpretation of the velocity–distance relation in terms of the uniform expansion of a homogeneous distribution of galaxies is entirely consistent with this principle. It can also be subjected to direct test by comparing the properties of the distribution of galaxies at different distances and in different regions of space. So far as we can tell, there is no reason to believe that the assumption that we live at a typical point in the Universe is incorrect. Indeed, if the principle were not to be correct and we are at a very special point in the Universe, we would be forced to return to a pre-Copernican, Ptolemaic view of the Universe in which we would occupy a privileged position.

We have already discussed the modern versions of Hubble's law, but we have yet to deal with the isotropy of the Universe and with the assumption that General Relativity can be used to describe the large-scale dynamics of the Universe. I will elevate the first of these observations to Fact 3 because of the quite spectacular accuracy with which this has now been established.

**Fact 3 The Universe is isotropic on very large scales to an accuracy of better than one part in 100 000.**

The Universe is obviously highly inhomogeneous on a small scale, with matter condensed into stars which are congregated into galaxies which are them-

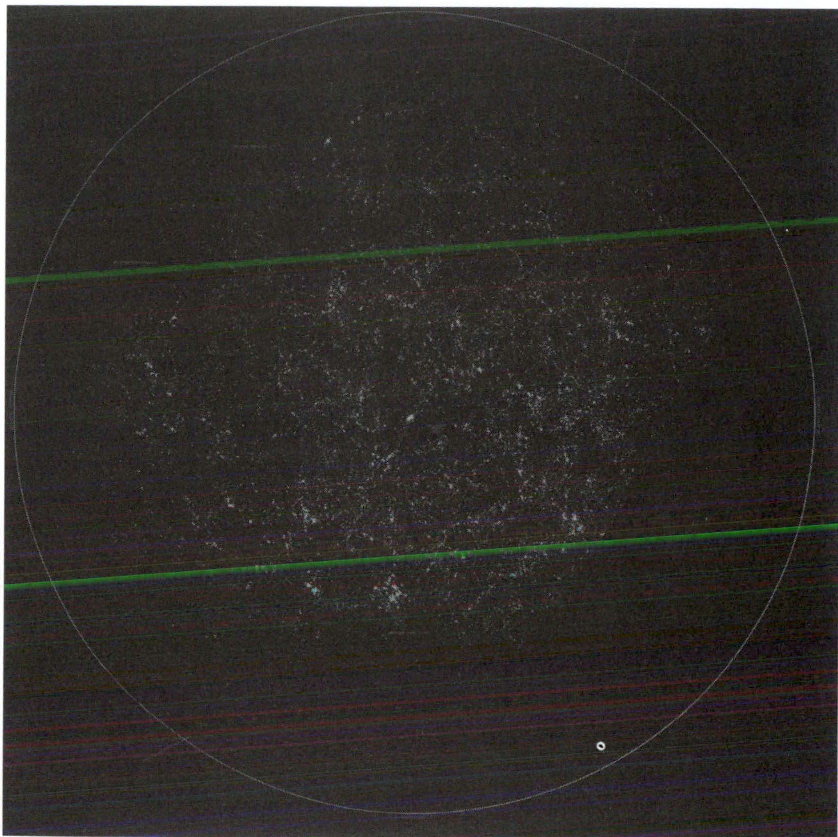

**Fig. 5** The distribution of galaxies in the northern galactic hemisphere derived from counts of galaxies undertaken by Shane, Wirtanen and their colleagues at the Lick Observatory in the 1960s. Over one million galaxies were counted in their survey. The northern galactic pole is at the centre of the picture and the galactic equator is represented by the solid circle bounding the diagram. The projection of the sky onto the plane of the picture is an equal area projection. This photographic representation of the galaxy counts was made by Peebles and his colleagues. The large sector missing from the bottom right-hand corner of the picture corresponds to an area in the southern celestial hemisphere which was not surveyed by the Lick workers. The decreasing surface density of galaxies towards the circumference of the picture, that is towards the galactic equator, is due to the obscuring effect of interstellar dust in the interstellar medium of our own Galaxy. The prominent cluster of galaxies close to the centre of the picture is the Coma cluster. (Seldner, Siebars, Groth and Peebles 1977).

selves clustered, the associations ranging from small groups to giant regular clusters of galaxies. If we take our averages on larger and larger scales, however, the inhomogeneity becomes less and less. Figure 5 shows the distribution of galaxies in the northern galactic hemisphere once all the stars of our own Galaxy have been removed. The large 'bite' out of the picture in the bottom right corresponds to an area of the sky which was not observed in the Lick survey and the decrease in the numbers of galaxies towards the edges of the picture is due to extinction by interstellar dust in our own Galaxy. Therefore, only in the central region of Fig. 5 do we obtain a reasonably clean picture of the large scale distribution of galaxies in the Universe. This is a picture of the distribution of galaxies on the grandest scale, a giant cluster such as the Coma cluster corresponding to the bright dot in the centre of the picture. We now know that much of the obvious clumping, the holes and the stringy structures are real features of the distribution of galaxies but, if we take averages over very large regions, one bit of Universe looks very much like another.

Even more impressive evidence comes from the distribution of extragalactic radio sources over the sky. It turns out that, when we make a survey of the radio sky, the objects which are easiest to observe are extragalactic radio sources associated with certain rare classes of galaxy at very great distances. Because they are rare objects, they sample the isotropy of the Universe on a large scale. Figure 6 shows the distribution of the brightest 3000 extragalactic radio sources in the northern hemisphere. There is a hole in the centre of the distribution corresponding to a region which was not observed as part of the Cambridge 4C survey, but otherwise the distribution is entirely consistent with the sources being distributed uniformly at random over the sky. The radio sources are ideal for probing the large-scale distribution of discrete objects since they are so readily observed at large distances. Similar maps will soon be available at X-ray wavelengths thanks to the ROSAT survey of the X-ray sky.

This is impressive enough, but it pales into insignificance compared with the recent results on the isotropy of the Cosmic Microwave Background Radiation. This background radiation was discovered in 1965 by Penzias and Wilson whilst commissioning a very sensitive telescope and receiver system for centimetre wavelengths. It was quickly established that this background radiation is remarkably uniform over the sky. The most recent results on the large scale distribution of this radiation over the sky have been obtained by the Cosmic Background Explorer (COBE) which was launched in November 1989. This satellite is dedicated to studies of the background radiation, not only in the millimetre waveband but throughout the infrared waveband as well.

The results in increasing levels of sensitivity are as follows. At sensitivity levels about one part in 1000 of the total intensity, there is a large scale anisotropy over the whole sky associated with the motion of the Earth through the frame of reference in which the radiation would be the same in all directions. This is no more than the result of Doppler effect due to the Earth's motion, and as a result the radiation is about one part in a thousand more intense in one direction and exactly the same amount less intense in the opposite direction. The intensity distribution has precisely the expected dipolar distribution and it turns out that the Earth is moving at about

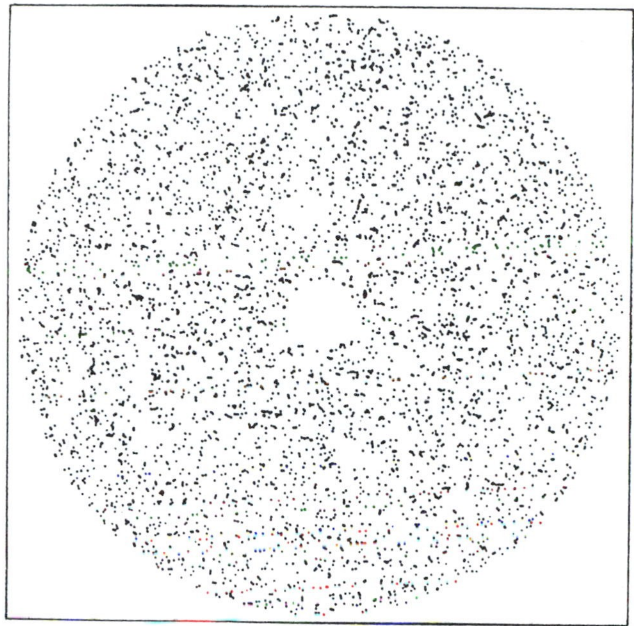

**Fig. 6** The distribution of radio sources in the fourth Cambridge (4C) catalogue in the northern celestial hemisphere. This part of the catalogue contains over 3000 sources of small angular diameter. In this equal area projection, the north celestial pole is in the centre of the diagram and the celestial equator around the solid circle. The area about the north celestial pole was not surveyed. The distribution does not display any significant departure from a random distribution. (Courtesy of Dr. M. Seldner)

350 $\mathrm{km\,s^{-1}}$ with respect to the frame of reference in which the radiation would be 100% isotropic. At about the same level of intensity the plane of our Galaxy can be observed as a faint band of emission over the sky.

In the most recent analyses of the isotropy of the Cosmic Microwave Background Radiation on angular scales 10° and greater by the COBE workers, sensitivity levels of only one part in 100 000 of the total intensity have been attained. At this level, the radiation from the plane of the Galaxy is intense but is confined to a relatively narrow strip centred on the galactic plane. Away from this region, the sky appears quite smooth on a large scale, but careful analysis of the variation of the intensity from point to point on the sky has found convincing evidence for tiny fluctuations in intensity over and above the instrumental noise. The signal amounts to only about 1 part in 100 000 of the total intensity and, when averaged over the clear region of sky, the significance of the result is at the $6\sigma$ level (Fig. 7). This is a very important result for cosmology as we will see later.

For the moment, however, our interest is in the isotropy of the Universe as a whole, and we can state that there is certainly no evidence for any anisotropy in the distribution of the Cosmic Microwave Background

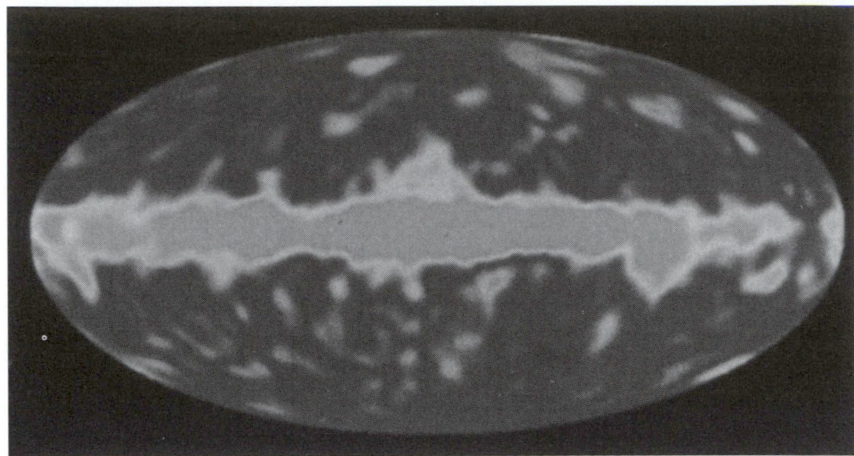

**Fig. 7** The map of the whole sky in galactic coordinates as observed in the millimetre waveband at a wavelength of 5.7 mm by the COBE satellite once the dipole component associated with the motion of the Earth through the background radiation has been removed. The residual radiation from the plane of the Galaxy can be seen as a bright band across the centre of the picture. The fluctuations seen at high galactic latitudes are noise from the telescope and the instruments, the rms value at each point being 36 μK but, when statistically averaged over the whole sky at high latitudes, an excess sky noise signal of 30 ± 5 μK is observed (Smoot *et al.* 1992).

Radiation at the level of one part in 100 000 when we look on the large scale. This is quite incredible precision for any cosmological experiment since one is normally lucky in cosmology if one knows anything within a factor of about 10. The obvious question is how the distribution of this radiation is related to the distribution of ordinary matter. The answer is not as straightforward as one would like and we need to understand the temperature history of the Universe to give the answer. In the standard picture of the evolution of the Hot Big Bang, when the Universe was squashed to only about one thousandth of its present size, the temperature of the Cosmic Background Radiation must have been about one thousand times greater than it is now, and that was sufficiently hot for all the hydrogen in the Universe to be ionised. When this occurs, there is very strong coupling between the Cosmic Background Radiation and the ionised matter. In fact, when we look back to these epochs, it is as if we were looking at the surface of a star surrounding us in all directions, but the temperature of the radiation we observe has been cooled by this factor of 1000 so that what we observe is redshifted into the millimetre waveband. This analogy also makes it clear that, because of the strong scattering of the radiation, we can only observe the very surface layers of the star and, in the same way, we can obtain no direct information about what was happening at earlier epochs as soon as we encounter the epoch at which the material of the Universe was ionised. This 'surface' at which the Universe becomes opaque to radiation is known as the last scattering surface and the fluctuations observed by COBE are believed to represent the very low intensity ripples present on that surface on angular scales of 10°. Thus, strictly speaking, in the standard interpretation, the COBE results provide information about the diffuse ionised intergalactic gas when the Universe was only about one thousandth of its present size. At that stage, the galaxies had not formed and so all the ordinary matter which was eventually to become galaxies as we know them was still in the form of a remarkably smooth intergalactic ionised gas. The extragalactic radio sources provide complementary information about the large-scale distribution of discrete objects such as galaxies at the present epoch once the galaxies had formed.

The upshot of this discussion is that the Cosmic Background Radiation provides us with information about the isotropy of the matter content of the Universe, not as it is now but as it was when the Universe was squashed by a factor of about 1000. None the less, this is the most powerful evidence we possess that the Universe is quite remarkably isotropic on a large scale.

The COBE mission has produced another remarkable result which we can add to our list of real facts about the Universe.

**Fact 4 The spectrum of the Cosmic Microwave Background Radiation has a pure black-body spectrum at a radiation temperature of 2.735 K.**

The evidence for this is the spectrum of the background radiation obtained by the Michelson interferometer on board the COBE satellite. The first published spectrum of the radiation is shown in Fig. 8. The boxes show the experimentally determined spectrum and the solid line is a black-body spectrum at a radiation temperature of 2.735 K. It can be seen that the line runs suspiciously precisely though the centres of all the error boxes which have been shown as 1% of the peak intensity. This means that the quoted errors are somewhat conservative estimates of the uncertainties. The most recent analysis of these magnificent data which I heard discussed only two months ago indicated that the spectrum is now known to be a black-body with an uncertainty of only 0.25% of the peak intensity at wavelengths longer than 500 μm. These are quite remarkable results. It is certainly by far the most precise black-body spectrum I know of in nature.

What is the significance of this observation? It was Planck who first showed in 1900 that the black-body spectrum is the unique radiation spectrum obtained when matter is in thermodynamic equilibrium with radiation at a single temperature. The implication of the observation of such a spectrum for cosmology is that the matter and radiation must have reached a state very closely approximating thermodynamic equilibrium at some time in the past. This occurs naturally in the hot early phases of the standard Big Bang model.

The last requirement of the Friedman models is the physics of the forces

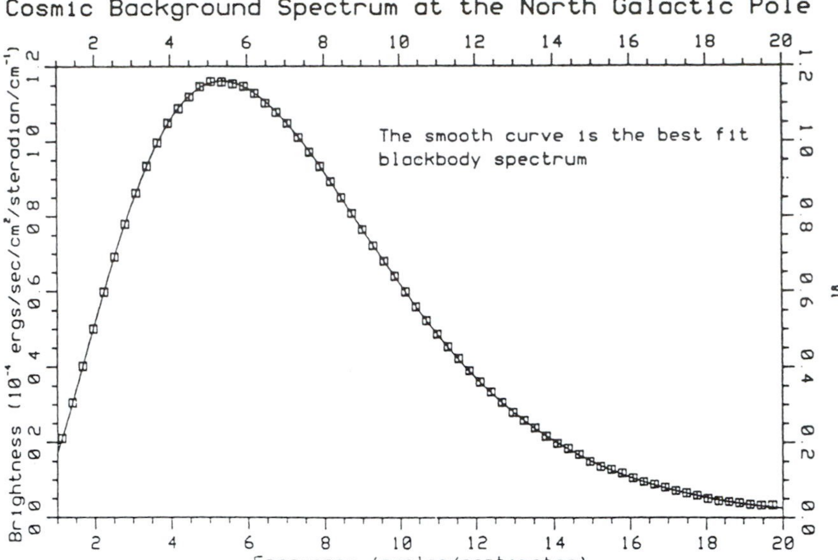


**Fig. 8** The spectrum of the Cosmic Microwave Background Radiation as measured by the COBE satellite in the direction of the North Galactic Pole. Within the quoted errors, the spectrum is that of a perfect black body at radiation temperature 2.735 ± 0.06 K (Mather *et al.* 1990).

which determine the large-scale dynamics of the Universe. Gravity is the only large-scale force we know of which acts upon all forms of matter and energy and General Relativity is the best theory of gravity we possess. When Friedman first solved the field equations of General Relativity for isotropically expanding Universes, the evidence for General Relativity was good but not perhaps overwhelming. The most remarkable result was the prediction of the exact perihelion shift of Mercury, which had remained an unsolved problem in the celestial mechanics of the Solar System since the time of its discovery by Leverrier in 1859. Modern tests of General Relativity, particularly those involving the use of pulsars as ideal astronomical clocks, are described in Professor Taylor's 1993 Milne Lecture. Suffice it to say that General Relativity has passed every test which has been made of the theory and we can have much greater confidence than in the past that it is an excellent description of the relativistic theory of gravity. I feel sufficiently confident that General Relativity is by far the best theory of gravity we possess that I will elevate it to Fact Number 5.

**Fact 5 Standard General Relativity has passed the most precise tests which have been devised so far and there is no astrophysical motivation for seeking any different theory.**

The upshot of this discussion is that we can have confidence in the basic assumptions behind the Friedman models and the next, and much more difficult, step is to determine which particular model, if any, provides the best description of the large-scale dynamics of the Universe.

## 4 The determination of cosmological parameters

Since the framework of the Friedman world models is now very well established, the next task is to pin down which of the models provides the best description of the large-scale dynamics of the Universe. It turns out that the models are defined by a very small number of parameters. These are:

1. The present rate of expansion of the Universe as defined by Hubble's constant $H_0$.
2. The present deceleration of the Universe as described by the deceleration parameter $q_0$.
3. The present average density of the Universe $\rho$. As discussed above, it is convenient to measure the density of the Universe relative to the critical density $\rho_{crit}$ and so define a density parameter $\Omega=\rho/\rho_{crit}$.

In the classical world models, the deceleration of the universal expansion is entirely due to the gravitational influence of the matter content of the Universe and, according to the Friedman models, $q_0 = \Omega/2$. Now the deceleration parameter and the mean density of matter in the Universe can be measured quite independently and so this prediction provides a test of the General Theory of Relativity on the scale of the Universe itself.

There is only one wrinkle in this story and that concerns the fact that, in his paper on the static Universe of 1917, Einstein introduced a further term into the field equations, the infamous cosmological constant $\Lambda$. Because of the attractive nature of gravity on the large scale, static Universes are not feasible unless a large-scale repulsive force is included in the equations to counteract the attractive influence of gravity. Once Hubble discovered that the Universe is not in fact static but expanding, there was no longer any reason to include this arbitrary term in the field equations and Einstein stated that the introduction of the $\Lambda$ term 'was the greatest blunder of my life' (Einstein, quoted by Gamow 1970). The presence of the cosmological constant changes the relation between $q_0$ and $\Omega$:

$$q_0 = \frac{\Omega}{2} - \frac{1}{3}\frac{\Lambda}{H_0^2}.$$

At the present time, there is absolutely no evidence from any astronomical observation that $\Lambda$ is not zero. As Zeldovich has remarked, however, 'the genie is out of the bottle and, once he is out, it is very difficult to put him back in again'. With great regularity, the cosmological constant has kept reappearing in the literature in response to some astronomical anomaly, only to be pushed back into the bottle until its next appearance. We will find that the cosmological constant continues to haunt the subject, but now in a completely different guise in the inflationary scenario for the very early Universe.

## 4.1 Hubble's constant

There are currently two schools of thought, one finding values of $H_0$ in the range 80–100 $\mathrm{km\,s^{-1}\,Mpc^{-1}}$ and the other values in the range 45–60 $\mathrm{km\,s^{-1}\,Mpc^{-1}}$. The big trouble is the difficulty of measuring accurate distances for the galaxies which sample the overall Hubble flow. The extragalactic distance scale has traditionally been determined by calibrating one set of distance indicators against another and so proceeding from the distance scale established in our Galaxy to extragalactic distances. This is an unresolved problem and my personal opinion is that the discrepancy simply reflects the difficulty of measuring extragalactic distances precisely. I have in mind a diagram I produced (for pedagogical purposes) of precise measurements of the speed of light between 1945 and 1960 (Fig. 9). It is noteworthy how often successive measurements lie outside the formal uncertainties of previous measurements and how many of the measurements are formally inconsistent with the present adopted value for the speed of light. Note also that these are laboratory measurements for which one might have thought that the errors could be precisely estimated. Almost certainly the problem lies in some unrecognised systematic error in the distance indicators used to measure the distances of the galaxies.

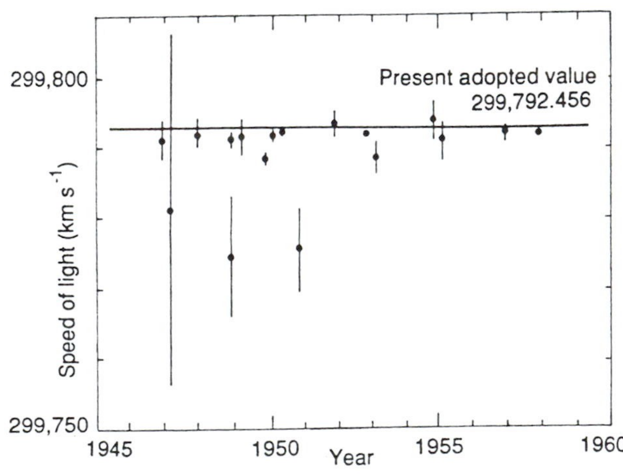


**Fig. 9** Precise measurements of the speed of light published in the period 1945 to 1960. (Data from *The American Handbook of Physics* (1969), 2nd edition. New York: McGraw Hill.)

I would make two comments about this problem. The first is that there is an urgent need to develop better physical methods of measuring extragalactic distances. Most of the steps in the traditional chain of arguments which lead to values of Hubble's constant involve assuming that the same types of astronomical object can be selected in different galaxies at different distances—this is why they are called distance indicators. In contrast, in the direct methods, the need to use distance indicators is eliminated by evaluating some physical dimension $d$ at the distant galaxy and then, by measuring its angular size $\theta$, the distance $D$ to the galaxy can be measured from $D = d/\theta$. The trick is to be able to find a method of measuring physical sizes of objects at extragalactic distances. I will mention only three of the more promising techniques. One method uses supernova explosions of Type II. In these explosions, the speed of expansion of the supernova as well as its spectrum and luminosity are measured as it decreases from maximum light. The physical rate of expansion is measured from the width of the emission lines and the change in angular size can be estimated from the change in surface brightness as the supernova expands. In the most recent observations, accuracies as good as those obtained by the traditional techniques are being found (Branch 1988, Kirshner and Schmidt (personal communication 1992)). A second method is to use the phenomenon of gravitational lenses to work out the geometry of the lensing galaxies and thus measure physical distances at the distance of the lensing galaxy. A third good example is the use of the Sunyaev–Zeldovich effect in clusters of galaxies which contain large amounts of hot X-ray emitting gas. In this effect, a decrement is observed in the intensity of the Cosmic Microwave Background Radiation in the direction of the hot gas cloud because of Compton scattering of the background photons by the electrons of the hot gas. Combining all the data on the properties of the hot gas cloud enables its physical size to be determined and hence, by observing its angular size, its distance can be measured.

My second comment concerns the use of more astrophysical methods of setting limits to Hubble's constant from the measurement of the ages of the oldest stellar systems in galaxies. In the standard world models, the age of the Universe is $t_0 = f(\Omega)/H_0$, where $f(\Omega)$ is less than or equal to 1. For the critical model, $\Omega = 1$, $f(\Omega) = 2/3$ and only for $\Omega = 1$, the Milne model, is $f(\Omega) = 0$. Thus, by measuring the ages of the oldest objects we can find in the Universe, we can find limits to both $H_0$ and $\Omega$. The best example of this approach is the beautiful work of Hesser and his colleagues in estimating the age of the globular cluster 47-Tucanae. The method involves measuring very precise colours and magnitudes for large numbers of stars in the cluster and then fitting the distribution of points in the resulting Hertzsprung–Russell diagram by models of the evolution of stars in old clusters. The type of analysis which can now be undertaken is illustrated in Fig. 10. According to Hesser *et al.*

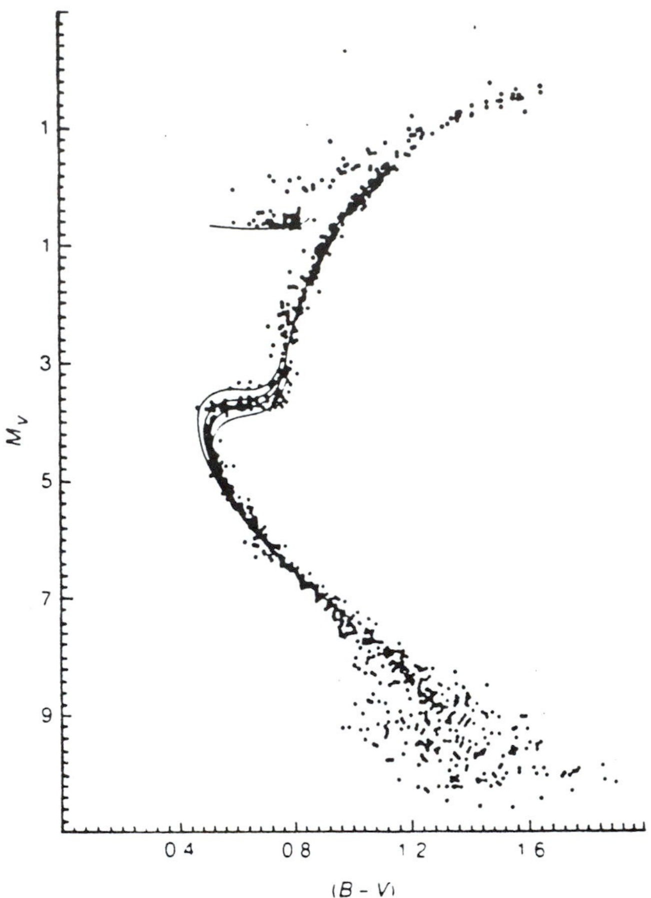


**Fig. 10** The H–R diagram for the globular cluster 47-Tucanae. The scatter in the points increases towards faint magnitudes because of the increase in observational error associated with the photometry of faint stars. The solid lines show best fits to the data using theoretical models for the evolution of stars from the main sequence onto the giant branch due to VandenBerg. For this cluster, the best-fit isochrones have ages between about 12 and 14 × $10^9$ years and the cluster is metal-rich relative to other globular clusters, the metal abundance corresponding to about 20% of the solar value. (From Hesser, Harris, VandenBerg, Allwright and Stetson (1989).)

(1989), the age of the globular cluster is probably about (12–14) × $10^9$ years old. If $\Omega$ were equal to 1, then $H_0$ would have to be less than or equal to 50 $\text{km s}^{-1}\,\text{Mpc}^{-1}$. In my opinion, this type of constraint upon $H_0$ is just as important as the traditional route through the calibration of the Hubble diagram.

There is clearly still about a factor of 2 uncertainty in the value of Hubble's constant. My instincts are to adopt a low value for Hubble's constant, but that is a working hypothesis rather than an established cosmological fact.

## 4.2 The deceleration parameter $q_0$

Effectively, in this test, we attempt to measure the rate of change of Hubble's constant as the Universe grows older. We therefore have to make observations of the Universe when it was significantly younger than it is now, that is, when the scale factor $R \sim 0.5$. It is a general property of the homogeneous, isotropic world models that the scale factor is directly related to the redshift of the galaxy $z$, through the relation $R = (1 + z)^{-1}$, if we normalise the scale factor so that it takes the value 1 at the present epoch. Thus, we have to measure galaxies with redshifts $z \sim 0.5$—1 in order to measure the deceleration of the Universe. The effects of the deceleration are reflected in variations in the properties of identical objects as they are observed at different redshifts. The method which has proved to be the most promising is to identify a class of galaxy with more or less standard properties and to find out how the observed intensity changes with increasing redshift. The problem is to find the classes of galaxy which can be observed at large redshifts and which have a narrow dispersion in their intrinsic properties.

Unfortunately, the options are quite limited. The quasars extend to very large redshifts, almost up to a redshift of 5, but they have a very wide dispersion in their intrinsic luminosities. Even worse, it appears that their mean luminosities have changed with cosmic epoch and so they are not very useful as standard candles. The brightest galaxies in clusters are probably the best standard extragalactic systems we have but at present the samples of clusters extend only out to redshifts of about 0.5, which is scarcely far enough to measure the deceleration parameter accurately without very large samples of clusters at that redshift which are not yet available (Fig. 1). Probably the best samples at the moment are the radio galaxies which are almost as good standard candles as the brightest galaxies in clusters but which can be observed out to redshifts of about 3. There is, however, a fundamental problem and that is that we have to take account of the evolution of the properties of the galaxies with cosmic epoch.

When galaxies are observed at large redshifts, they are observed at significantly earlier stages in their evolution. For example, in the critical world model ($\Omega = 1$, $q_0 = 0.5$), the relation between cosmic time, the scale factor and redshift is very simple:

$$R = \frac{1}{(1+z)} = \left(\frac{t}{t_0}\right)^{2/3},$$

where $t_0$ is the age of the Universe which for this model is $t_0 = (2/3)H_0^{-1}$. Thus, for the critical model, the Universe was only 35% of its present age at a redshift of 1, 19% of its present age at a redshift of 2 and so on. It is evident that we cannot assume that the galaxies have remained unchanged over such

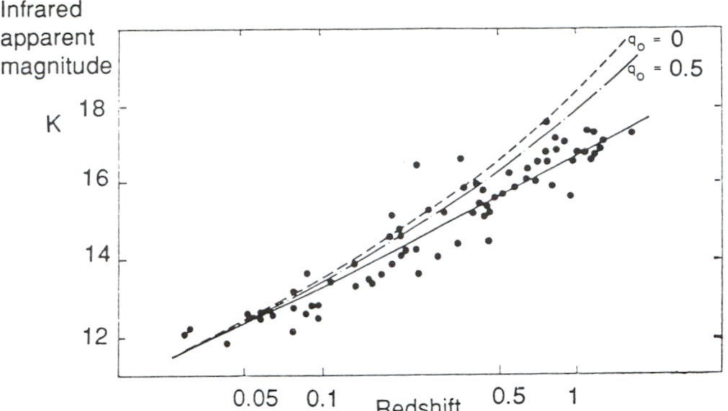


**Fig. 11** The infrared redshift–magnitude relation for a complete sample of radio galaxies selected from radio sources in the revised 3C catalogue. The dashed lines show the expectations of world models having $q_0 = 0$ and 0.5. The solid line shows the expected relation when account is taken of the evolution of the stellar populations of the radio galaxies. (Lilly and Longair 1984)

time-scales. Indeed, when we look at the redshift–magnitude relation for the radio galaxies, there is clear evidence that they were considerably brighter in the past. In Fig. 11, the redshift–magnitude relation for a complete sample of bright radio galaxies extending out to redshifts of almost 2 is compared with the expectations of the standard world models. One of the important aspects of this diagram is that the observed intensities are infrared magnitudes measured at 2.2 μm rather than optical magnitudes. This has many advantages over working in the optical waveband, principally because of the absence of the obscuring effect of dust and because the evolutionary effects are much easier to estimate that in the optical waveband. Recalling that, when intensities are measured in magnitudes, the greater the apparent magnitude, the fainter the object, it can be seen that the galaxies at redshifts about 1 are brighter than is expected for standard world models for which $q_0 = 0$ and $q_0 = 0.5$.

Simon Lilly and I interpreted these results as indicating that the radio galaxies at a redshift of about 1 were brighter by about 1 magnitude as compared with their luminosities at the present epoch. It turns out that this is almost exactly the change in luminosity expected if the galaxies at redshifts of 1 have the same numbers of stars as similar objects nearby but that the stellar population was only half its present age or less. The galaxies were brighter in the past than they are now because there were more bright stars populating the giant branch at that time. It turns out that there are great advantages in carrying out this type of analysis in the near-infrared waveband because the evolution corrections are remarkably model-independent (Lilly and Longair 1984). The upshot of these studies is that world models with $\Omega \sim 0$–1, $q_0 \sim 0$–1 are consistent with the redshift–magnitude relation for these galaxies once the corrections for the effects of the stellar evolution of the populations of the galaxies are taken into account.

Thus, in this case, there is strong evidence for the evolution of the properties of the radio galaxies with cosmic epoch. There is, however, much more evidence for changes in the populations of various classes of object with cosmic epoch. The half fact which Peter Scheuer told me about in 1963 has now become part of what I will call Fact 6.

**Fact 6 Many different classes of extragalactic systems show changes in their average properties with cosmic epoch.**
Let me summarise some of the evidence for these changes.

1. The counts of galaxies to faint optical magnitudes show an excess of faint blue galaxies. The natural interpretation of these results is that there were more blue galaxies in the past than there are now. The problem with this interpretation is that, when the redshifts of the galaxies responsible for the excess are measured, they do not seem to be any more distant than would be expected if there were no blue excess. In other words, there simply seem to be more blue galaxies at redshifts of about 0.5 than there are at the present epoch (Ellis 1992).
2. The populations of extragalactic radio sources and radio quasars show very strong evolutionary changes with cosmic epoch. The simplest way of accounting for these changes quantitatively is to assume that the luminosities of all the quasars were on average about an order of magnitude greater at redshifts $z \sim 2$–3 than they are at the present day. According to Peacock (1992), these evolutionary changes can be described by functions of the following form:

$$L(z) = L(0)(1+z)^3 \qquad \text{for } 0 < z < 2$$
$$L(z) = \text{constant} = 27L(0) \qquad \text{for } z > 2,$$

   where $L(0)$ is the luminosity of the quasar at zero redshift.
3. Optically selected samples of quasars show similar evolutionary changes with cosmic epoch to those of the radio quasars and radio galaxies (Boyle *et al.* 1987). Unfortunately, no one knows exactly what the significance of this result is. Apparently, the epoch when the quasars were at their most active is not the present epoch but the time when the Universe was about 20% of its present age. The behaviour of the population at even larger redshifts is not understood but the increase in luminosity does not continue beyond redshifts of about 2–3. The statistics are limited and it is not certain whether or not the population overall decreases at the largest redshifts. The problem is that there are fewer and fewer quasars at very large redshifts and it requires a huge effort to find the largest redshift quasars.

Nowadays, however, there are about 20 quasars known with redshifts greater than 4 and so eventually we may know the answer, but it is a very long and difficult task.

4. The physical sizes of double radio sources are smaller at redshifts $z \sim 1$ than they are at the present epoch, $z \sim 0$.
5. In the spectra of distant quasars, absorption lines are observed associated with intergalactic clouds along the line of sight to the quasar. Studies of the space distribution of these clouds with cosmic epoch have shown that the clouds which show only hydrogen absorption lines increase in number with increasing redshift, while those systems which possess metals increase in number as the redshift decreases. Again, the cause of these changes is not understood but the reality of the changes is clearly demonstrated.

In one way or another, these observations provide information about the sequence of events which must have taken place as the Universe evolved to its present state. Unfortunately, the evidence is still too sparse to build these separate pieces of evidence into a convincing self-consistent picture. Part of the trouble is that there are only limited classes of objects which can be observed at large redshifts. It is still an impossible task to study normal galaxies at the same distances as the most distant quasars and radio galaxies. Optimistically, we may be able to extend these studies for normal galaxies to redshifts of about 1 with the next generation of 10-m class telescopes and this is an essential step before we are able to tackle the evolution of normal systems at the very largest redshifts.

## 4.3 The density parameter $\Omega$

It might seem that we should be able to make better estimates of the density parameter $\Omega$ since we only have to count up all the galaxies in the Universe and work out the total amount of mass associated with them. If we do this calculation, it is found that the average density of matter in the Universe corresponds to only about 1% of the critical density. We know, however, that this is a serious underestimate of the total amount of mass present. If we make observations of the outer regions of giant spiral and elliptical galaxies, it is found that there must be about ten times as much mass present as would be inferred from the optical light. The reason is that with increasing distance from the centre of a galaxy, the light falls off much more rapidly than the mass so that the visible galaxies are surrounded by dark haloes. The same problem is found in clusters of galaxies. The total mass of clusters of galaxies can be estimated from their internal velocity dispersions and this exceeds by a factor of about 10–20 the mass which would be inferred from the optical light alone. These are aspects of the famous dark matter problem and it is so pervasive in extragalactic astronomy that I will elevate it to the status of a fact.

**Fact 7 Most of the mass of the Universe is in some dark form and it exceeds the amount of visible matter by a factor of at least ten.**
It will be noted that I have been careful not to make any statement about:

(a) How much there actually is?

(b) What is it?

We only possess firm lower limits to the amount of dark matter present. It is likely that there is at least 10–20 times the amount of visible matter, but a key issue is whether or not there could be about a factor of 100 times more matter so that the Universe approaches its critical density. In recent work, it is claimed that, when the Universe is observed on scales much greater than clusters of galaxies, more dark matter is found and that the density parameter approaches 1. It is somewhat disturbing that, if this really is the case, most of the dark matter must reside in the regions in which it is most difficult to observe it—namely, in the regions between clusters of galaxies. There remains a great deal to be done to establish precisely how much dark matter there really is. A personal concern I have is that, in their enthusiasm for the critical model of the Universe which is one of the predictions of the inflationary model of the early Universe, some of the interpreters have been more than anxious to show that our Universe does indeed have the critical density.

If estimating the total density of the Universe is difficult, determining what the dark matter is turns out to be even more uncertain. The basic problem is that, if some constituent of the Universe does not emit or absorb much radiation, it is very difficult to detect and so it is remarkably easy to hide dark matter. The example which I like to quote is that, if the critical density were made up of standard bricks, we would not know about it from any observation because they would not emit detectable radiation and would not absorb background radiation either.

One form which I am sure must exist is what might be called *ordinary (or baryonic) dark matter* in the form of cold dust, rocks, cool solid bodies, all the way up to planet-sized objects and brown dwarfs. Objects such as brown dwarfs are expected to be cool but none of them has yet been definitively found in infrared surveys despite intensive efforts. I am sure these searches must continue because this issue is crucial for cosmology. Another form of dark matter consists of the black holes, either massive black holes, solar mass black holes or mini-black holes. One very beautiful idea, which is the subject of a very large survey at the present time, is to search for dark matter candidates by searching for the rare gravitational lensing effects expected when a black hole, brown dwarf or planet passes in front of a background star. When this occurs, there is a characteristic brightening and dimming of the light from the star and many of the properties of the intervening dark object can be found from the characteristic signature of the changing brightness of the star. This is a very demanding survey because lensing events are very rare but they must occur at some level if the dark matter in our Galaxy is in some discrete form (Alcock 1992).

The forms of dark matter which have caused the greatest excitement in the theoretical community are ultraweakly interacting particles, as yet unknown to science. These are predicted to exist according to different versions of those theories which seek to unify the forces of nature—grand unified theories, supersymmetry, superstring theories and so on. Examples of these types of hypothetical particle include photinos, gravitinos, axions and so on. None of these has yet been observed in laboratory experiments. According to current speculation, the early Universe was hot enough for these particles to be produced and so the particle physicists invert the whole process and use the very early Universe as a laboratory within which to test theories of elementary processes. The methodological problem is that cosmology and particle physics then boot-strap their way to a self-consistent solution and there is no independent way of constraining the theories by observation or experiment.

One important possibility is that the least massive of these hypothetical particles is stable and so might well be present in the Universe now as relics of the very hot early phases. Any such particles would have to be rather massive. It is now possible to search for these dark matter particles by laboratory searches. The idea is that, if the dark matter in our own Galaxy were made up of these exotic particles, we know roughly what their velocities would have to be because they have to form a bound self-gravitating halo about the Galaxy. Therefore, a considerable number of them would be passing through terrestrial laboratories each second. The new generation of very low temperature crystal detectors can detect the very rare collisions between one of these particles and the crystal lattice resulting in a tiny but measurable temperature increase in the crystal. Such experiments are now underway in a number of countries and they should enable constraints to be placed upon a wide range of possible candidates for the exotic dark matter. The optimists argue that this approach is no different from that of Newton and Einstein, in that astronomical discoveries and problems result in new physical concepts and suggest experiments which lead to the discovery of phenomena which could not be detected by purely laboratory experiments. We shall have to wait and see. I will return to this interface between particle physics and cosmology later.

What can we say about the value of the density parameter $\Omega$? I believe it is best to treat the statement in three parts. It is certain that $\Omega$ is greater than about 0.01 because that corresponds to the amount of visible matter present in the Universe. Probably $\Omega$ is at least 0.1 because of the presence of the dark haloes about giant galaxies and the dark matter in clusters of galaxies. Possibly $\Omega$ is about 1 from the peculiar velocities of galaxies on a large scale but, in my view, this is not established with any real certainty.

### 4.4 The comparison of $\Omega$ and $q_0$

As for our comparison of the deceleration parameter $q_0$ and the density parameter $\Omega$, it is probable that they are equal within a factor of about 10 although I believe the measurement of $q_0$ is quite uncertain in the range 0 to 1. If this is the case, some cosmologists argue that they are very likely to satisfy the equality $q_0 = \Omega/2$, but we would really like to know this with much more certainty from improved measurements of both $\Omega$ and $q_0$.

Another way of looking at these numbers is to note that $\Omega$ is probably within a factor of 10 of the critical value $\Omega = 1$. The cosmologist can argue that this surely cannot be a coincidence. There is nothing in the standard model of the Universe which determines what the value of $\Omega$ should be but there is one important aspect of the evolution of the standard models which leads to one of the basic problems of cosmology. It can be shown very easily that, if the Universe has a density which is different from the critical world model at any epoch, then the value of $\Omega$ diverges rapidly so that after a long enough time, the Universe would have density parameter either very much greater or very much less than unity. Since the Universe is within a factor of ten of the critical model now, this means that the Universe must have been very close indeed to the critical density in the very distant past. The cosmologist argues that, since it is very unlikely that the Universe was set up with $\Omega$ just infinitesimally different from 1 in the very early Universe, the only stable value for the density parameter is exactly one. To many theorists, this is a very persuasive argument and they take the point of view that there is no other value which $\Omega$ can take other than 1. It is not clear to me how strongly this argument has influenced theorists, either consciously or unconsciously, in their analyses of cosmological problems.

## 5 The hot early universe

Despite the problems of determining the cosmological parameters described in the last section, we can have considerable confidence that the basic physical picture is correct. The case for the essential correctness of the Big Bang comes from the study of the early evolution of these models. It turns out that radiation was the dominant form of 'mass' which determined the early dynamics of the Universe—the Universe was radiation-dominated and the

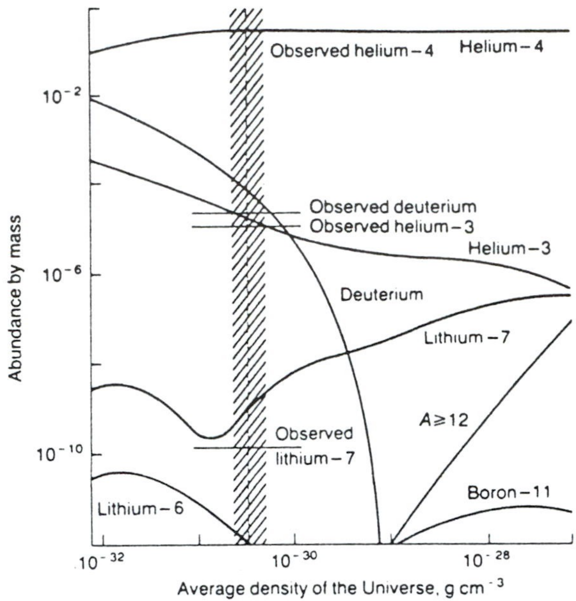


**Fig. 12** The predicted abundances of the light elements synthesised in the standard Big Bang model of the early Universe. In the standard model, it is assumed that the lepton number is zero and that the model evolves from an equilibrium state at a temperature of about $10^{12}$ K. The predicted abundances depend only upon the mean density in ordinary matter now. The abundance of deuterium is a sensitive measure of the density of the Universe now since, if the mean density is high, all the synthesised deuterium is converted into helium. (After Audouze (1982))

Cosmic Background Radiation we observe today is the cooled remnant of these early stages. As a consequence, we can work out rather precisely the early dynamical evolution of the Universe, in fact, with greater accuracy than we can work out its present dynamics. One of the remarkable results of these studies is that, as the Universe cooled down through temperatures of about $10^8$–$10^9$ K, the light elements helium, deuterium and lithium were synthesised from the primordial plasma which at that stage consisted mostly of protons and neutrons. The results are dramatic. It turns out that we can account for the observed abundances of deuterium, lithium and the isotopes of helium for a single value of the ordinary matter density in the Universe now (Fig. 12). This is a great triumph for the Big Bang picture since it has been a great astrophysical problem to account for the observed abundances of these elements by nucleosynthesis in stars. These are very fragile elements and they are destroyed rather than synthesised in stars.

We obtain two essentially independent pieces of information about the dynamics of the Universe during the period when the light elements were synthesised. The amount of helium produced is remarkably independent of the density of matter at that time, because the ratio of the numbers of neutrons to protons depends only upon the temperature of the Universe when the neutrinos decoupled from the reactions which maintained them in thermal equilibrium with the neutrons and protons. Subsequently, almost all the neutrons combine with protons to create helium nuclei. Thus, the helium abundance is essentially a thermometer for the early Universe. This fact enables us to set further limits on physical processes in the early Universe. If the early expansion of the Universe were any faster than it is in the standard model, the neutron-to-proton ratio would freeze out at a higher temperature resulting in the over-production of helium. This argument enables us to rule out models in which the gravitational constant was stronger in the past and also excludes the possibility that there are more than three neutrino species. In the latter case, the energy density of the early Universe would have been greater, resulting in a more rapid expansion than is permitted by the observed abundance of helium. The recent experiments at the Large Electron–Positron collider (LEP) at CERN have confirmed that there are only three neutrino species. This same experiment also shows that there cannot be any unknown weakly interacting particles with rest masses less than about 40 GeV, and so the cold dark matter cannot be any neutrino-like particle with rest mass less than about 40 GeV.

The second aspect of these calculations is that the light elements, deuterium and helium-3, are density probes of the early Universe. This is because these elements are by-products of the synthesis of helium from protons and neutrons. In the case of deuterium, the amount produced depends upon the density of the matter content of the Universe. The first step in the synthesis of helium is the formation of deuterium nuclei by combining a proton and a neutron and these deuterons then combine successively with protons and other helium-3 nuclei to form helium. Thus, if there is a high density of protons and neutrons, almost all of the deuterium is converted into helium but, if the density is low, fewer deuterium nuclei are used up in the production of helium-4. Thus, the greater the density, the lower the expected abundance of deuterium.

One of the important results of this study is that it provides an upper limit to the mean density of ordinary matter in the Universe. This turns out to be less than about 10% of the critical density. Now, as discussed above, the theorists have a strong preference for models of the Universe which have exactly the critical density. Therefore, the hard-line inflationist has to assume that most of the mass in the Universe is in some extraordinary form which does not disturb the excellent agreement between the observed abundances of the light elements and the predictions of the standard Big Bang—the matter would have to be in some exotic form of dark matter—black holes, ultraweakly interacting particles, axions, photinos, gravitinos and all the other exotica discussed by the particle physicist. These must not change the early dynamics of the Universe significantly from the canonical picture.

I find the agreement between the observed light element abundances and the predictions of the standard Big Bang picture so impressive that I will elevate these results to the status of Fact 8.

**Fact 8 The light elements, helium, deuterium, helium-3 and possibly lithium, were created primordially.**

It is the combination of the observational Facts 2, 3, 4 and 8 and the success of General Relativity (Fact 5) which gives us such confidence that the Big Bang model can provide an excellent description of the evolution of the global properties of the Universe. It is therefore the best framework within which to study the more difficult problems of astrophysical cosmology. Let me emphasise two points. The first is that the Facts 2, 3, 4 and 8 are independent pieces of observational evidence, all of which find a natural explanation within the context of Big Bang models of the Universe, although which of them is the best description is not uniquely defined.

The second point is that the reason we can be confident about the success of the model is that we do not need to extrapolate the physics beyond what has been tested in the laboratory. Thus, we are still dealing with known physics from the time the Universe was about 1 millisecond, or even less, old to the present epoch. When we extrapolate to earlier times, we rapidly run out of known physics, the maximum energies for which laboratory experiments have been carried out corresponding to about 80 GeV. Thus, it is only safe to regard many of the inferences about the physics of the Universe earlier than about 1 millisecond as speculative. On the other hand, it is a very respectable aspect of theoretical physics to use the early evolution of the Universe as a constraint upon possible theories of elementary processes. In the strict Popperian sense, we can use the class of Big Bang models as a constraint upon physical theories in that, if the latter result in Universes which bear no resemblance to our Universe, they can be discarded. Thus, I believe a healthy approach to speculations about the very early Universe is to regard them as hypotheses which can be disproved, but to be wary of taking them seriously as real physics until there is some form of independent experimental or observational validation.

For me, the real significance of these remarkable developments is that we can ask meaningful questions about the very early evolution of the Universe. How far one is prepared to extrapolate our present understanding of particle physics is, in my view, a matter of taste. What is unquestionable is the fact that, if we accept the essential correctness of the Big Bang picture, there are some basic problems which we cannot avoid.

*Basic Problem 1*

This is often known as the Horizon Problem. As we go further and further back in time, the distance over which information can be communicated gets smaller and smaller. For example, when we observe the Microwave Background Radiation in opposite directions on the sky, we detect radiation from the last scattering surface when the Universe was contracted by a factor of about 1000 compared to its present size. We can work out easily how far a light ray could have travelled since the beginning of the Universe and convert that into an angular scale on the sky. This turns out to correspond to an angle of only a few degrees. This means that there is no way in which regions in opposite directions on the sky could have communicated. In other words, according to the standard picture, it is a puzzle why regions in opposite directions on the sky are so precisely the same. How could the different regions of the Universe know that they had to end up looking the same in all directions?

*Basic problem 2*

This is often referred to as the Flatness Problem and we have already discussed it in Section 4.4. Why is the Universe so close to its critical density $\Omega = 1$ when, *a priori* it could have taken any value at all? Furthermore, all values of $\Omega \neq 1$ are unstable.

*Basic Problem 3*

The third problem is called the Asymmetry Problem. In the Universe now, there are about $10^9$ photons of the Cosmic Microwave Background Radiation for every proton. In the very early Universe, at temperatures greater than about $10^{12}$ K, particle–antiparticle pair production flooded the Universe with protons and antiprotons, neutrons and antineutrons and so on, with one particle–antiparticle pair for each pair of photons. Therefore, when the clocks are run forward, a slight asymmetry in favour of matter as opposed to antimatter has to be built in at the level of one part in $10^9$, or the present observed ratio of photons to particles is not obtained. Why was there this very slight asymmetry in the initial conditions in favour of matter?

*Basic Problem 4*

Finally, what was the origin of the fluctuations from which galaxies and the large-scale structure of the Universe formed? The problem is that density perturbations grow so slowly in the expanding Universe that there have to be some 'seed' fluctuations introduced in the very early Universe to produce the large-scale features of the Universe we see now. Where did these 'seeds' come from? The big problem in understanding the origin of galaxies is to reconcile the quite remarkable smoothness of the Cosmic Background Radiation (Fig. 7) with the gross irregularity in the distribution of galaxies, most vividly portrayed in the three-dimensional picture of their large-scale

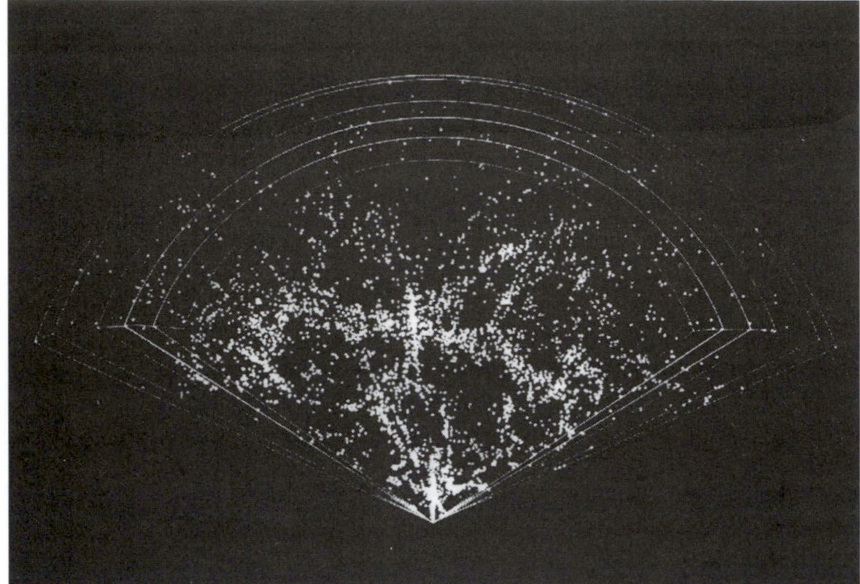

**Fig. 13** The distribution of galaxies in the nearby Universe as derived from the Harvard–Smithsonian Center for Astrophysics survey of galaxies. This picture is a projection of the three-dimensional distribution of galaxies. Our Galaxy is located at the apex of the segments in which the distances of the galaxies have been measured. If the distribution of galaxies were uniform, the points would be distributed uniformly and at random within these segments. In fact, the distribution is grossly non-uniform with huge sheets, filaments and voids in the distribution of galaxies. (Courtesy of Margaret Geller, John Huchra and the Harvard–Smithsonian Center for Astrophysics.)

distribution (Fig. 13). If the distribution of galaxies were uniform, the sectors would be uniformly filled with points. It can be seen that the distribution of galaxies is very non-uniform with connected structures on scales much greater than clusters of galaxies. There are huge two-dimensional sheets of galaxies as well as filaments and great 'voids'. These features have to be explained by any satisfactory theory of galaxy formation.

These large-scale features of our Universe pose a major challenge for theories of galaxy formation. They are unquestionably there and so I designate them Fact 9:

**Fact 9 The distribution of galaxies on large scales in the Universe, although uniform on the cosmological scale, possesses large-scale irregularities on a scale much greater than that of clusters of galaxies.**

Within the framework of the standard Hot Big Bang model of the Universe, the four basic problems require *ad hoc* initial conditions which are introduced arbitrarily in order to 'explain' the large scale properties of the Universe now. Let me suggest five ways of attempting to solve these problems.

1. That is just how the Universe is—the initial conditions were set up that way.
2. There are only certain classes of Universe in which intelligent life can evolve. For example, the fundamental constants of nature should not be too different from the values we observe them to have or else there would be no chance of life ever forming as we know it. This approach is known as the Anthropic Principle and it asserts that the Universe is the way it is because we are here to observe it (see Barrow and Tipler 1987, Gribbin and Rees 1991).
3. Seek clues from particle physics and extrapolate the understanding that has been gained, beyond what has been confirmed by experiments with large accelerators, to the earliest phases of the Universe.
4. Adopt the inflationary scenario for the early Universe.
5. Something else is required which we have not yet thought of. This will certainly involve some new physics.

There is some merit in each of these positions. In approach (1), it might just be too hard a problem to decipher what it was that set up the initial conditions from which our Universe has evolved. How can we possibly check that the physics adopted for the very early Universe is correct? In approach (2), there is certainly truth in the statement that the mere fact that we can ask questions about the origin of the Universe must say something about the sort of Universe we live in. Whilst the Cosmological Principle asserts that we do not live in any special location in the Universe, we are certainly privileged in that we are able to ask the question at all. I do not like this line of reasoning, however, because it means that we could never seek any physical reason for the relations between the fundamental constants of nature. I regard the Anthropic Principle as the last resort if all other physical approaches fail.

The third approach provides many important clues to possible physical solutions to the basic problems, one of them being approach (4), the inflationary scenario for the early Universe. I like to think of the inflationary model in three stages. The *first* is inflation without physics, in the sense that, if the Universe expanded exponentially by an enormous factor in its very early phases, for whatever reason, we are able to eliminate the first and third of the basic problems. This occurs for two reasons. First, if the scale factor of the Universe expanded exponentially in its very earliest phases, regions which were originally very close together are separated exponentially rapidly by the inflationary expansion. Thus, causally connected regions are swept beyond their local horizons by the inflationary expansion. The second effect is that the very rapid expansion has the effect of straightening out the geometry of the early Universe, however complicated it was to begin with. The geometry of the Universe is driven towards flat Euclidean geometry and so, when the inflationary expansion ceases and the Universe transforms over to the 'normal' Universe, the geometry is flat and consequently the Universe must readjust to the critical density. These ideas can be formulated without specific reference to any particular physical realisation of the process of inflation. Ironically, this description is no more than de Sitter's model of the dynamics of an empty Universe, which he developed to show that there can exist solutions of Einstein's field equations including the cosmological constant even if there is no matter present (de Sitter 1917).

In the second stage, if we put in a little bit of physics, similar to that which accounts for the asymmetry of elementary processes in the 'low-temperature limit', specifically the charge asymmetry of $K^0$ decay, we can explain the matter–antimatter asymmetry. In the third stage, if one is much bolder, one uses the best theories we have of elementary particles to identify forces which could cause the exponential expansion of the early Universe. The intriguing discovery has been that processes required by the theories of elementary particles bear a close resemblance to what is needed to produce the exponential expansion of the very early Universe. Specifically, the equation of state at very high energies has to be a negative energy equation of state $p = -\rho c^2$ and this is a property of the scalar fields needed to account for the masses of elementary particles. Similar processes are assumed to have taken place in the very early Universe. All the inflationary action is supposed to take place before the Universe was $10^{-35}$ seconds old at extremely high energies (Guth and Steinhardt 1989).

It will be noted that I have referred to the fourth possibility as the inflationary scenario since it has been designed to solve specifically problems (1) and (3) in the above list. There is certainly no other evidence for the inflationary picture beyond the need to solve these four problems, and the physical realisations of the physics of inflation cannot be tested in the laboratory. I therefore applaud the endeavours of the theorists to give a proper physical basis for the inflationary scenario, but it is not clear how we are to find independent evidence that the physics is along the right lines.

Part of the concern is tied up with the fifth approach—the need for new physics. I have shown schematically a popular representation of the evolution of the Universe from the Planck era, when the Universe was only $10^{-44}$ seconds old, to the present epoch (Fig. 14). Halfway up the diagram, from the time when the Universe was only about a millisecond old, to the present epoch, we can be reasonably confident that we have the correct picture for the Big Bang despite the four basic problems described above. However, it will be seen that, at times earlier than about 1 millisecond, we very quickly run out of known physics. Indeed, the models of the very early Universe suppose that we can extrapolate across that huge gap from $10^{-3}$ s to $10^{-44}$ s

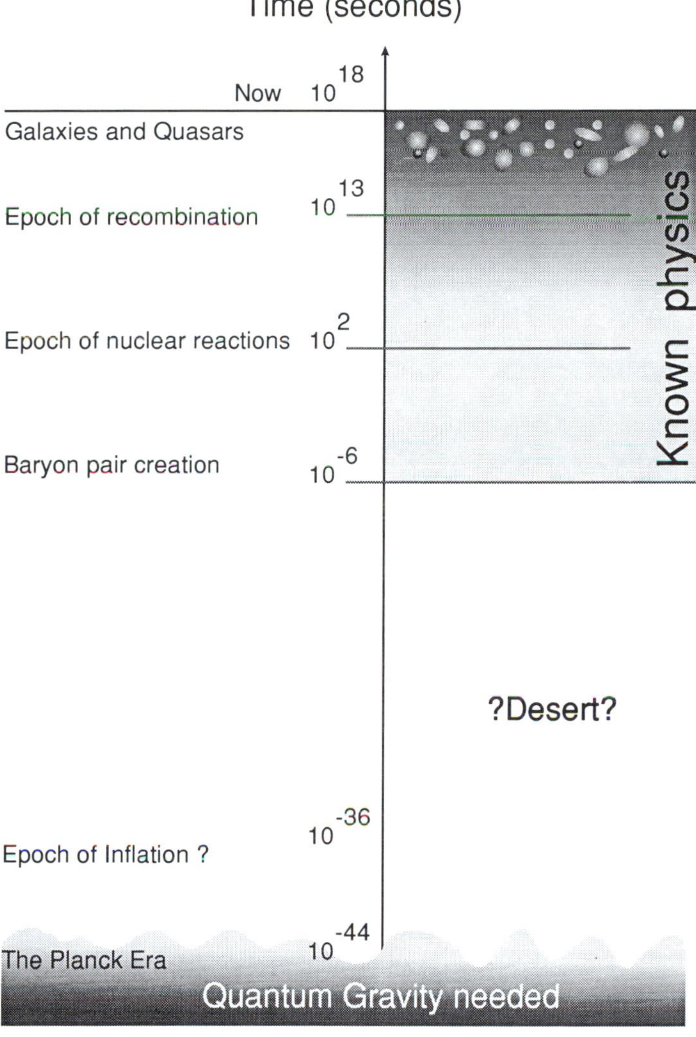


**Fig. 14** A schematic diagram illustrating the evolution of the Universe from the Planck era to the present time. The shaded area to the right of the diagram indicates the regions of known physics.

using our understanding of laboratory physics. The theorists may be correct but one must have some concern that there may be some fundamentally new physics to be understood at higher and higher energies before we reach the Planck era, $t \sim 10^{-44}$ seconds.

The one thing which is certain is that at some stage we will have to understand how to quantise gravity. The singularity theorems of Penrose and Hawking show that, according to classical theories of gravity under very general conditions, there is inevitably a physical singularity at the origin of the Big Bang (see Hawking and Elllis 1973). One of the possible ways of eliminating this problem may be to find a proper quantum theory of gravity. This remains an unsolved problem and we can be certain that our understanding of the very earliest stages will remain seriously incomplete until it is solved. Thus, there is no question but that there is some new physics needed if we are to develop a serious physical picture of the very early Universe.

## 7 Conclusions

I have described 9 facts about the large scale properties of the Universe which I believe are now secure. This represents remarkable progress since 1963 when there were only 2½ facts. In my opinion, the best bits of the story of modern cosmology are those in which laboratory and cosmological physics come together to form a unique and convincing synthesis.

I have also described four major problems of cosmology and it is a matter of speculation, in my opinion, how far we will be able to understand the solution to them in convincing physical terms. I have described 5 ways of approaching the problems, some of which are more physical than others. I will defend to the last the right of theorists to probe more deeply into the structure of the theories of particle physics and quantum gravity, and the hope is that these studies will cast new light on these problems. I would not be surprised if they remained difficult problems for some time.

My own position is that there is so much to be done in consolidating the observational and experimental foundations of the subject that I hope a very major effort can be made to advance the discipline of observational cosmology over the next thirty years. It would be a wonderful thing if the next thirty years were to contribute as many new real facts about the Universe as the last thirty.

## Addendum

In preparing these notes, it is an enormous advantage that my 1992 Milne Lecture, 'Modern cosmology—a critical assessment', appears well down the batting order and so has not dated too badly over the last four years. I have no reason to change any of my broad conclusions about the current state of cosmology—the standard Hot Big Bang is the most convincing framework for astrophysical cosmology, and the great problems remain—but there have been a number of important advances.

## The isotropy of the Cosmic Microwave Background Radiation

Three further years' worth of data have been added to the observations which I illustrated in Fig. 8 and this final data set will be published by the COBE team in 1996. The intensity fluctuations on an angular scale of 10°, first reported by the COBE observers in 1992, have been confirmed by the subsequent years' observations (see Mather (1995), Smoot (1995) for a review of the first two years' data). Estimates have been made of the spectral index of the perturbations associated with these fluctuations, $n = 1.1 \pm 0.3$, and they correspond to the standard scale-free index, $n = 1$, which is preferred by theorists. In addition, there have been positive detections from ground-based observations, some of them on similar angular scales to the COBE observations; for example, the Tenerife experiment by Hancock *et al.* (1994) (see also Lasenby and Hancock (1995)), and others on the crucial angular scale of about 1° (see, for example, White, Scott and Silk (1994), Mather (1995), Partridge (1995)). An angular scale of 1° is of great importance for galaxy formation since it corresponds to the sizes of the huge 'voids' observed in the distribution of visible matter today. The detection of fluctuations in this angular size-range is a sensitive test of different models of the early evolution of the density fluctuations from which galaxies eventually formed.

## The spectrum of the Cosmic Microwave Background Radiation

Just as spectacular as the isotropy of the Cosmic Microwave Background Radiation is the precision with which it is now known that the spectrum of the radiation is of blackbody form (Kogut 1995, Mather 1995). In the wavelength interval $2.5 \geqslant \lambda \geqslant 0.5$ mm, which includes the maximum of the background spectrum, the intensity spectrum is now known to be of precisely blackbody form at a temperature of 2.725 K with a precision of 0.03% of its maximum intensity. This provides very powerful limits to the amout of energy which could have been dissipated in intergalactic space at any time from a redshift of about $10^6$ to the present epoch. For example, it provides direct evidence that the Universe cannot be filled with hot gas which might account for the intensity of the hard X-ray background radiation. It can also be used to set limits to the amount of star formation which could have taken place in distant star-forming galaxies, since these are known to be intense sources of sub-millimetre and far-infrared emission (Blain and Longair 1993 *a* and *b*).

## The value of Hubble's constant and the age of the oldest stars

One of the most important developments for classical cosmology over the last 10 years has been that the distances of nearby galaxies have become much better known. Among the best distance measures are those derived using the period–luminosity relation for Cepheid variable stars as observed in galaxies out to about 20 Mpc by the Hubble Space Telescope. Good examples of the precision now possible are provided by the papers by Freedman and her colleagues (1994) and by Tanvir and his colleagues (1995). They find values of Hubble's constant of $(80 \pm 17)\ \mathrm{km\,s^{-1}\,Mpc^{-1}}$ and $(69 \pm 8)\ \mathrm{km\,s^{-1}\,Mpc^{-1}}$ respectively. The importance of these recent determinations is that the error analyses have been carried out very carefully. Notice that these are $1\sigma$ errors. There are good prospects that these values will be improved over the lifetime of the HST. At the same time, estimates of the ages of the oldest stars in our Galaxy remain about $(16 \pm 2)$ Gyr (Maeder 1994). For comparison, Sandage (1995) estimates that the ages of the oldest globular clusters are about $(14 \pm 1.5)$ Gyr. As a result, there is a potential discrepancy between the values of Hubble's constant and the age of the Universe, if $\Omega = 1$, at about the $(1 - 2)\sigma$ level. If this discrepancy were to persist in the light of improved estimates of these cosmological parameters, it might be necessary to reintroduce the cosmological constant $\Lambda$.

## The Density Parameter $\Omega$

There has been a great deal of effort in the area of bulk streaming motions of galaxies, the objective being to estimate the total amount of gravitating matter present in the Universe which drives these motions. A good example of this procedure is discussed in Dekel's review (1994) in which he describes the results of the POTENT programme which attempts to derive in a self-consistent way the local density and velocity field for galaxies. Dekel's analysis suggests that the density parameter is close to unity but other estimates from the typical mass-to-luminosity ratios of galaxies in clusters have suggested lower values. This issue is far from closed. Another aspect of the dark matter problem concerns the issue of whether or not the dark halo of our Galaxy can be bound by massive compact halo objects. The MACHO project has indeed found a number of gravitational lensing events associated with the passage of dark halo objects in front of background stars, but there have not been enough of them to account for the total amountof mass needed to bind our Galaxy in the from of dark objects with masses in the range $10^{-7} < M < 0.1\ M_{\odot}$ (see, for example, Evans 1996). It is inferred that the dark halo of our Galaxy may be in some other, quite possibly non-baryonic, form.

## Observations of distant galaxies

Since writing the 1992 review, there have been concerted efforts to understand the nature of the excess of faint blue galaxies. Deep counts by the

Hubble Space Telescope have shown convincingly that the excess of blue galaxies is primarily associated with interacting/irregular/merging systems (Glazebrook *et al.* 1995*b*, Abrahams *et al.* 1996). At first, it seemed that the excess blue galaxies were mostly dwarf galaxies (Glazebrook *et al.* 1995*a*), but, as deeper surveys have been carried out, it is becoming clear that among the fainter samples of galaxies, there are large redshift blue galaxies. For example, in the deep Canada–France Redshift Survey, the blue galaxies have redshifts between about 0.4 and 1.2 (Schade *et al.* 1995). The deep surveys carried out by Cowie and his colleagues (1995) have shown that, in their deep fields, the redshifts of the galaxies extend beyond 1 and that the star formation rates are about an order of magnitude greater than those at redshifts $z \sim 0.5$. The first result of the surveys carried out in the Keck Deep Survey Fields have found a similar result, namely that the mean redshift of the galaxies studied continues to increase as fainter and fainter samples are studied. In addition to star-forming galaxies, there also seem to be present in these samples galaxies which are as red as elliptical galaxies at the present day—they must have formed their old stellar populations at reshifts $z > 2$ (Koo 1996). These studies provide a flavour of the types of astrophysical cosmological studies which are becoming possible with the coming generation of 8-metre optical-infrared telescopes.

Searches for star-forming galaxies at large redshifts have been undertaken by searching for the characteristic signature of a region of star formation which has a flat spectrum with an abrupt cut-off at the Lyman limit. This procedure, pioneered by Steidel and Hamilton (1992), has been used by Macchetto and Giavalisco (1995) to find a number of star-forming candidates at redshift $z \sim 3.2$. They suggest that these objects are foming stars at such a great rate that the bulk of the stars in these galxies could be formed at that epoch. Early formation of the metals in galaxies is also suggested by studies of the abundances of the heavy elements in the absorption lines systems observed in Lyman-$\alpha$ forest in large redshift quasars (Pettini *et al.* 1994). The sum of these observations suggests that a considerable amount of star and metal formation must have occurred at redshifts about 3 to 4. It may not be a coincidence that this is also the redshift at which the maximum of quasar activity is also found from studies of the evolution of the luminosity functions of quasars and active galaxies with cosmic epoch.

## References

Alcock, C. (1993). In *Sky surveys: protostars to protogalaxies*, (ed. T. Soifer), p. 29. ASP conference series.

Audouze, J. (1982). In *Astrophysical cosmology*, (eds. H. A. Bruck, G. V.

Coyne and M. S. Longair), 409. Vatican City: Pontifica Academia, Scientiarum Scripta Varia.

Barrow, J. D. and Tipler, F. J. (1987). *The anthropic cosmological principle*. Oxford: Clarendon Press.

Bondi, H. (1952). *Cosmology*. Cambridge University Press.

Boyle, B. J., Shanks, T., Fong, R. and Peterson, B. A. (1987). *Observational cosmology* (eds. A. Hewett, G. R. Burbidge and L. Z. Fang), IAU Symposium No. 124, 643. Dordrecht: D. Reidel Publishing Co.

Branch, D. (1988). *The extragalactic distance scale* (eds. S. van den Bergh and C. J. Pritchet), 146. San Francisco: Astron. Soc. Pacific.

de Sitter, W. (1917). *Mon. Not. R. astr. Soc.*, **78**, 3.

Einstein, A. (1917). *Sitz. Preuss. Akad. d. Wiss*, **142**.

Ellis, R. S. (1992). In *Sky surveys: protostars to protogalaxies*, (ed. T. Soifer), p. 165. ASP conference series.

Gamow, G. (1970). *My world line*. New York: Viking Press.

Gribbin, J. and Rees, M. J. (1991). *Cosmic coincidences: dark matter, mankind and anthropic cosmology*. London: Black Swan.

Guth, A. H. and Steinhardt, P. J. (1989). *The new physics* (ed. P. C. W. Davies), p. 34. Cambridge University Press.

Hawking, S. W. and Ellis, G. F. R. (1973). *The large scale structure of space–time*. Cambridge University Press.

Hesser, J. E., Harris, W. E., VandenBerg, D. A., Allwright, J. W. B., Schott, P. and Stetson, P. (1989). *Publ. Astron. Soc. Pacific*, **99**, 739.

Lilly, S. J. and Longair, M. S. (1984). *Mon. Not. R. astr. Soc.*, **211**, 833.

Longair, M. S. (1989). In *The new physics*, (ed. P. C. W. Davies), p. 94. Cambridge University Press.

Longair, M. S. (1992). *Theoretical concepts in physics*, p. 345. Cambridge University Press.

Mather, J. C., Cheng, E. S., Eplee, R. E. Jr., Isaacman, R. B., Meyer, S. S., Shafer, R. A., Weiss, R., Wright, E. L., Bennett, C. L., Boggess, N. W., Dwek, E., Gulkis, S., Hauser, M. G., Janssen, M., Kelsall, T., Lubin, P. M., Moseley, S. H. Jr., Murdock, T. L., Silverberg, R. F., Smoot, G. F. and Wilkinson, D. T. (1990). *Astrophys. J.*, **354**, L37.

Milne, A. E. and McCrea, W. H. (1934). *Q.J. Math.*, **5**, 73.

Peacock, J. A. (1994). In *The nature of compact objects in active galactic nuclei*, (eds A. Robinson and A. Terlevich) NATO Advanced Study Institute, p. 101. Cambridge University Press.

Sandage, A. R. (1968). *Observatory*, **88**, 99.

Seldner, M., Siebars, B., Groth, E. J. and Peebles, P. J. E. (1977). *Astron. J.*, **82**, 249.

Smoot, G. F., Bennett, C. L., Kogut, A., Wright, E. L., Aymon, J., Boggess, N. W., Cheng, E. S., DeAmici, G., Gulkis, S., Hauser, M. G., Hinshaw, G.,

Lineweaver, C., Loewenstein, K., Jackson, P. D., Janssen, M., Kaita, E., Kelsall, T., Keegstra, P., Lubin, P., Mather, J. C., Meyer, S. S., Moseley, S. H., Murdock, T. L., Rokke, L., Silverberg, R. F., Tenorio, L., Weiss, R. and Wilkinson, D. T. (1992). *Astrophys. J. Letts*, **396**, L1.

## References (Addendum)

Abrahams, R. G., van den Bergh, S., Glazebrook, K., Ellis, R. S., Santiago, B. X., Surma, P. and Griffiths, R. E. (1996). *Mon. Not. R. Astron. Soc.*, in press.

Blain, A. W. and Longair, M. S. (1993*a*). *Mon. Not. R. Astron. Soc.*, **264**, 509.

Blain, A. W. and Longair, M. S. (1993*b*). *Mon. Not. R. Astron. Soc.*, **265**, L21.

Cowie, L. L., Hu, E. M. and Songaila, A. (1995). *Nature*, **337**, 603.

Dekel, A. (1994). *Ann. Rev. Astron. Astrophys*, **32**, 371.

Evans, N. W. (1996). *Mon. Not. R. Astron. Soc.*, **278**, L5.

Freedman, W. L., Madore, B. F., Mould, J. R., Hill, R., Ferrarese, L., Kennicutt, R. C. Jr., Saha, A., Stetson, P. B., Graham, J. A., Ford, H., Hoessel, J. G., Huchra, J., Hughes, S. M. and Illingworth, G. D. (1994). *Nature*, **371**, 757.

Glazebrook, K., Ellis, R., Colless, M., Broadhurst, T., Allington-Smith, J. and Tanvir, N. (1995*a*). *Mon. Not. R. Astron. Soc.*, **273**, 157.

Glazebrook, K., Ellis, R., Santiago, B. and Griffiths, R. (1995*b*). *Mon. Not. R. Astron. Soc.*, **275**, L 19.

Hancock, S., Davies, R. D., Lasenby, A. N., Gutiérrez de la Cruz, C. M., Watson, R. A., Rebolo, R. and Beckman, J. E. (1994) *Nature*, **367**, 333.

Kogut, A. (1995). In *Current topics in astrofundamental physics: the early universe* (eds. N. Sánchez and A. Zichichi), p. 277. Dordrecht: Kluwer Academic Publishers.

Koo, D. (1996). In *HST2*, Proceedings of the Second European Conference on the Hubble Space Telescope (eds. P. Benvenuti, D. Macchetto and E. Schreier), in press.

Lasenby, A. and Hancock, S. (1995). In *Current topics in astrofundamental physics: the early universe* (eds. N. Sánchez and A. Zichichi), p. 327. Dordrecht: Kluwer Academic Publishers.

Macchetto, D. and Giavalisco, M. (1995). *ESO Messenger*, No. 81, September 1995, p. 14.

Mather, J. (1995). In *Current topics in astrofundamental physics: the early universe* (eds. N. Sánchez and A. Zichichi), p. 257. Dordrecht: Kluwer Academic Publishers.

Maeder, A. (1994). In *Frontiers of space and ground-based astronomy* (eds.

W. Wamsteker, M. S. Longair and Y Kondo), p. 177. Dordrecht: Kluwer Academic Publishers.

Partridge (1995). In *Current topics in astrofundamental physics: the early universe* (eds. N. Sánchez and A. Zichichi), p. 357. Dordrecht: Kluwer Academic Publishers.

Pettini, M., Smith, L. J., Hunstead, R. W. and King, D. L. (1994). *Astrophys. J.*, **426**, 79.

Sandage, A. R. (1995). In Sandage, A. R., Kron, R. and Longair, M. S. *The deep universe*, vol. 1. Berlin: Springer-Verlag.

Schade, D., Lilly, S. J., Crampton, D., Hammer, F., Le Fèvre, O. and Tresse, L. (1995). *Astrophys. J.*, **451**, L1.

Smoot, G. (1995). In *Current topics in astrofundamental physics: the early universe* (eds. N. Sánchez and A. Zichichi), p. 301. Dordrecht: Kluwer Academic Publishers.

Steidel, C. C. and Hamilton, D. (1992). *Astron. J.*, **104**, 941.

Tanvir, N. R., Shanks, T., Ferguson, H. C. and Robinson, D. R. T. (1995). *Nature*, **377**, 27.

White, M., Scott, D. and Silk, J. (1994). *Ann. Rev. Astron Astrophys*, **32**, 319.

# Binary pulsars and relativistic gravity*

Joseph H. Taylor, Jr.

## 1 Search and discovery

Work leading to the discovery of the first pulsar in a binary system began more than twenty years ago, so it seems reasonable to begin with a bit of history. Pulsars burst onto the scene [1] in February 1968, about a month after I completed my PhD at Harvard University. Having accepted an offer to remain there on a post-doctoral fellowship, I was looking for an interesting new project in radio astronomy. When *Nature* announced the discovery of a strange new rapidly pulsating radio source, I immediately drafted a proposal, together with Harvard colleagues, to observe it with the 92 m radio telescope of the National Radio Astronomy Observatory. By late spring we had detected and studied all four of the pulsars which by then had been discovered by the Cambridge group, and I began thinking about how to find further examples of these fascinating objects, which were already thought likely to be neutron stars. Pulsar signals are generally quite weak, but have some unique characteristics that suggest effective search strategies. Their otherwise noise-like signals are modulated by periodic, impulsive waveforms; as a consequence, dispersive propagation through the interstellar medium makes the narrow pulses appear to sweep rapidly downward in frequency. I devised a computer algorithm for recognizing such periodic, dispersed signals in the inevitable background noise, and in June 1968 we used it to discover the fifth known pulsar [2].

Since pulsar emissions exhibited a wide variety of new and unexpected phenomena, we observers put considerable effort into recording and studying their details and peculiarities. A pulsar model based on strongly magnetized, rapidly spinning neutron stars was soon established as consistent with most of the known facts [3]. The model was strongly supported by the



The substance of this paper was first delivered as the Milne Lecture on 4 November 1993.

discovery of pulsars inside the glowing, gaseous remnants of two supernova explosions, where neutron stars should be created [4, 5], and also by an observed gradual lengthening of pulsar periods [6] and polarization measurements that clearly suggested a rotating source [7]. The electrodynamical properties of a spinning, magnetized neutron star were studied theoretically [8] and shown to be plausibly capable of generating broadband radio noise detectable over interstellar distances. However, the rich diversity of the observed radio pulses suggested magnetospheric complexities far beyond those readily incorporated in theoretical models. Many of us suspected that detailed understanding of the pulsar emission mechanism might be a long time coming—and that, in any case, the details might not turn out to be fundamentally illuminating.

In September 1969, I joined the faculty at the University of Massachusetts, where a small group of us planned to build a large, cheap radio telescope especially for observing pulsars. Our telescope took several years to build, and during this time it became clear that whatever the significance of their magnetospheric physics, pulsars were interesting and potentially important to study for quite different reasons. As the collapsed remnants of supernova explosions, they could provide unique experimental data on the final stages of stellar evolution, as well as an opportunity to study the properties of nuclear matter in bulk. Moreover, many pulsars had been shown to be remarkably stable natural clocks [9], thus providing an alluring challenge to the experimenter, with consequences and applications about which we could only speculate at the time. For such reasons as these, by the summer of 1972 I was devoting a large portion of my research time to the pursuit of accurate timing measurements of known pulsars, using our new telescope in western Massachusetts, and to planning a large-scale pulsar search that would use bigger telescopes at the national facilities.

I suspect it is not unusual for an experiment's motivation to depend, at least in part, on private thoughts quite unrelated to avowed scientific goals. The challenge of a good intellectual puzzle, and the quiet satisfaction of finding a clever solution, must certainly rank highly among my own incentives and rewards. If an experiment seems difficult to do, but plausibly has interesting consequences, one feels compelled to give it a try. Pulsar searching is the perfect example: it's clear that there must be lots of pulsars out there, and once identified, they are not so very hard to observe. But finding each one for the first time is a formidable task, one that can become a sort of detective game. To play the game you invent an efficient way of gathering clues, sorting, and assessing them, hoping to discover the identities and celestial locations of all the guilty parties.

Most of the several dozen pulsars known in early 1972 were discovered by examination of strip-chart records, without benefit of further signal processing. Nevertheless, it was clear that digital computer techniques would be essential parts of more sensitive surveys. Detecting new pulsars is necessarily a multi-dimensional process; in addition to the usual variables of two spatial coordinates, one must also search thoroughly over wide ranges of period and dispersion measure. Our first pulsar survey, in 1968, sought evidence of pulsar signals by computing the discrete Fourier transforms of long sequences of intensity samples, allowing for the expected narrow pulse shapes by summing the amplitudes of a dozen or more harmonically related frequency components. I first described this basic algorithm [10] as part of a discussion of pulsar search techniques, in 1969. An efficient dispersion-compensating algorithm was conceived and implemented soon afterward [11, 12], permitting extension of the method to two dimensions. Computerized searches over period and dispersion measure, using these basic algorithms, have by now accounted for discovery of the vast majority of nearly 600 known pulsars, including forty in binary systems [13, 14].

In addition to private stimuli related to 'the thrill of the chase', my outwardly expressed scientific motivation for planning an extensive pulsar survey in 1972 was a desire to double or triple the number of known pulsars. I had in mind the need for a more solid statistical basis for drawing conclusions about the total number of pulsars in the Galaxy, their spatial distribution, how they fit into the scheme of stellar evolution, and so on. I also realized [15] that it would be highly desirable 'to find even *one* example of a pulsar in a binary system, for measurement of its parameters could yield the pulsar mass, an extremely important number'. Little did I suspect that just such a discovery would be made, or that it would have much greater significance that anyone had foreseen! In addition to its own importance, the binary pulsar PSR 1913 + 16 is now recognized as the harbinger of a new class of unusually short-period pulsars with numerous important applications.

An up-to-date map of known pulsars on the celestial sphere is shown in Fig. 1. The binary pulsar PSR 1913 + 16 is found in a clump of objects close to the Galactic plane around longitude 50°, a part of the sky that passes directly overhead at the latitude of Puerto Rico. Forty of these pulsars, including PSR 1913 + 16, were discovered in the survey that Russell Hulse and I carried out with the 305 m Arecibo telescope [16–18]. Figure 2 illustrates the periods and spin-down rates of known pulsars, with those in binary systems marked by larger circles around the dots. All radio pulsars slow down gradually in their own rest frames, but the slowdown rates vary over nine orders of magnitude. Figure 2 makes it clear that binary pulsars are special in this regard. With few exceptions, they have unusually small values of both period and period derivative—an important fact which helps to make them especially suitable for high-precision timing measurements.

Much of the detailed implementation and execution of our 1973–1974

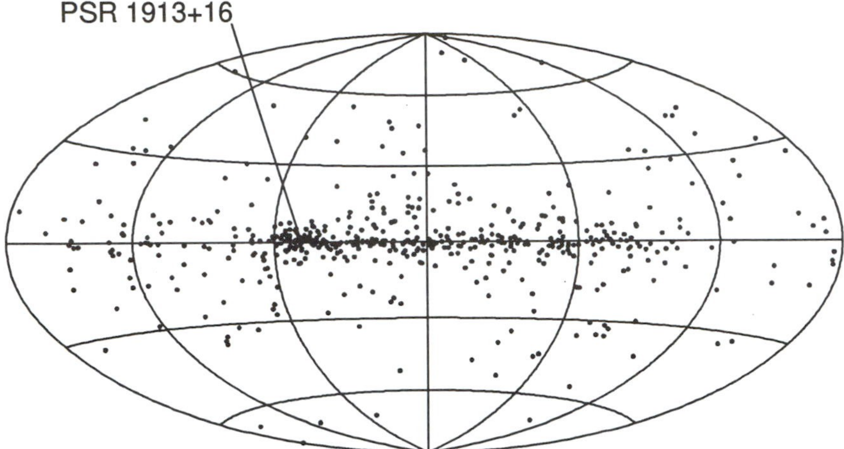


**Fig. 1** Distribution of 558 pulsars in Galactic coordinates. The Galactic center is in the middle, and longitude increases to the left.

Arecibo survey was carried out by Russell Hulse. He describes that work, and particularly the discovery of PSR 1913 + 16, in his accompanying lecture [19]. The significant consequences of our discovery have required accurate timing measurements extending over many years, and since 1974–1976 I have pursued these with a number of other collaborators. I shall now turn to a description of these observations.

## 2 Clock-comparison experiments

Pulsar timing experiments are straightforward in concept: one measures pulse times of arrival (TOAs) at the telescope, and compares them with time kept by a stable reference clock. A remarkable wealth of information about a pulsar's spin, location in space, and orbital motion can be obtained from such simple measurements. For binary pulsars, especially, the task of analyzing a sequence of TOAs often assumes the guise of another intricate detective game. Principal clues in this game are the recorded TOAs. The first and most difficult objective is the assignment of unambiguous pulse numbers to each TOA, despite the fact that some of the observations may be separated by months or even years from their nearest neighbors. During such inevitable gaps in the data, a pulsar may have rotated through as many as $10^7$–$10^{10}$ turns, and in order to extract the maximum information content from the data, these integers must be recovered *exactly*. Fortunately, the correct sequence of pulse numbers is easily recognized, once attained, so you can tell when the game has been 'won'.

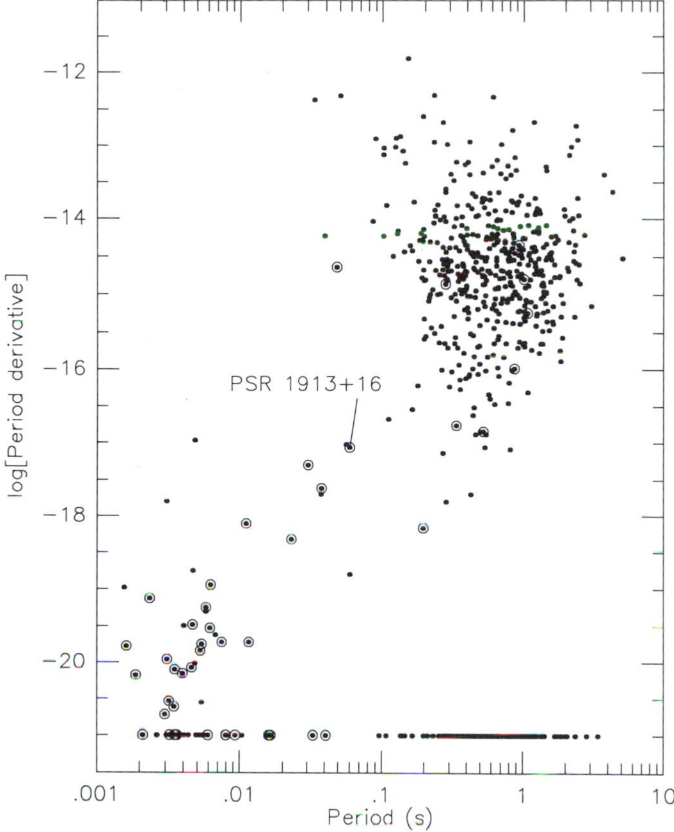


**Fig. 2** Periods and period derivatives of known pulsars. Binary pulsars, denoted by larger circles around the dots, generally have short periods and small derivatives. Symbols aligned near the bottom represent pulsars for which the slow-down rate has not yet been measured.

A block diagram of equipment used for recent pulsar timing observations [20] at Arecibo is shown in Fig. 3. Incoming radio-frequency signals from the antenna are amplified, converted to intermediate frequency, and passed through a multichannel spectrometer equipped with square-law detectors. A bank of digital signal averagers accumulates estimates of a pulsar's periodic waveform in each spectral channel, using a pre-computed digital ephemeris and circuitry synchronized with the observatory's master clock. A programmable synthesizer, its output frequency adjusted once a second in a phase-continuous manner, compensates for changing Doppler shifts caused by accelerations of the pulsar and the telescope. Average profiles are recorded once every few minutes, together with appropriate time tags. A log is kept of small measured offsets (typically of order 1 $\mu$s) between the observatory

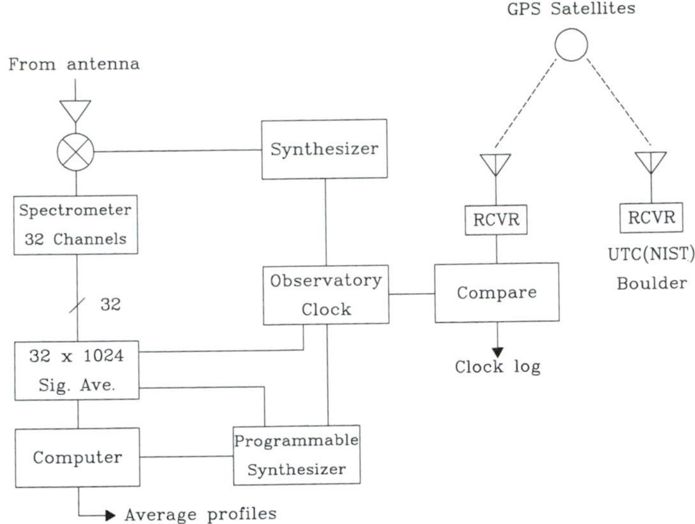


**Fig. 3** Simplified block diagram of equipment used for timing pulsars at Arecibo.

clock and the best available standards at national time-keeping laboratories, with time transfer accomplished via satellites in the Global Positioning System.

An example of pulse profiles recorded during timing observations of PSR 1913 + 16 is presented in Fig. 4, which shows intensity profiles for 32 spectral channels spanning the frequency range 1383–1423 MHz, followed by a 'de-dispersed' profile at the bottom. In a five-minute observation such as this, the signal-to-noise ratio is just high enough for the double-peaked pulse shape of PSR 1913 + 16 to be evident in the individual channels. Pulse arrival times are determined by measuring the phase offset between each observed profile and a long-term average with much higher signal-to-noise ratio. Differential dispersive delays are removed, the adjusted offsets are averaged over all channels, and the resulting mean value is added to the time-tag to obtain an equivalent TOA. Nearly 5 000 such five-minute measurements have been obtained for PSR 1913 + 16 since 1974, using essentially this technique. Through a number of improvements in the data-taking systems [21–26], the typical uncertainties have been reduced from around 300 μs in 1974 to 15–20 μs since 1981.

## 3 Model fitting

In the process of data analysis, each measured topocentric TOA, say $t_{\text{obs}}$, must be transformed to a corresponding proper time of emission $T$ in the pul-

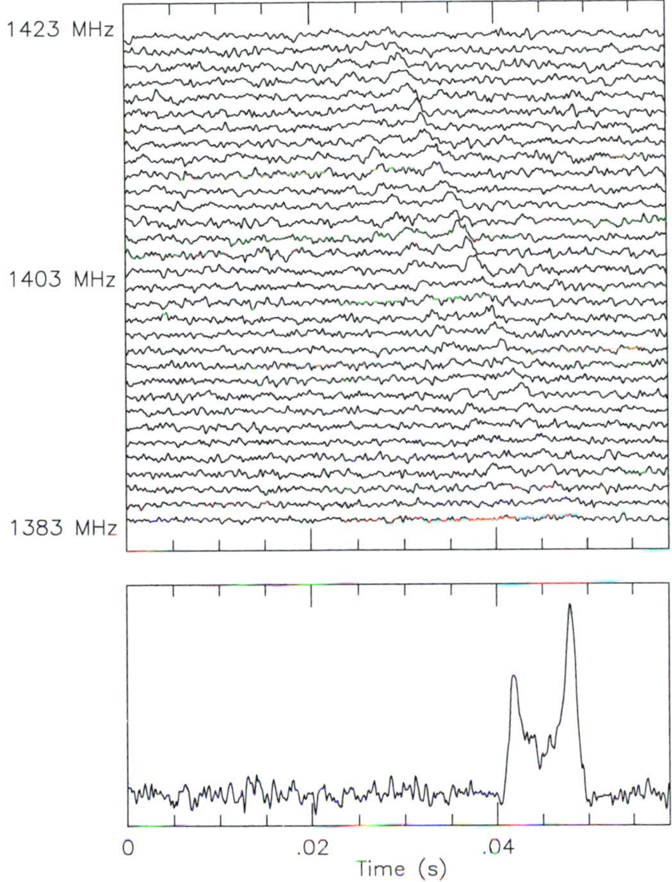


**Fig. 4** Pulse profiles obtained on 24 April 1992 during a five-minute observation of PSR 1913 + 16. The characteristic double-peaked shape, clearly seen in the de-dispersed profile at the bottom, is also discernible in the 32 individual spectral channels.

sar frame. Under the assumption of a deterministic spin-down law, the rotational phase of the pulsar is given by

$$\phi(T) = \nu T + \tfrac{1}{2}\dot{\nu}T^2, \tag{1}$$

where $\phi$ is measured in cycles, $\dot{\nu} \equiv 1/P$ is the rotation frequency, $P$ the period, $\dot{\nu}$ the slowdown rate. Since a topocentric TOA is a relativistic space–time event, it must be transformed as a four-vector. The telescope's location at the time of a measurement is obtained from a numerically integrated solar-system model, together with published data on the Earth's unpredictable rotational variations. As a first step one normally transforms to the solar-system barycenter, using the weak-field, slow-motion limit of general relativity. The necessary equations include terms depending on the positions, velocities, and masses of all significant solar-system bodies. Next, one accounts for propagation effects in the interstellar medium; and finally, for the orbital motion of the pulsar itself.

With presently achievable accuracies, all significant terms in the relativistic transformation can be summarized in the single equation

$$T = t_{obs} - t_0 + \Delta_C - D/f^2 + \Delta_{R\odot}(\alpha, \delta, \mu_\alpha, \mu_\delta, \pi) + \Delta_{E\odot} - \Delta_{S\odot}(\alpha, \delta) - \Delta_R(x, e, P_b, T_0, \omega, \dot{\omega}, \dot{P}_b) - \Delta_E(\gamma) - \Delta_S(r, s). \quad (2)$$

Here $t_0$ is a nominal equivalent TOA at the solar system barycenter; $\Delta_C$ represents measured clock offsets; $D/f^2$ is the dispersive delay for propagation at frequency $f$ through the interstellar medium; $\Delta_{R\odot}$, $\Delta_{E\odot}$, and $\Delta_{S\odot}$ are propagation delays and relativistic time adjustments within the solar system; and $\Delta_R$, $\Delta_E$, and $\Delta_S$ are similar terms for effects within a binary pulsar's orbit. Subscripts on the various $\Delta$'s indicate the nature of the time-dependent delays, which include 'Römer', 'Einstein', and 'Shapiro' delays in the solar system and in the pulsar orbit. The Römer terms have amplitudes comparable to the orbital periods times $v/2\pi c$, where $v$ is the orbital velocity and $c$ the speed of light. The Einstein terms, representing the integrated effects of gravitational redshift and time dilation, are smaller by another factor $ev/c$, where $e$ is the orbital eccentricity. The Shapiro time delay is a result of reduced velocities that accompany the well-known bending of light rays propagating close to a massive object. The delay amounts to about 120 μs for one-way lines of sight grazing the Sun, and the magnitude depends logarithmically on the angular impact parameter. The corresponding delay within a binary pulsar orbit depends on the companion star's mass, the orbital phase, and the inclination $i$ between the orbital angular momentum and the line of sight.

Figure 5 illustrates the combined orbital delay $\Delta_R + \Delta_E + \Delta_S$ for PSR 1913 + 16, plotted as a function of orbital phase. Despite the fact that the Einstein and Shapiro effects are orders of magnitude smaller than the Römer delay, they can still be measured separately if the precision of available TOAs is high enough. In fact, the available precision is very high indeed, as one can see from the lone data point shown in Fig. 5 with 50 000$\sigma$ error bars.

Equations (1) and (2) have been written to show explicitly the most significant dependences of pulsar phase on as many as nineteen *a priori* unknowns. In addition to the rotational frequency $\nu$ and spin-down rate $\dot{\nu}$, these phenomenological parameters include a reference arrival time $t_0$, the dispersion constant $D$, celestial coordinates $\alpha$ and $\delta$, proper-motion terms $\mu_\alpha$ and $\mu_\delta$, and annual parallax $\pi$. For binary pulsars the terms on the second line of eqn (2), with as many as ten significant orbital parameters, are also required. The additional parameters include five that would be necessary even in a purely Keplerian analysis of orbital motion: the projected semi-major axis

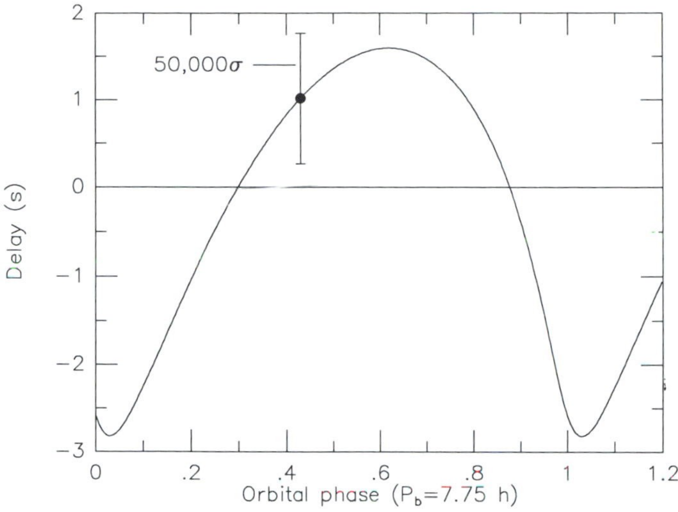


**Fig. 5** Orbital delays observed for PSR 1913 + 16 during July 1988. The uncertainty of an individual five-minute measurement is typically 50 000 times smaller than the error bar shown.

$x \equiv a_1 \sin i/c$, eccentricity $e$, binary period $P_b$, longitude of periastron $\omega$, and time of periastron $T_0$. If the experimental precision is high enough, relativistic effects can yield the values of five further 'post-Keplerian' parameters: the secular derivatives $\dot{\omega}$ and $\dot{P}_b$, the Einstein parameter $\gamma$, and the range and shape of the orbital Shapiro delay, $r$ and $s \equiv \sin i$. Several earlier versions of this formalism for treating timing measurements of binary pulsars exist [27–29], and have been historically important to our progress with the PSR 1913 + 16 experiment. The elegant framework outlined here was derived during 1985–1986 by Damour and Deruelle [30, 31].

Model parameters are extracted from a set of TOAs by calculating the pulsar phases $\phi(T)$ from eqn (1) and minimizing the weighted sum of squared residuals,

$$\chi^2 = \sum_{i=1}^{N} \left( \frac{\phi(T_i) - n_i}{\sigma_i / P} \right)^2, \tag{3}$$

with respect to each parameter to be determined. In this equation, $n_i$ is the closest integer to $\phi(T_i)$, and $\sigma_i$ is the estimated uncertainty of the $i$th TOA. In a valid and reliable solution the value of $\chi^2$ will be close to the number of degrees of freedom, i.e. the number of measurements $N$ minus the number of adjustable parameters. Parameter errors so large that the closest integer to $\phi(T_i)$ may not be the correct pulse number are invariably accompanied by huge increases in $\chi^2$; this is the reason for my earlier statement that correct pulse numbering is easily recognizable, once attained. In addition to providing a list of fitted parameter values and their estimated uncertainties, the

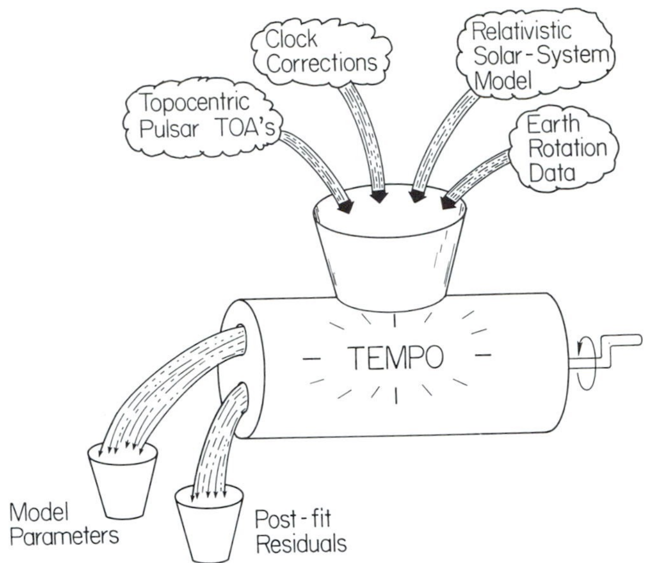


**Fig. 6** Schematic diagram of the analysis of pulsar timing measurements carried out by the computer program TEMPO. The essential functions are all described in the text.

least-square solution produces a set of post-fit residuals, or differences between measured TOAs and those predicted by the model (see Fig. 6). The post-fit residuals are carefully examined for evidence of systematic trends that might suggest experimental errors, or some inadequacy in the astrophysical model, or perhaps deep physical truths about the nature of gravity.

Necessarily, some model parameters will be easier to measure than others. When many TOAs are available, spaced over many months or years, it generally follows that at least the pulsar's celestial coordinates, spin parameters, and Keplerian orbital elements will be measurable with high precision, often to as many as 6–14 significant digits. As we will see, the relativistic parameters of binary pulsar orbits are generally much more difficult to measure—but the potential rewards for doing so are substantial.

## 4 The Newtonian limit

Thirty-five binary pulsar systems have now been studied well enough to determine their basic parameters, including the Keplerian orbital elements, with good accuracy. For each system the orbital period $P_b$ and projected semi-major axis $x$ can be combined to give the mass function,

$$f_1(m_1, m_2, s) = \frac{(m_2\, s)^3}{(m_1 + m_2)^2} = \frac{x^3}{T_\odot (P_b/2\pi)^2}. \qquad (4)$$

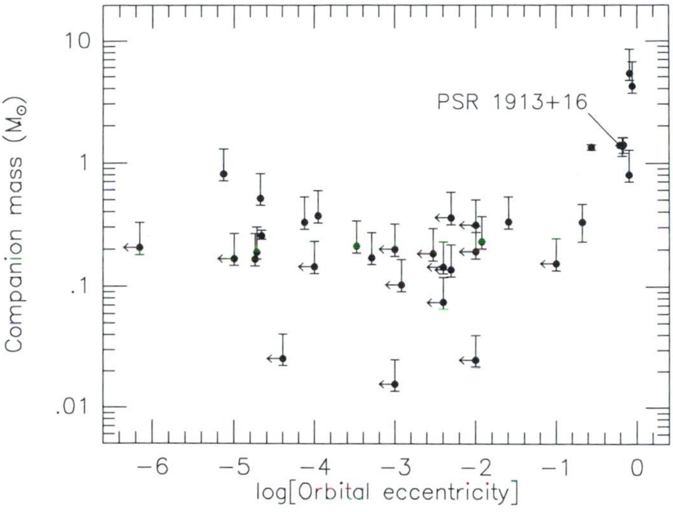


**Fig. 7** Masses of the companions of binary pulsars, plotted as a function of orbital eccentricity. Near the marked location of PSR 1913 + 16, three distinct symbols have merged into one; these three binary systems, as well as their two nearest neighbors in the graph, are thought to be pairs of neutron stars. The two pulsars at the upper right are accompanied by high-mass main-sequence stars, while the remainder are believed to have white-dwarf companions.

Here $m_1$ and $m_2$ are the masses of the pulsar and companion in units of the Sun's mass, $M_\odot$; I use the shorthand notations $s \equiv \sin i$, $T_\odot \equiv GM_\odot/c^3 = 4.925490947 \cdot 10^{-6}$ s, where $G$ is the Newtonian constant of gravity. In the absence of other information, the mass function cannot provide unique solutions for $m_1$, $m_2$, or $s$. Nevertheless, likely values of $m_2$ can be estimated by assuming a pulsar mass close to 1.4 $M_\odot$ (the Chandrasekhar limit for white dwarfs) and the median value $\cos i = 0.5$, which implies $s = 0.87$. With this approach one can distinguish three categories of binary pulsars, which I shall discuss by reference to Fig. 7: a plot of binary-pulsar companion masses versus orbital eccentricities.

Twenty-eight of the binary systems in Fig. 7 have orbital eccentricities $e < 0.25$ and low-mass companions likely to be degenerate dwarfs. Most of these have nearly circular orbits; indeed, the only ones with eccentricities more than a few percent are located in globular clusters, and their orbits have probably been perturbed by near collisions with other stars. Five of the binaries have much larger eccentricities and likely companion masses of $0.8M_\odot$ or more; these systems are thought to be pairs of neutron stars, one of which is the detectable pulsar. Their large orbital eccentricities are almost certainly the result of rapid ejection of mass in the supernova explosion. Unlike the binary pulsars with compact companions, these two systems have orbits that could be significantly modified by complications such as tidal forces or mass loss.

## 5 General relativity as a tool

As Russell Hulse and I suggested [17] in the discovery paper for PSR 1913 + 16, it should be possible to combine measurements of relativistic orbital parameters with the mass function, thereby determining masses of both stars and the orbital inclination. In the post-Keplerian (PK) framework outlined above, each measured PK parameter defines a unique curve in the $(m_1, m_2)$ plane, valid within a specified theory of gravity. Experimental values for any two PK parameters (say $\dot{\omega}$ and $\gamma$, or perhaps $r$ and $s$) establish the values of $m_1$, $m_2$, and $s$ unambiguously. In general relativity the equations for the five most significant PK parameters are as follows [25, 31, 32]:

$$\dot{\omega} = 3\left(\frac{P_b}{2\pi}\right)^{-5/3} (T_\odot M)^{2/3} (1 - e^2)^{-1}, \tag{5}$$

$$\gamma = e\left(\frac{P_b}{2\pi}\right)^{1/3} T_\odot^{2/3} M^{-4/3} m_2 (m_1 + 2m_2), \tag{6}$$

$$\dot{P}_b = -\frac{192\pi}{5}\left(\frac{P_b}{2\pi}\right)^{-5/3}\left(1 + \frac{73}{24}e^2 + \frac{37}{96}e^4\right) \times (1 - e^2)^{-7/2} T_\odot^{5/3} m_1 m_2 M^{-1/3}, \tag{7}$$

$$r = T_\odot m_2, \tag{8}$$

$$s = x\left(\frac{P_b}{2\pi}\right)^{-1/3} T_\odot^{-1/3} M^{2/3} m_2^{-1}, \tag{9}$$

Again the masses $m_1$, $m_2$, and $M \equiv m_1 + m_2$ are expressed in solar units. I emphasize that the left-hand sides of eqns (5–9) represent directly measurable quantities, at least in principle. Any two such measurements, together with the well-determined values of $e$ and $P_b$, will yield solutions for $m_1$ and $m_2$ as well as explicit predictions for the remaining PK parameters.

The binary systems most likely to yield measurable PK parameters are those with large masses and high eccentricities and which are astrophysically 'clean', so that their orbits are overwhelmingly dominated by the gravitational interactions between two compact masses. The five pulsars clustered near PSR 1913 + 16 in Fig. 7 would seem to be especially good candidates, and this has been borne out in practice. In the most favorable circumstances, even binary pulsars with low-mass companions and nearly circular orbits can yield significant post-Keplerian measurements. The best present example is PSR 1855 + 09: its orbital plane is nearly parallel to the line of sight, greatly magnifying the orbital Shapiro delay. The relevant measurements [33–35] are illustrated in Fig. 8, together with the fitted function $\Delta_S(r, s)$, in this case closely approximated by

$$\Delta_S = -2r \log(1 - s\cos[2\pi(\phi - \phi_0]) \tag{10}$$

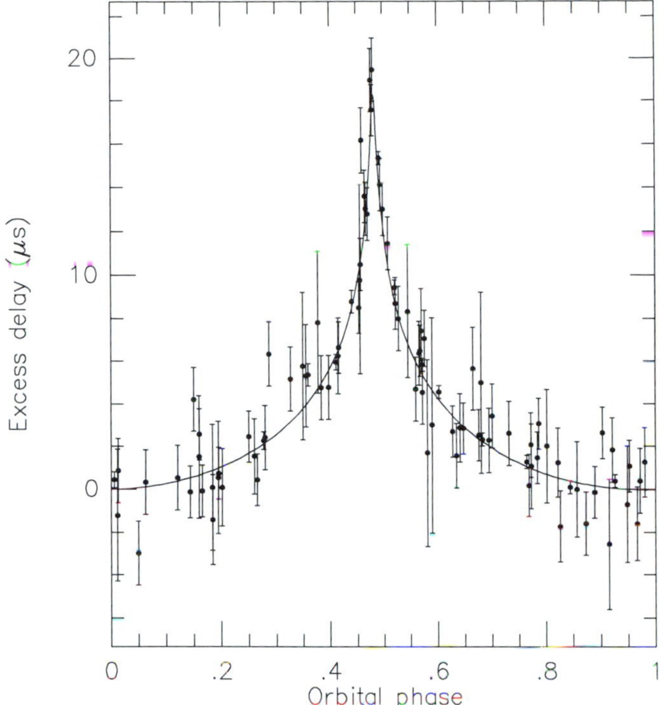


**Fig. 8** Measurements of the Shapiro time delay in the PSR 1855 + 09 system. The theoretical curve corresponds to eqn (10), and the fitted values of *r* and *s* can be used to determine the masses of the pulsar and companion star.

where $\phi$ is the orbital phase in cycles and $\phi_0 = 0.4823$ is the phase of superior conjunction. The fitted values of *r* and *s* yield the masses $m_1 = 1.50^{+0.26}_{-0.14}$, $m2 = 0.258^{+0.028}_{-0.016}$. In a similar way, all binary pulsars with two measurable PK parameters yield solutions for their component masses. At present, most of the experimental data on the masses of neutron stars (see Fig. 9) come from such timing analyses of binary pulsar systems (see [36, 37], and references therein).

## 6 Testing for gravitational waves

If three or more post-Keplerian parameters can be measured for a particular pulsar, the system becomes over-determined, and the extra experimental degrees of freedom transform it into a calibrated laboratory for testing relativistic gravity. Each measurable PK parameter beyond the first two provides an explicit, quantitative test. Because the velocities and gravitational

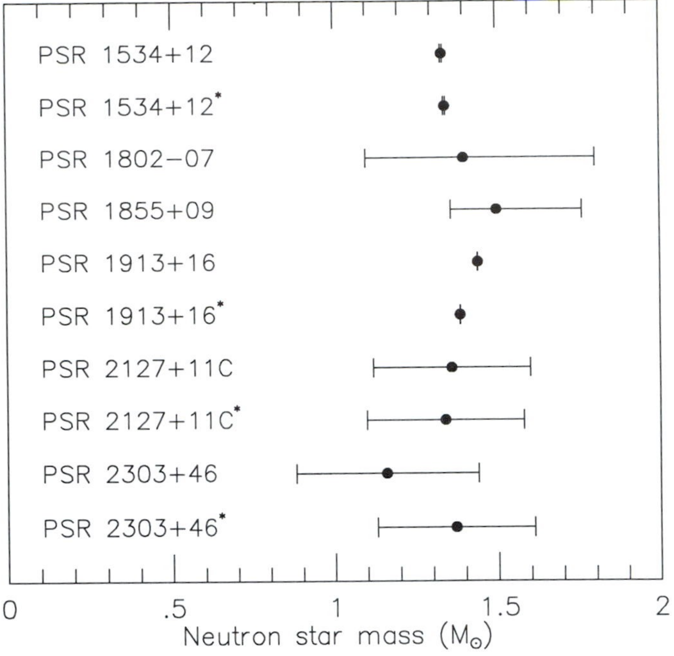


**Fig. 9** The masses of ten neutron stars, measured by observing relativistic effects in binary pulsar orbits. Asterisks after pulsar names denote companions to the observed pulsars.

energies in a high-mass binary pulsar system can be significantly relativistic, strong-field and radiative effects come into play. Two binary pulsars, PSRs 1913 + 16 and 1534 + 12, have now been timed well enough and long enough to yield three or more PK parameters. Each one provides significant tests of gravitation beyond the weak-field, slow-motion limit [32, 38].

PSR 1913 + 16 has an orbital period $P_b \approx 7.8$ h, eccentricity $e \approx 0.62$ and mass function $f_1 \approx 0.13\ M_\odot$. With the available data quality and time span, the Keplerian orbital parameters are actually determined with fractional accuracies of a few parts per million, or better. In addition, the PK parameters $\dot{\omega}$, $\gamma$, and $\dot{P}_b$ are determined with fractional accuracies better than $3 \times 10^{-6}$, $5 \times 10^{-4}$, and $4 \times 10^{-3}$, respectively [25, 39]. Within any viable relativistic theory of gravity, the values of $\dot{\omega}$ and $\gamma$ yield the values of $m_1$ and $m_2$ and a corresponding prediction for $\dot{P}_b$ arising from the damping effects of gravitational radiation. At present levels of accuracy, a small kinematic correction (approximately 0.5% of the observed $\dot{P}_b$) must be included to account for accelerations of the solar system and the binary pulsar system in the Galactic gravitational field [40]. After doing so, we find that Einstein's theory passes this extraordinarily stringent test with a fractional accuracy better than 0.4% (see Figs 10 and 11). The clock-comparison experiment for PSR 1913 + 16

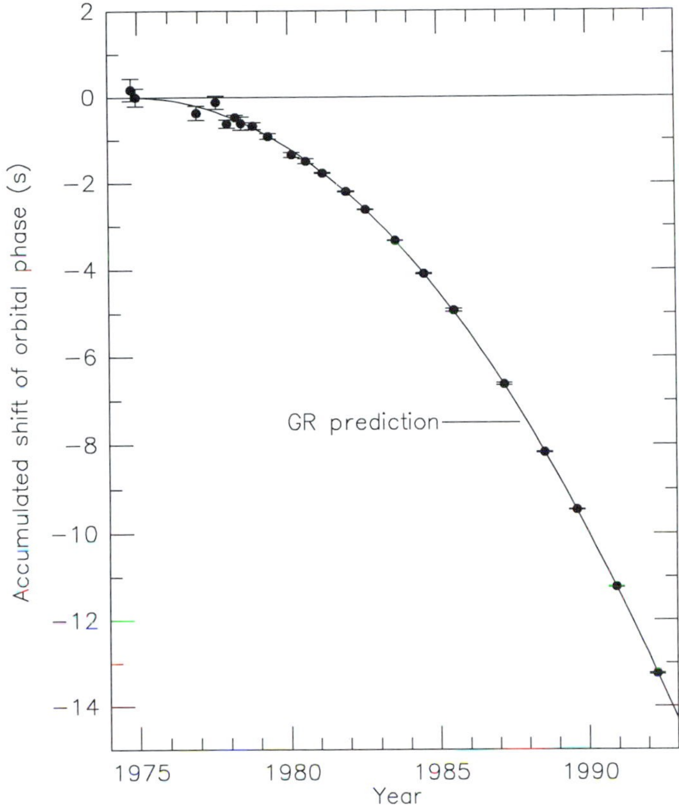


**Fig. 10** Accumulated shift of the times of periastron in the PSR 1913 + 16 system, relative to an assumed orbit with constant period. The parabolic curve represents the general relativistic prediction for energy losses from gravitational radiation.

thus provides direct experimental proof that changes in gravity propagate at the speed of light, thereby creating a dissipative mechanism in an orbiting system. It necessarily follows that gravitational radiation exists and has a quadrupolar nature.

PSR 1534 + 12 was discovered just three years ago, in a survey by Aleksander Wolszczan [41] that again used the huge Arecibo telescope to good advantage. This pulsar promises eventually to surpass the results now available from PSR 1913 + 16. It has orbital period $P_b \approx 10.1$ h, eccentricity $e \approx 0.27$, and mass function $f_1 \approx 0.31\ M_\odot$. Moreover, with a stronger signal and narrower pulse than PSR 1913 + 16, its TOAs have considerably smaller measurement uncertainties, around 3 μs for five-minute observations. Results based on 15 months of data [39] have already produced significant measurements of four PK parameters: $\dot{\omega}$, $\gamma$, $r$, and $s$. In recent work not yet published, Wolszczan and I have measured the orbital decay rate, $\dot{P}_b$, and

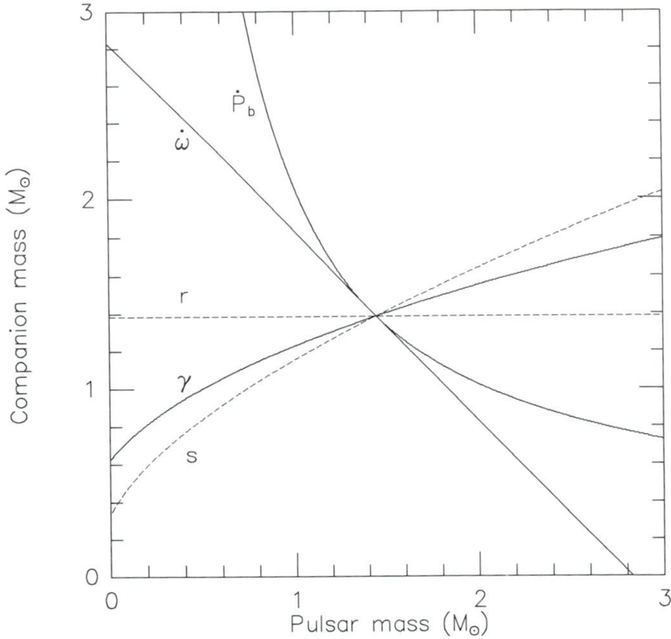


**Fig. 11** Solid curves correspond to eqns (5–7) together with the measured values of $\dot{\omega}$, $\gamma$, and $\dot{P}_b$. Their intersection at a single point (within the experimental uncertainty of about 0.35% in $\dot{P}_b$ establishes the existence of gravitational waves. Dashed curves correspond to the *predicted* values of parameters $r$ and $s$; these quantities should become measurable with a modest improvement in data quality.

found it to be in accord with general relativity at about the 20% level. In fact, *all* measured parameters of the PSR 1534 + 12 system are consistent within general relativity, and it appears that when the full experimental analysis is complete, Einstein's theory will have passed three more very stringent tests under strong-field and radiative conditions.

I do not believe that general relativity necessarily contains the last valid words to be written about the nature of gravity. The theory is not, of course, a quantum theory, and at its most fundamental level the universe appears to obey quantum-mechanical rules. Nevertheless, our experiments with binary pulsars show that, whatever the precise directions of future theoretical work may be, the correct theory of gravity must make predictions that are asymptotically close to those of general relativity over a vast range of classical circumstances.

## Acknowledgements

Russell Hulse and I have many individuals to thank for their important work, both experimental and theoretical, without which our discovery of PSR 1913

+ 16 could not have borne fruit so quickly or so fully. Most notable among these are Roger Blandford, Thibault Damour, Lee Fowler, Peter McCulloch, Joel Weisberg, and the skilled and dedicated technical staff of the Arecibo Observatory.

## References

1. A. Hewish, S. J. Bell, J. D. H. Pilkington, P. F. Scott, and R. A. Collins. Observation of a rapidly pulsating radio source. *Nature*, 217:709–713, 1968.
2. G. R. Huguenin, J. H. Taylor, L. E. Goad, A. Hartai, G. S. F. Orsten, and A. K. Rodman. New pulsating radio source. *Nature*, 219:576, 1968.
3. T. Gold. Rotating neutron stars as the origin of the pulsating radio sources. *Nature*, 218:731–732, 1968.
4. D. H. Staelin and E. C. Reifenstein, III. Pulsating radio sources near the Crab Nebula. *Science*, 162:1481–1483, 1968.
5. M. I. Large, A. E. Vaughan, and B. Y. Mills. A pulsar supernova association. *Nature*, 220:340–341, 1968.
6. D. W. Richards and J. M. Comella. The period of pulsar NP 0532. *Nature*, 222:551–552, 1969.
7. V. Radhakrishnan and D. J. Cooke. Magnetic poles and the polarization structure of pulsar radiation. *Astrophys. Lett.*, 3:225–229, 1969.
8. P. Goldreich and W. H. Julian. Pulsar electrodynamics. *Astrophys. J.*, 157:869–880, 1969.
9. R. N. Manchester and W. L. Peters. Pulsar parameters from timing observations. *Astrophys. J.*, 173:221–226, 1972.
10. W. R. Burns and B. G. Clark. Pulsar search techniques. *Astron. Astrophys.*, 2:280–287, 1969.
11. R. N. Manchester, J. H. Taylor, and G. R. Huguenin. New and improved parameters for twenty-two pulsars. *Nature Phys. Sci.*, 240:74, 1972.
12. J. H. Taylor. A sensitive method for detecting dispersed radio emission. *Astron. Astrophys. Supp. Ser.*, 15:367, 1974.
13. J. H. Taylor, R. N. Manchester, and A. G. Lyne. Catalog of 558 pulsars. *Astrophys. J. Supp. Ser.*, 88:529–568, 1993.
14. F. Camilo. Millisecond pulsar searches. In A. Alpar (ed), *Lives of the neutron stars, (NATO ASI Series)*, Dordrecht, 1994. Kluwer.
15. J. H. Taylor. A high sensitivity survey to detect new pulsars. Research proposal submitted to the US National Science Foundation, September, 1972.
16. R. A. Hulse and J. H. Taylor. A high sensitivity pulsar survey. *Astrophys. J. (Letters)*, 191:L59–L61, 1974.

17. R. A. Hulse and J. H. Taylor. Discovery of a pulsar in a binary system. *Astrophys. J.*, 195:L51–L53, 1975.
18. R. A. Hulse and J. H. Taylor. A deep sample of new pulsars and their spatial extent in the galaxy. *Astrophys. J. (Letters)*, 201:L55–L59, 1975.
19. R. A. Hulse. The discovery of the binary pulsar. *Les prix Nobel 1993*, 58–79. The Nobel Foundation, 1994.
20. J. H. Taylor. Millisecond pulsars: nature's most stable clocks. *Proc. I. E. E. E.*, 79:1054–1062, 1991.
21. J. H. Taylor, R. A. Hulse, L. A. Fowler, G. E. Gullahorn, and J. M. Rankin. Further observations of the binary pulsar PSR 1913 + 16. *Astrophys. J.*, 206:L53–L58, 1976.
22. P. M. McCulloch, J. H. Taylor, and J. M. Weisberg. Tests of a new dispersion-removing radiometer on binary pulsar PSR 1913 + 16. *Astrophys. J. (Letters)*, 227:L133–L137, 1979.
23. J. H. Taylor, L. A. Fowler, and P. M. McCulloch. Measurements of general relativistic effects in the binary pulsar PSR 1913 + 16. *Nature*, 277:437, 1979.
24. J. H. Taylor and J. M. Weisberg. A new test of general relativity: Gravitational radiation and the binary pulsar PSR 1913 + 16. *Astrophys. J.*, 253:908–920, 1982.
25. J. H. Taylor and J. M. Weisberg. Further experimental tests of relativistic gravity using the binary pulsar PSR 1913 + 16. *Astrophys. J.*, 345:434–450, 1989.
26. D. R. Stinebring, V. M. Kaspi, D. J. Nice, M. F. Ryba, J. H. Taylor, S. E. Thorsett, and T. H. Hankins. A flexible data acquisition system for timing pulsars. *Rev. Sci. Instrum.*, 63:3551–3555, 1992.
27. R. Blandford and S. A. Teukolsky. Arrival-time analysis for a pulsar in a binary system. *Astrophys. J.*, 205:580–591, 1976.
28. R. Epstein. The binary pulsar: post-Newtonian timing effects. *Astrophys. J.*, 216:92–100, 1977.
29. M. P. Haugan. Post-Newtonian arrival-time analysis for a pulsar in a binary system. *Astrophys. J.*, 296:1–12, 1985.
30. T. Damour and N. Deruelle. General relativistic celestial mechanics of binary systems. I. The post-Newtonian motion. *Ann. Inst. H. Poincaré (Physique Théorique)*, 43:107–132, 1985.
31. T. Damour and N. Deruelle. General relativistic celestial mechanics of binary systems. II. The post-Newtonian timing formula. *Ann. Inst. H. Poincaré (Physique Théorique)*, 44:263–292, 1986.
32. T. Damour and J. H. Taylor. Strong-field tests of relativistic gravity and binary pulsars. *Phys. Rev. D*, 45:1840–1868, 1992.
33. L. A. Rawley, J. H. Taylor, and M. M. Davis. Fundamental astrometry and millisecond pulsars. *Astrophys. J.*, 326:947–953, 1988.

34. M. F. Ryba and J. H. Taylor. High precision timing of millisecond pulsars. I. Astrometry and masses of the PSR 1855 + 09 system. *Astrophys. J.*, 371:739–748, 1991.
35. V. M. Kaspi, J. H. Taylor, and M. Ryba. High-precision timing of millisecond pulsars. III. Long-term monitoring of PSRs B1855 + 09 and B1937 + 21. *Astrophys. J.*, **428**, 713–728, 1994.
36. J. H. Taylor and R. J. Dewey. Improved parameters for four binary pulsars. *Astrophys. J.*, 332:770–776, 1988.
37. S. E. Thorsett, Z. Arzoumanian, M. M. McKinnon, and J. H. Taylor. The masses of two binary neutron star systems. *Astrophys. J. (Letters)*, 405:L29–L32, 1993.
38. J. H. Taylor, A. Wolszczan, T. Damour, and J. M. Weisberg. Experimental constraints on strong-field relativistic gravity. *Nature*, 355:132–136, 1992.
39. J. H. Taylor. Testing relativistic gravity with binary and millisecond pulsars. In R. J. Gleiser, C. N. Kozameh, and O. M. Moreschi (eds), *General relativity and gravitation 1992*, pp. 287–294, Bristol, 1993. Institute of Physics Publishing.
40. T. Damour and J. H. Taylor. On the orbital period change of the binary pulsar PSR 1913 + 16. *Astrophys. J.*, 366:501–511, 1991.
41. A. Wolszczan. A nearby 37.9 ms radio pulsar in a relativistic binary system. *Nature*, 350:688–690, 1991.

# Taking the measure of the universe

Robert P. Kirshner

## Small brains in a big universe

Despite the limitations of our small brains, brief lives, and limited experience, humans have embarked on an adventurous exploration of the dimensions of the universe. A frank appraisal of our equipment for this expedition is not encouraging: we have brains with a volume of about one liter and we're trying to encompass a volume $10^{52}$ times larger. At best we'll have to leave out some details. Our lives are brief, too. We're trying to understand the evolution over time of something which is 100 000 000 times older than a very old person. That's the ratio of one second to the duration of an undergraduate degree. I caution my students not to blink. Even the most patient and diligent observation over a lifetime is unlikely to reveal a local change due to cosmic evolution. It's a little like seeing the last scene of *Hamlet* and trying to reconstruct the plot, the characters, and the forces that act between them from the bodies on the stage and the play's last word ['shoot'].

Most cumbersome is our common sense—developed on the surface of an 12 800 km diameter planet around a middleweight star which, though congenial to us as a place with the delights of warmth, light, and major league baseball (said to be a refined form of rounders), is far from the cold emptiness that characterizes most of the universe. What we find typical is actually very unusual. Our common sense tells us some very misleading things: the Earth looks flat, but that's just because we are very short. If we were *big* compared to the Earth (like le Petit Prince on his asteroid), nobody would have been surprised to see Magellan's ships stagger back into port after circumnavigating the globe. Of course, if we were that big, we might not need the ships. And there might not be room enough for you, me, and Magellan.

Despite these handicaps, the case is not altogether desperate. We have tools of physical measurement to construct a picture of the universe that spans space and time from the smooth, hot, dense Big Bang about 15 billion years ago to the lumpy and elaborate universe of stars and galaxies we see today. How have we been able to go beyond the limitations of our brief and local lives? One reason is that our information about the universe comes to us in the form of light. Light travels about a foot (a unit of distance used in the USA, Liberia, and few other jurisdictions, but possibly familiar to older readers in the UK) in a nanosecond—a billionth of a second. That means we don't see objects as they are, but as they were when the light was emitted. So in a lecture hall, the professor sees the eager students in the front row as they were a few nanoseconds into the past, and those snoozing in the back row perhaps a hundred nanoseconds ago. Other things being equal, the students in the back will appear *younger*. What is true to a negligible degree here on Earth is a deep fact of nature on the cosmic scale. We see familiar stars in the night-time sky as they were a few hundred years in the past, and even the nearest galaxies as they were millions of years ago. For the most distant galaxies we can image, a telescope is a true time machine, showing us what the universe was like several billion years into the past. Common sense separates time and space—but they are actually woven together. As we look out from an observatory, our view scans a curious surface of the 'present' for nearby objects and the distant past for remote ones. So despite our evanescent lives, we have the chance to probe cosmic evolution.

Another aspect of the world allows us to escape our parochial view. Matter on Earth is identical to matter elsewhere in the solar system, in other stars, and in distant galaxies. For example, atoms of calcium in an earthbound laboratory emit and absorb light at definite and well-defined wavelengths or colors. Identical atoms in the sun impress the same pattern of lines on the solar spectrum: calcium is present in the outer layers of the sun. Knowledge obtained *here* on the Earth and *now* in the latest phase of cosmic change works to reveal what was happening *there* in distant quasars and *then* 10 billion years ago. This powerful fact lets us measure the chemical composition of distant (and ancient) matter by decoding the clues imprinted by atoms at the source, or along the intervening path. Spectra also show the motion of the source through a systematic shift of the pattern of atomic emission or absorption. Objects which are moving away have their spectra shifted toward longer wavelengths, toward the red, and approaching objects are shifted toward shorter wavelengths, toward the blue.

Our ability to see into the distant past, to identify the atoms that make up distant stars and galaxies and to detect their motion, has led to the picture of an evolving universe whose properties we now seek to measure quantitatively. How big is it? How old? How do we know?

## Candles in the darkness

The chain of reasoning that culminates in the scale of the universe begins with the size of the Earth's orbit and the small parallax shift our annual travel around the sun produces in the apparent positions of nearby stars. This shift is so subtle that it eluded detection for even the nearest stars until the technology of telescopes was up to the task. Once distances to a handful of stars were established by parallax, distances to similar stars could be estimated from their apparent brightness. Since the surface area of a sphere (at least for the ordinary geometry of flat space) increases as the square of the distance, the light from a distant star flowing through each square centimeter of your eye (or a telescope) diminishes as the square of the distance, so the brightness of a standard object diminishes as the square of the distance. Navigating up a coastline by judging the brightness of lighthouses is a certain way to drown, but the penalties for error in cosmology are less severe.

This approach to measuring distances demands that we recognize stars of the same intrinsic brightness by some other observable property. One type of star has proved especially useful: the Cepheid variables. Cepheids are very luminous stars (10 000 times the power of the sun) that we recognize by their rhythmic pulsations which cycle from dim to bright to dim again in a convenient period of a few days or weeks. In the opening decade of this century, Henrietta Swan Leavitt, working at the Harvard College Observatory, studied variable stars in the Magellanic Clouds, small galaxies that orbit our own Milky Way Galaxy. All the stars in the Magellanic Clouds are at about the same distance from us, so their apparent brightness signals their true relative luminosity. The stars in the Magellanic Clouds which *appear* bright really *are* brighter than the stars that look dim. Henrietta Leavitt noticed that the dim Cepheids had short periods while the bright stars changed more slowly. The 'period–luminosity relation' for Cepheids connects a property that doesn't depend on distance (the period) with the luminosity of the star. Then if you measure the apparent brightness you can use the inverse square law to infer the distance to the galaxy which hosts that Cepheid. Edwin Hubble used the world's largest telescope, which in the 1920s was the 100-inch (2.5 m) reflector at Mount Wilson, to discover Cepheid variable stars in the spiral galaxy M31. Hubble was operating at the edge of technology, measuring stars which had the same period as stars in the Magellanic Clouds, but which were about 100 times fainter—so they must be located 10 times as far from us, since the brightness goes down as the square of the distance.

Hubble's work created a new picture for the universe: the spiral nebulae were not part of our own galaxy, but separate islands of stars equivalent to our own system—100 000 light years across, and each as bright as 100 billion suns. By gauging the relative brightness of the most luminous stars in galaxies, and then the properties of galaxies themselves, Hubble estimated distances out to the nearby Virgo Cluster of galaxies. Hubble combined his distances with physical measurements of galaxy velocities derived from the shifts of spectral features. He found a remarkable fact that lies at the center of our view of the universe: almost all galaxies have *red*shifts—they are moving *away* from us. More precisely, Hubble found that the measured recession velocity was just proportional to the distance:

$$v = H \times d,$$

$v$ is the recession velocity (in km $s^{-1}$) and $d$ is the distance (in astronomers' units of 'megaparsecs'—a megaparsec is about 3 million light years). The number that connects the velocity and the distances is $H$, the Hubble Constant, which has the units of kilometers per second per megaparsec. Hubble himself modestly used the symbol '$K$' for this constant, a usage which I personally prefer, but the eponymous '$H$' has become standard.

## Cosmic expansion and cosmic time

What does it mean to have the other galaxies rushing away from us? Do they know something of human nature? One lesson from the history of our slow discovery of our place in the large scheme of things, is that our vantage point is more typical than special. The Earth is *not* at the center of things. The sun, promoted in the sixteenth century to a central role, is just one of 100 billion serviceable stars in the Milky Way Galaxy. It would require a particularly obtuse reading of this trend to infer from Hubble's Law that *we* have the central view of the universe.

Instead we take the opposite and democratic view that *everyone* should see the same overall properties of the universe—and no observer should have a unique and privileged position. This is not a political statement, but a mathematical one, and it is surprisingly easy to construct a situation where all observers see Hubble's Law. Imagine a lecture hall full of attentive students (dream on!) and then imagine what that would look like if you were to double the size of the place by stretching the space between each scholar. What would this look like to one of the students in the middle of the room? All his neighbors would be receding; somebody one seat away would now be at twice the distance. The speed of recession would be proportional to the distance: somebody ten seats away would now have moved to twenty in the same time that his nearest neighbor moved from one to two. The student would observe exactly Hubble's Law. Everyone would be moving away from him and the velocity would be proportional to the distance. But the same would be true for everyone else—it has nothing to do with you. Each observer sees an expanding lecture hall, just as Hubble (and his possible counterparts on distant galaxies) observed an expanding universe with velocity proportional to distance.

The interesting part of this picture comes from taking it seriously and evaluating the expansion quantitatively. We take the cxpanding universe seriously because it is a simple idea that accounts for a very large number of facts. The expansion is the large scale residue of cosmic evolution from the hot, dense phase of the Big Bang. We also see (with radio receivers) the light from that early fireball, now cooled by expansion to just 2.73 degrees above absolute zero, and we can touch helium atoms which were synthesized in that hot cauldron. But we seek a quantitative understanding of the Big Bang: *when* did it happen? Can we check that date by other methods? What will be the future of cosmic expansion?

Hubble's Law says that the velocity is proportional to the distance. We're interested in knowing how long this has been going on. Think about two galaxies, whose present-day separation is $d$ megaparsecs. Since they are receding from one another at $v$ kilometers per second, the time elapsed since they were cheek-by-jowl is just given by $t = d/v$. In a universe where Hubble's Law operates, we get *the same* c-b-j time for all galaxies. That's because $v = H \times d$ so

$$t = d/v = \frac{d}{H \times d} = 1/H.$$

In words, the time it would take for a galaxy to get where we see it is the same whether the galaxy is near or far because the distant galaxies are receding faster by just the right amount. So we can tell how long the universe has been expanding by measuring $H$: Hubble's Constant.

In fact, when you put in the number of kilometers in a megaparsec and the number of seconds in a year, the relation between the expansion time and the Hubble Constant is fairly simple:

$$t_H = 1/H = 10 \times 10^9 \text{ years} \left( \frac{100 \text{ km s}^{-1} \text{ Mpc}^{-1}}{H} \right)$$

As we shall see, plausible modern values range from about 50 to 80 in these units: for $H = 50$ km s$^{-1}$ Mpc$^{-1}$, $t_H = 20$ billion years, and for $H = 80$ km s$^{-1}$ Mpc$^{-1}$, $t_H = 12$ billion years.

It is interesting to note that Hubble's own original measurement of the expansion rate was $H = 500$ km s$^{-1}$ Mpc$^{-1}$ which led to an expansion time of only 2 billion years, shorter than the age of the Earth as known in Hubble's time. This was not a good sign for taking the Hubble time seriously, though the situation today is much better. The quantitative agreement of the expansion time with the age of known objects remains a powerful test for our understanding.

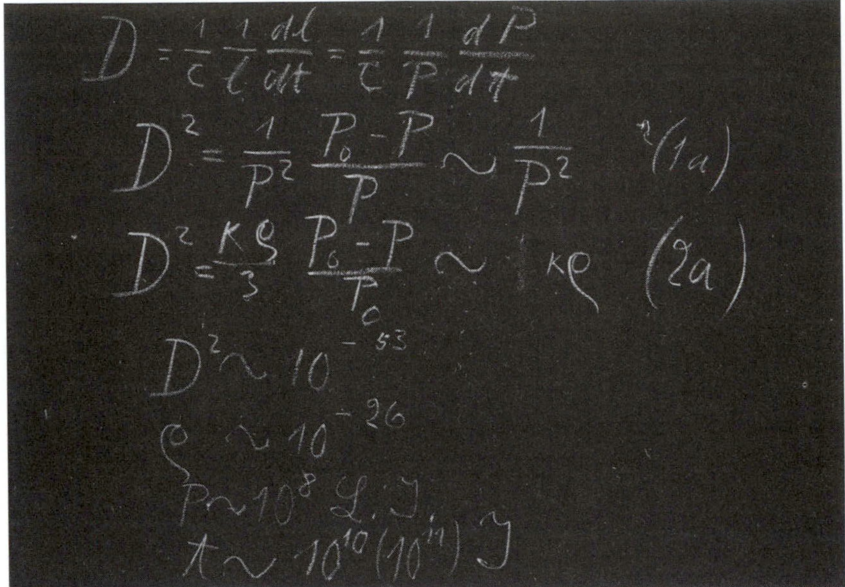

**Fig. 1** An early attempt to compute the age of the universe, from the Oxford Museum of the History of Science, where the blackboard from Einstein's 1931 talk on relativity is preserved. The final line gives an estimated age of the universe of $10^{10}$–$10^{11}$ years. Einstein's work is not so closely linked to recent observational developments as this lecture: it may be fortuitous that the numbers are so similar.

Measuring the Hubble Constant with good precision has been an important goal for observational astronomy. This requires measuring the redshifts and the distances to galaxies. The problem is not usually in measuring the redshift: though it can be laborious, the measurement of the shift in spectrum lines is straightforward. All the problems come from the distances. Hubble's original distances suffered serious errors because some of the objects he identified as luminous stars turned out to be clumps of many stars when observed with modern equipment. The present uncertainty comes from the difficulty in measuring the really good distance indicators (Cepheids) out at the distances where the cosmic expansion dominates the less orderly velocities that individual galaxies acquire from the gravitational pulls of their neighbors.

Even though there is a lively discussion about the correct *value* of the Hubble Constant, the evidence is overwhelming that the Hubble Law holds over a large range in distance. You can tell that this is true if you have a set of luminous objects with a small spread in intrinsic properties: even if you don't know how bright they really are, they will follow the inverse square law. This means they can be used to check that the apparent brightness goes down as the square of the redshift (which is the square of the distance if Hubble's Law holds).

We have so much confidence that the Hubble Law is correct (even though the Hubble Constant is still under debate) that we often use the redshift of a galaxy as a surrogate for its distance. On the night sky, or in a photograph, galaxies are confusing; some are in the foreground, some are in the background. Because there is a very broad range in the true luminosity and diameter of galaxies, you cannot tell just from their apparent brightness and size which are nearby and which are distant. But a redshift measurement can sort this out, and large-scale redshift surveys provide the best way to reveal the true texture of the galaxy distribution: the galaxy distance is proportional to its redshift. Measure the redshift and you can use it as an indicator of distance.

From Hubble's day to the present, the technology for measuring redshifts has improved dramatically. Most of the change is very recent, so that the number of galaxies with measured redshifts has zoomed from about 1000 when I started graduate school in 1970 to well over 100000 today. For example, results from the Las Campanas Redshift Survey (see Fig. 2) which I have helped carry out were obtained with a system that measures 100 galaxies simultaneously. My graduate student, Huan Lin had the unique opportunity to measure 25000 galaxies whose typical redshift of 30 000 km $s^{-1}$ places them at a distance of about a billion light years. This massive redshift survey extends our understanding that galaxies are distributed very unevenly through space; they form large sheets of high galaxy density that separate huge voids where galaxies are rare. This high-contrast and inhomogeneous universe has grown through the action of gravity from very slight variations in density at early epochs. The early seeds of present-day structure are so subtle that they show up as fluctuations in the light from the hot opaque phase of the universe of only 1 part to $10^5$: ten times smoother than the proverbial baby's bottom. Yet the peaks have grown at the expense of the valleys to give a universe with voids 300 million light years across bordered by dense walls of galaxies stretching for 500 million light years.

Modern measurements of the Hubble Constant exploit improvements in technology to extend old methods to new distances. For example, the Hubble Space Telescope (named after Edwin H.) is no larger than the telescope Hubble used at Mount Wilson, but it is at a much better site. Without the smog and city lights or the atmosphere to warp its images, HST detects individual stars in galaxies that cannot be resolved from the ground. Direct measurement of Cepheid variable stars in galaxies of the Virgo Cluster provides an accurate distance to those galaxies by comparing them to Cepheids in the LMC with the same period of pulsation. This is exactly analogous to Hubble's own comparison of M31 Cepheids with the stars Henrietta Leavitt studied. The stars in question are roughly another factor of 100 times fainter than those that Hubble measured, and the Virgo Cluster galaxies are again

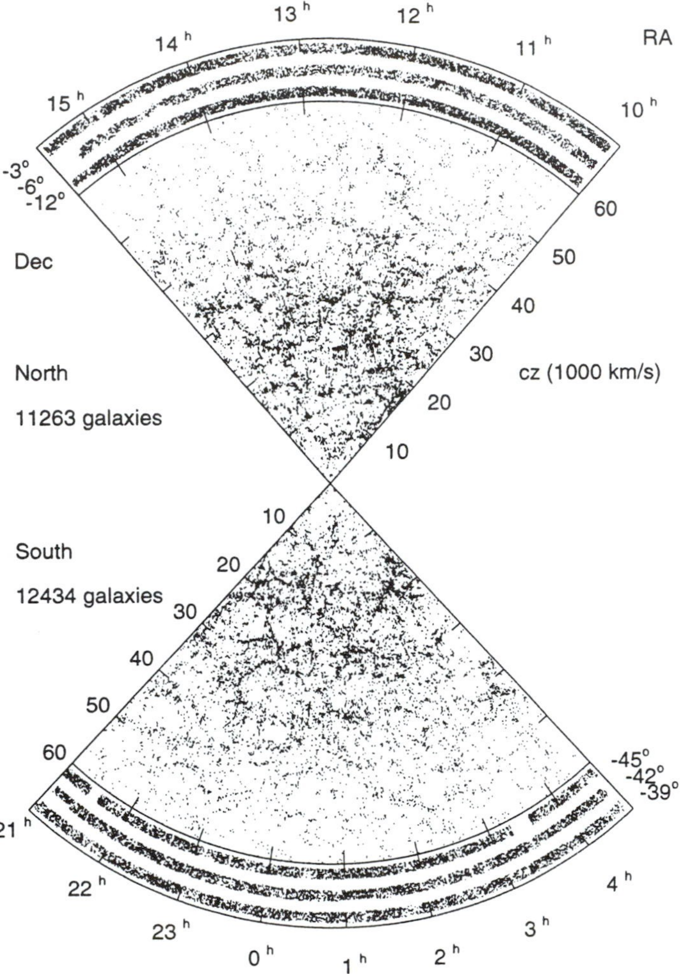


**Fig. 2** Distribution of galaxies measured in the Las Campanas Redshift Survey. This shows the two-dimensional location for each of 23 697 galaxies in six thin strips on the sky. The observed galaxy distribution shows a pattern of voids and walls with a typical size of about 200 million light years. This figure resembles that shown in the chapter by Malcom Longair except that the *depth* of this sample is much larger—extending out to about 2 billion light years.

about 10 times more distant than nearby galaxies of our local neighborhood. Even in this case, the distance is still too small to trust that the velocity measured is solely the expansion velocity in Hubble's Law. At the distance of Virgo, the motion of our own Milky Way, or the idiosyncratic path of an individual galaxy, provides a significant uncertainty to the velocity. Still, the best estimate from this approach gives a Hubble Constant of about 80 km $s^{-1}$ $Mpc^{-1}$, corresponding to an expansion time of about 12 billion years.

## Gravitational effects

So far, I have talked about the expansion without considering the effects of gravitation on the overall behavior of the universe. The modern theory of gravitation that provides a framework for cosmology is Einstein's General Theory of Relativity. Formulated in 1916, General Relativity is a geometric theory that tells how matter and energy produce curvature in space and time. It has proved to be a phenomenally successful theory—accurately predicting several geometric effects that were not correctly accounted for in Newton's gravitation, and now extended by precision timing of pulsars to include radiation of energy by gravitational waves, and by astrophysical measurements to include black holes orbiting around observed stars.

However, the early application of General Relativity to cosmology came too soon for the astronomers to give Einstein sound guidance. Early theoretical investigation showed that General Relativity could account for an expanding universe or for a contracting universe, but not for one that was standing still. Since this work predated Hubble's discovery of expansion, to agree with known 'facts' Einstein introduced a new term into his equations, the 'Cosmological Constant', whose effect was to create a long nearly stationary phase in the cosmic expansion. When the expanding universe was later established by observation, Einstein referred to the Cosmological Constant as his 'greatest blunder', since it kept him from *predicting* the expanding universe.

If we set aside the Cosmological Constant (holding it in reserve for truly desperate circumstances), there are three possibilities for the balance of expansion and gravitation. One is that the density of matter in the universe is so low that it has negligible effect on the expansion. Another is that the density of matter is high enough so that the expansion we see today will eventually slow, reverse, and become a contraction leading to a gnab gib (a backwards big bang—or big crunch). The third possibility is that the universe lies exactly on the border between these two—it has just enough matter to balance the present expansion and will expand forever at an ever-decreasing rate. Each of these gravitational states has a geometrical consequence for the global curvature of the universe which can be found by applying Einstein's theory.

Will the universe expand forever? This is a quantitative question, not a philosophical one, which we can try to answer through observation of the cosmic density. The big redshift surveys enable us to measure the number of galaxies per cubic megaparsec. If we knew the mass associated with each galaxy, we could compute the density. We have perfectly good ways to measure a galaxy's mass by measuring the gravitational acceleration the mass produces on the galaxy's own stars and gas, yet galaxies are not quite what they appear to be in pictures. The measurements show that the mass in galaxies is not confined to the bright inner regions where the stars are most plentiful. Instead, the mass keeps rising even in the outer parts of galaxies where there are few luminous stars. Each visible galaxy must be embedded in an invisible penumbra of dark matter, whose mass we detect, but whose light we do not.

It is embarrassing, but well established, that at least 90% of the universe is in the form of this 'dark matter' whose nature has not been determined. It might be in the form of very low-mass stars or planets. It might be in the form of subatomic particles—and if so, those particles might be very different from the ordinary protons, neutrons, and electrons that make up our everyday world. Astronomers are very good at detecting the effects of invisible objects like planets, black holes, and the dark matter, yet we would feel much more comfortable if we had a clear idea of the nature of the dark matter. Even without understanding its nature, we detect the mass by its gravitational effects and we can estimate its amount. The dark matter associated with galaxies adds up to about 1/3 of the density needed to balance the cosmic expansion.

For many investigators, usually theorists with a strongly developed aesthetic sense, the just-balanced picture is so appealing and our inventory of the contents of the universe so appalling that they forge ahead with most of their computations in the just-balanced case. The effect of gravitation is to provide a continuous slowing down of the cosmic expansion. How does this affect the estimate of cosmic age?

If you took a train from Oxford to London, you could estimate the time elapsed from the start of your journey by knowing the distance travelled and your present rate of speed. The time since the Big Goodbye would be just the distance divided by your speed. This would be precisely right if the train travelled at constant speed. But you would get the wrong elapsed time if the train ran very fast at first, then decelerated, and then crept into town through a mess of construction and repairs. For example, if you woke up after travelling 60 miles and noticed you were going into the station at 10 miles per hour, you might compute that you'd been travelling for 6 hours, when in fact you'd only been riding for one. Deceleration means you overestimate the duration of your journey by using the distance and your present velocity. Maybe you should wear a watch.

Deceleration of the universal expansion by gravitation means you would overestimate the age of the universe by using today's Hubble Constant. Quantitatively, the actual age, for a universe that is just balanced between expansion and gravitation, is 2/3 of the expansion time computed for the present Hubble Constant. That means that a Hubble Constant of 80 km $s^{-1}$ $Mpc^{-1}$, which corresponds to an age of 12 billion years in a low-density universe, would have an elapsed time since the Big Bang of only $2/3 \times 12 = 8$

billion years if the density is high enough to balance the expansion. In the case of the train from Oxford to London, you could compare the elapsed time you compute from your speed and the distance with the much more certain time on your watch to see whether you've been decelerating, or at least to check if the whole picture makes quantitative sense. What about checking our cosmological timescale derived from cosmic distances and motions against some clocks with a different physical basis?

## Quelle heure est-il?

Do we have any chronometers suitable for charting this cosmic voyage? Yes. Our own lives last 100 years and human chronology goes back 10 000, but we have natural clocks that have been running steadily since the formation of the chemical elements. The radioactive decay of long-lived isotopes provides a nuclear clock that ticks at a rate that's independent of temperature, barometric pressure, or violent shaking. From this type of evidence, we know that the Earth is about 4.5 billion years old. However, the Earth and the solar system are certainly not the oldest objects in our galaxy and we can estimate the ages of the older stars by understanding the way stars generate energy. We know that the sun generates its energy from nuclear fusion of hydrogen into helium. The sun will simmer along in this way for about 10 billion years, then swell up to become a red giant star, and finally spend a long retirement as a white dwarf star. Less massive stars than the Sun take longer to reach the red giant stage, and we can compute how long they need from understanding the way they generate energy. More massive stars than the sun rapidly cook the elements up to iron in steady burning and synthesize even heavier elements in the violent blast of a supernova explosion. The generations of stars that preceded the sun created the heavy nuclei that are in our star, our planet, and in us. But the oldest stars formed too early to inherit the legacy of earlier generations. We know from their spectra that the old stars have only 1/1000 the abundance of heavy elements, such as iron, seen in the sun. By looking at the globular clusters and these old anaemic stars in our galaxy, we can measure which of these stars is just now swelling up to become a red giant. We can compute the ages of those stars because we understand the way stars generate their energy through fusion. The oldest stars in our galaxy have an age of about 15 billion years.

The alert reader will note a discrepancy here. The expansion time for a Hubble Constant of 80 km $s^{-1}$ $Mpc^{-1}$ is 12 billion years. If the universe is dense enough for gravitation to balance expansion, that time is an overestimate and the real age is closer to 8 billion years. If the oldest stars in our galaxy are 15 billion years old, even making allowance for some uncertainty in the calculation, there could be a problem. Although I said at the outset that our common sense is not always a reliable guide to understanding the universe, you really shouldn't be older than your mother, and the universe should not be younger than its stars.

How should we regard this disagreement? Does it mean that the picture of an expanding universe is fundamentally flawed? Should we insist that the Hubble Constant must be set to 50 km $s^{-1}$ $Mpc^{-1}$ to allow enough time for stars to bloom? Or should we reintroduce the Cosmological Constant which can produce a long period of slow expansion to reconcile these ages?

My own view is that we have come a long way toward turning observational cosmology into an experimental science. There are a few facts obtained at great effort and it would be a mistake to abandon the quest for better measurements because of an apparent paradox. This means we need to look for better ways to measure astronomical distances and to devise tests to see if the present methods have hidden errors.

## Supernovae illuminate the distant universe

Fortunately, we have tools for this job. My own work has centered on the use of supernova explosions as cosmic yardsticks. Supernovae have the advantage that they are a million times brighter than Cepheids, so they can (at least in principle) be observed a thousand times as far away. Supernovae mark the violent collapse of a massive star or the thorough thermonuclear explosion of a white dwarf. They play a key role in the formation of the chemical elements out of which the Earth, and we, are made, but it is their utility for the distance problem that concerns us here.

The brightest supernova seen on Earth since 1604 was supernova 1987A (SN1987A), first observed in Chile in February 1987. SN1987A was the moment of destruction for a star of about 20 solar masses, Sanduleak-69 202 by name, which is no longer present in the Large Magellanic Cloud (LMC). The Magellanic Clouds have a central role in the web of distance measuring methods: it is by comparing Cepheids in distant galaxies to Cepheids in the LMC that Hubble himself measured the distances to nearby spirals, and the same type of calibration is used with the Hubble telescope to stretch the Cepheid scale out to the Virgo Cluster. We observed SN1987A in 1987 and 1988 with the International Ultraviolet Explorer (IUE) satellite (the Hubble Space Telescope was still on the ground in Sunnyvale, California). One interesting phenomenon we observed was the appearance, about 100 days after the explosion, of emission from very slow-moving nitrogen gas. These narrow lines grew stronger, finally reaching a well-defined maximum about 400 days after the explosion. We reasoned that this might be emission from a shell of slow-moving gas surrounding the supernova that was lit up by the explosion. The 400-day delay in reaching a maximum would be the time it took light from the supernova to reach the most distant part of the shell plus the time it took the fluourescent emission from the shell to reach us. From the timing of this delay, we knew the shell was about 200 light-days across.

When the Hubble Space telescope (HST) was launched, one of the early targets was SN1987A (see Plate VIII). The images showed that the shell we had inferred is actually there, but is more interesting and complex than we dared to think. There is a bright ring of gas, which emitted the light we measured in 1987 and 1988, and subsequent pictures taken after the repair of the Hubble's vision problems show there are also fainter outer rings which were similarly lit up by the explosion. A plausible source for these rings is gas exhaled from the pre-supernova star in the 30 000 years just before the explosion.

In any case, the rings provide a unique opportunity to check the extragalactic distance scale. From the 1987–1988 timing measurements with IUE, we know the true size of the inner ring. From HST images, we measure its angular size. An elementary calculation gives the distance at which a ring of that size would cover the observed angle. The result is about 150 000 light years. So the star Sk-69 202 exploded about 150 000 years ago, and the news of that outburst arrived in Chile on a Tuesday in February 1987. The importance of this distance measurement is that it is completely *independent* of the long chain of reasoning used to derive a distance to the Cepheids in the LMC from conventional methods. The pleasant fact is that the two distances agree. So this supernova-based distance suggests that the basis of the Cepheid distance scale is in good shape and that the problem, if there is one, must lie farther out.

Over the last twenty years, I have been working on another method for using supernovae to measure cosmic distances. Progress on this has come in the last few years through the efforts of my excellent graduate students, Ronald Eastman and Brian Schmidt. What we've done is to use the light from supernova explosions in massive stars to measure the distances to these objects directly, without any use of Cepheid variables near or far. In this way, our investigation provides another independent check on the distance scale. We measure the temperature of a supernova from its color, the velocity of its rapidly expanding atmosphere from its spectrum, and the flux from the supernova detected here at Earth. By repeated measurements of these expanding photospheres, a few simple assumptions, and some serious computer models for the way light escapes from a supernova atmosphere, we can derive the distance to a supernova without using any other astronomically determined distance. The results are very gratifying. We get a distance to SN 1987A of 152 000 light years, in good agreement with both the geometric distance to the ring and conventional measurements from Cepheids. More importantly, we've extended this out to distant galaxies where not even HST can see the Cepheids. Based on 18 objects that stretch out to a distance of 180 Mpc (about 10 times the distance to the Virgo Cluster) we find a good confirmation of Hubble's Law and a Hubble Constant of $H_0 = 73 \pm 6$ km s$^{-1}$ Mpc$^{-1}$. This corresponds to an expansion time of just under 14 billion years. Of course, there's the possibility that our modelling of the supernova atmospheres is not quite perfect, so further work needs to be done, but this method agrees well with the Cepheids in the galaxies where they have been measured and reaches much deeper into the expanding universe where the particular motions of individual galaxies are dominated by the overall cosmological expansion.

Finally, I've been working with my colleague William Press and another graduate student, Adam Riess, to improve the use of supernovae as 'standard candles'. It has long been recognized that a particular type of supernova, the 'Type Ia' (SNIa) events might be harnessed for cosmological distance measurements. Since these explosions are, for about a month, as bright as a billion stars like the sun, they can be seen and measured to great distances. However, recent measurements of individual SNIa have shown that they simply do *not* all have the same intrinsic brightness. This introduces errors into the distances you might compute based on the assumption that they all had the same light output. Of course, the same is true of Cepheids. If you lump all Cepheids together, they cover a wide range in brightness. But Henrietta Leavitt showed how to avoid this error by taking into account the relation between the luminosity of a Cepheid variable and its period of vibration. In a similar way, following the ideas introduced by Mark Phillips and his colleagues at the Cerro Tololo Inter-American Observatory in Chile, we have developed a method that uses the shape of the light curve for a SNIa (see Fig. 3) to sharpen up our knowledge of its brightness. Just as the slow-changing Cepheids are brighter than those that change rapidly, the slower fading SNIa are brighter than those that decline rapidly. By calibrating this effect, and accounting precisely for the variation in brightness as revealed by the shape of the light curve, we have been able to turn SNIa into *the best* tools for measuring distances to galaxies and for confirming Hubble's Law out to a recession velocity of over 30 000 km s$^{-1}$. To derive a value of Hubble's Constant, we need to know the distances to some galaxies that have had well-observed SNIa. Abi Saha, working with Allan Sandage and others, has measured Cepheid variable stars with HST to obtain this calibration. Based on their distances to modern and well-observed supernovae SN1972E, SN1981B, and SN1990N we find a Hubble Constant of $H_0 = 66 \pm 3$ km s$^{-1}$ Mpc$^{-1}$. Here, the very small uncertainty reflects the fact that the light-curve-shape method turns SNIa into extremely reliable distance measuring tools. It does *not* reflect the fact that the underlying calibration of the Cepheids them-

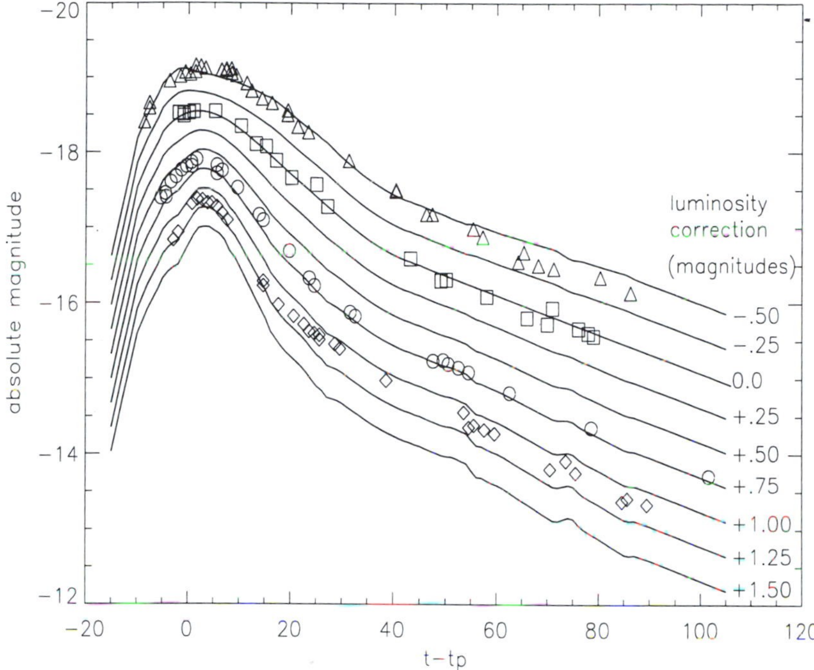


**Fig. 3** The light curves of a variety of Type Ia supernova. The brightest supernovae have a slower rise and fall than the dimmer supernovae, and this fact can be used to improve their use as distance measuring tools.

selves still has some uncertainties and the whole scale could conceivably slide up or down by 10% based on that investigation. So perhaps a more modest way to compare this value with others is to say $H_0 = 66 \pm 6$ km s$^{-1}$ Mpc$^{-1}$.

The convergence of these supernova methods is encouraging. First, the Cepheid scale seems to agree with the geometric method of the SN1987A ring. Second, the expanding photosphere method is completely independent of the Cepheids and gives a distance scale that corresponds to $H_0 = 73 \pm 6$ km s$^{-1}$ Mpc$^{-1}$. Finally, the method based on light-curve shapes for SNIa gives a phenomenally precise way to account for the variation in the luminosity of these explosions. When it is placed on the Cepheid distance scale, we find $H_0 = 66 \pm 6$ km s$^{-1}$ Mpc$^{-1}$, including in our error-estimate the unresolved uncertainties in the Cepheid distance scale.

## Synchronize your watching!

Where does that leave the problem of the age of the universe? If the Hubble Constant is in the neighborhood of 70 km s$^{-1}$ Mpc$^{-1}$, that corresponds to an expansion time of 14 billion years, in tolerable agreement with the ages of the oldest stars in our galaxy. However, if the density is high enough to balance expansion, then deceleration shrinks the age to about 10 billion years, which is not in good agreement with the globular cluster ages. Is this enough evidence to say that the aesthetically-pleasing balance is incorrect? My own view is that we should be very wary of arguments contrived to achieve consistency. Since the universe is certainly indifferent to human preferences based on aesthetic ideas, we should always try to make a measurement rather than rely on argument or taste.

Is there any measurement that could show whether the universe has been decelerating due to its unseen dark matter? Yes. In General Relativity the effect of matter is to curve space. This means that the density of the universe affects its geometry. Observers can tell whether the universe will expand forever or not by measuring whether the apparent brightness of objects drops off exactly as the inverse square of the distance, because the geometrical effect of gravitation alters that relation in an observable way. All you need are redshifts and apparent brightnesses for objects you know have the same intrinsic brightness across a fair fraction of the universe. Just as the effects of curvature only show up on the Earth when you consider navigating across oceans, the curvature of the universe is only seen when we use telescopes to voyage deep into the past. SNIa are bright enough to be seen 1/3 or more of the way across the universe with today's telescopes, and with the method of light-curve shapes, they are precise enough to give us a sporting chance at measuring not just the size, but the shape and the fate of the universe. Our brains are small and our lives are brief, but we have some hope of taking the measure of the universe.

## Addendum (1996)

The program of measuring very distant SNIa to determine the cosmic deceleration is underway and has met with some success. Two groups, one based in Berkeley and another in which I am participating, have found over 20 very faint and very distant supernovae at redshifts from 0.2 to 0.5, including 5 new ones this month. Spectra show that many of these objects are Type Ia supernovae, much like those we observe nearby. If the light-curve shapes of these objects can be accurately determined, we can apply the same powerful methods based on the light-curve shape to these distant supernovae (see Fig. 4). A measurement of the deceleration will result from the careful construction of a diagram of redshift against apparent brightness, out in the range where curvature effects become large enough to see. This will show whether the age problem is resolved and can provide a quantitative limit on the Cosmological

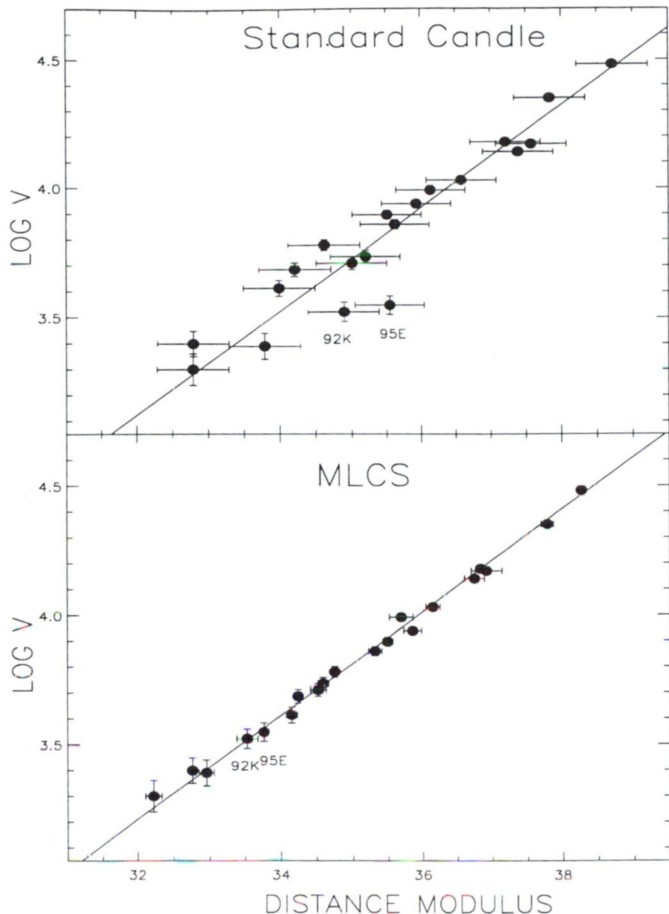


**Fig. 4** Hubble diagrams, plotting redshift against distance for type Ia supernovae. The top panel shows the relation without correcting for the light curve shape, while the bottom panel shows how powerful this technique is in reducing the scatter. The value of the Hubble Constant based on this diagram is 65 km $s^{-1}$ $Mpc^{-1}$, corresponding to an expansion time of about 15 billion years.

Constant. These measurements are very difficult, but they hold genuine promise for resolving our uncertainties about the size and shape of the universe through observation, not just through debate.

# Observing the Big Bang

John C. Mather

## 1 Introduction

It is a wonderful honor to be invited to give the Milne Lecture, and I thank the Society for the opportunity to show these wonderful new results. As I turn down the lights in the room so I can show the transparencies, I am reminded of the classic joke about the NASA astronomer who reached the Pearly Gates, and Saint Peter asked him what he had done in his life. The astronomer said, 'let me have my first viewgraph.' I'm going to show you the results we've measured, the baby pictures of the infant universe, but to give you a context I'll start with the reason we think there was a Big Bang.

## 2 Ways of measuring the universe

The key principle in studying the big bang is that we can look back in time by looking at things far away. Light travels fast, but not infinitely fast. It takes 25 000 years for light from the middle of our own Milky Way galaxy to reach us, and maybe 15 billion years from the Big Bang. We can always tell how far back in time we are looking if we know the distance.

Now we need to know distances. Astronomers have two basic methods: surveying according to the ordinary rules of geometry, and the standard candle method. The trigonometric method works very well, but only for very close-up things, within a few hundred light years of the Earth. The standard candle method also works very well, but there's a problem. It's very difficult to prove that the distant candles have the same intrinsic brightness as the known ones close by. An immense fraction of the international effort of astronomers for centuries has gone into this question of determining standard candles.

When the Palomar observatory was built, one of the main reasons was to understand the shape of the universe by studying the brightness of galaxies as a function of distance. It's never worked. The problem is that galaxies evolve with time, so we can never find a distant galaxy that is just like a nearby one. On the other hand, policy makers should take note that one of the most famous and productive telescopes in the world was justified for wrong reasons, but turned out brilliantly. Even Columbus was wrong about the distance he would have to go across the Atlantic to find the Indies, and the scientific review committees of the day no doubt said so. Fortunately he was not deterred entirely by true but incomplete facts.

Astronomers measure velocity in two parts, tangential and radial. The first part comes from watching an object going across the sky. Multiplying the rate of change of angle by the distance of the object gives the part of the velocity that is perpendicular to the line of sight. The second part is radial, along the line of sight, and is relatively easy to measure with the Doppler shift. This is a change of the measured wavelength of light from the object, caused by its motion towards or away from us. Fortunately nature has given us standard wavelengths to use. If you form the spectrum of the Sun with a prism and spread it out quite far, you can find some fine lines across the spectrum that come from particular atoms and molecules in the atmosphere of the Sun. These lines are at wavelengths that are fixed by the laws of physics, so we know what they are even for objects that are far away.

To find out how much stuff there is in the whole universe, we add up the masses of all the parts we can find. That's a little tricky if we fail to find an important part, like some kind of dark matter. We can measure masses of planets, stars, galaxies, and clusters of galaxies whenever we can measure the velocities of things orbiting around them, and the spacing between them. Then we just apply Newton's law of gravity to calculate the masses. This works very well when we can see complete orbits, as we can for planets or for binary stars, but it's less precise when we have to guess the shape of the orbits, as we do for clusters of galaxies.

We have two main ways to measure the ages of very old things. One is based on the natural radioactivity of certain elements like uranium or carbon-14. These atomic nuclei disintegrate at a known rate and their decomposition products accumulate. For example, if we start with a piece of pure uranium ore, after a while it will have some lead in it as a result of the radioactive decay. These little clocks have been used to get very precise ages for rocks on the earth, including some meteorites that come from elsewhere in the solar system. Our other main way of getting ages comes from the theory of the structure of stars. We know that they produce energy by nuclear reactions deep inside, and we think we know the physical laws and the constants of nature that govern them. Every star has a certain lifetime before it uses up its nuclear fuel, after which it must change into a dead star or else explode. Massive stars live very short lives, only a few million years, while small ones can live for many billions. If we find a cluster of stars that has no stars with masses greater than a certain maximum, then we can compute the age of the cluster.

## 3 Gravity and the expanding universe

Now we would like to see how much we can know of the nature of the whole universe, simply from pure thought. Suppose that the universe has 7 galaxies in it, and that they are just sitting still. This is an unstable situation, because they are all attracting one another with the gravitational force. The particular arrangement in the picture would collapse in a definite period of time, all the galaxies falling in to the middle one and colliding with each other. Since we don't see that happening in the real universe, something is wrong with this picture. It took centuries to find out what it was.

Suppose that the universe is infinite in extent, and filled uniformly with galaxies. Now what will happen? A simple argument says that our imaginary infinite universe has no center, so there is no reason for the anything to fall toward a center. Therefore the penny in the middle of this viewgraph, representing a single galaxy in the infinite uniform universe, will not go anywhere. To be more precise, Newton's law of gravity shows that the gravitational force inside a uniform spherical shell of matter is zero. We imagine the whole universe divided into uniform shells centered on our penny, and none of the shells do anything to the penny.

However, this argument is misleading. Imagine two galaxies, A and B, both inside this collection of spherical shells. According to the previous argument, there is no gravitational force inside the shells from the infinite universe outside them. Nevertheless, A and B do attract each other, just as in my picture with 7 galaxies. They cannot sit still. They must eventually move, either in, out, or around each other. The problem with the argument that there is no center is that there does not need to be a center to have expansion or contraction.

There is an exactly equivalent set of arguments using Einstein's Theory of General Relativity instead of Newton's theory of gravity. It is astonishing that Einstein was so firmly convinced that the universe should be static that he actually changed his equations, to allow a repulsive force to oppose the gravitational force. As other astronomers and mathematicians like Alexander Friedmann and Georges Lemaitre did these calculations using General Relativity, and showed that the universe could not be static, Einstein became quite defensive. At one point, Einstein told Lemaitre that his equations were correct, but his physics was abominable. The dispute was extremely heated.

The question was not resolved until 1929, when Edwin Hubble used the

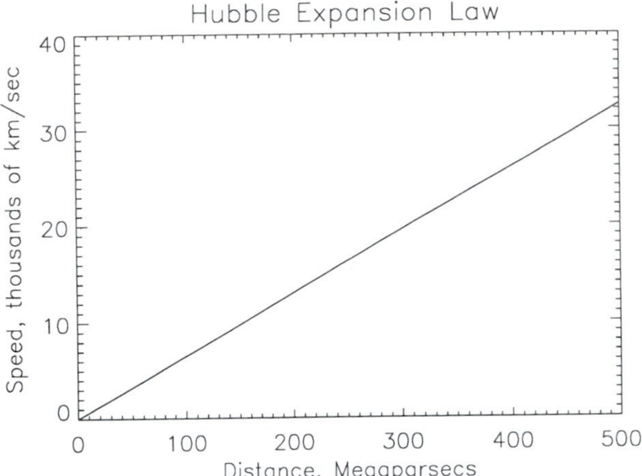


**Fig. 1** Hubble's Law, velocity of galaxy proportional to distance. Actual observations deviate from the trend by several hundred km $s^{-1}$.

great 100 inch telescope at Mount Wilson to find that distant galaxies are receding from us with speeds proportional to their distances (Fig. 1). This is exactly what would happen if the whole universe exploded from a very compressed state a few billion years ago. Einstein was wrong and Lemaitre was right; the universe was not static. On the other hand, not everyone was happy with the result. The calculated age of the universe, the distance of a galaxy divided by its speed, was much too small. It was already known that stars and even the Earth were older than this calculated age of the universe. This problem was not cleared up until the 1950s, when Walter Baade found that the standard candles used by Hubble were not the same as the ones he had measured nearby to establish the distance scale.

This is a good time to mention the definition of redshift. The redshift $z$ of a moving object is the fractional change in wavelength, $z = (\lambda_{obs} - \lambda_{emit})/\lambda_{emit}$. In ordinary laboratory experiments it can be used to calculate velocities, just as Hubble did. In the expanding universe picture, it has another simple and startling and equally good interpretation. According to the Einstein picture of curved space–time, the wavelengths of light expand with the expansion of the universe. Hence, we can also say that the redshift $z$ is $z = (R_{now} - R_{then})/R_{then}$. Equivalently, $1 + z = R_{now}/R_{then}$, where $R$ is a typical distance between two objects in the expanding universe.

Hubble's law has a remarkable consequence: there does not need to be any center of the universe. Any astronomer in any galaxy would find that other galaxies are receding from him, and he would measure the same rate of expansion as we see here. This result is purely mathematical, and results from the exact proportionality between speed and distance. If the speeds

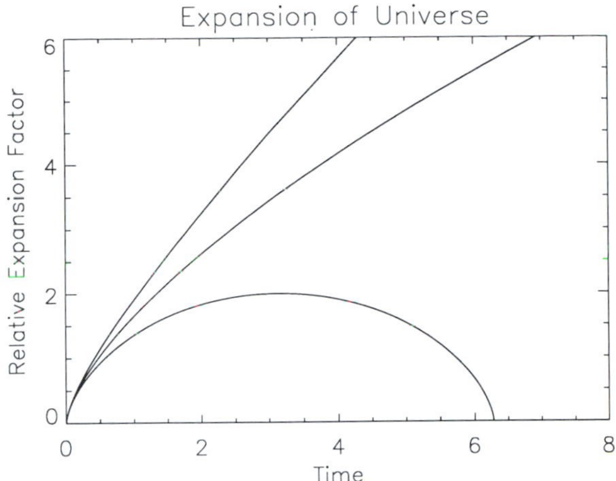


**Fig. 2** Possible fates of the expanding universe. Scales are arbitrary units.

were not the same in every direction, or did not follow the straight line on the graph, then the situation would be different. Similarly, if there were an edge of the matter distribution in one direction, a place beyond which there were no more galaxies, then we could imagine that there is a center. However, no observations have ever shown us any of these exceptions. There is no detectable center or edge, and we are not in a special spot.

Now that we know that Friedmann and Lemaitre were right, we can classify the possible fates of the expanding universe (Fig. 2). Either the universe will keep on expanding forever, or its gravity will stop the expansion and pull it back in, leading to a cosmic crunch at the end of time. If it does keep expanding forever, there are two choices too: either it will continue at high speed, or it will slow down gradually so that it nearly stops. That would be the case if the kinetic energy of the expansion just neatly balances the attractive force of gravity. Theoretical physicists tend to think that this case is more natural or more beautiful, but in truth we don't know. The velocity of expansion is easy enough to measure, but the calculation of the gravitational attraction is not. It depends on adding up all the masses of all the parts, and this is very difficult.

In the Einstein theory of gravity, space–time is curved by the gravitational field. The spatial part of the curvature turns out to be positive for the universe which will collapse again, negative for the universe that keeps on expanding forever, and exactly zero for the universe that expands but tends towards zero velocity. In other words, the shape of the whole universe depends on the amount of gravitating matter in it. If there is enough gravity, then a light ray can be bent so that it goes around a circle, and the whole universe has a finite volume, even though it has no boundary.

A long time ago, two astronomers counted the number of galaxies and made an enormous table. When the computer age arrived, these numbers could be plotted on a map, and a great surprise occurred. The galaxies are not randomly distributed at all, but are arranged in huge stringy structures that stretch across vast expanses of the sky. It was already known that they are clustered into huge groups, but the stringy appearance had not been recognized. This general appearance is one of the great mysteries of modern cosmology. If we assume that cosmology is as simple as possible, then the only force available to act on such cosmic scales is gravity. Although these clouds and strings might be reminiscent of the weather on the Earth, there is an important difference. On the Earth, the weather changes completely in a few weeks, and there is no way to predict the detailed motion of the air for more than a few days. On cosmic scales, the expansion of the universe continually weakens the gravitational forces between distant galaxies, so the cosmic weather moves more and more slowly. When the calculations are made, we find that these cosmic structures must come directly from pre-existing seeds that were left from the Big Bang. This is one of the strongest motivations for trying to find those seeds in the cosmic background radiation.

The cosmic background radiation is the heat radiation left after the Big Bang. It still contains 99% of all the known radiant energy in the whole universe, so it is very bright. Unfortunately, it is also difficult to measure because it occurs at wavelengths where the Earth and its atmosphere also glow brightly.

With a lot more effort, astronomers have mapped the locations of the galaxies in three dimensions. Using the redshift from the cosmic expansion as a measure of distance, we can place each galaxy on a map. There are the huge stringy structures seen on the map of a million galaxies, but there is something new as well. There are huge empty regions, called cosmic voids, that stretch up to 300 million light years across. These voids also must be the result of some pre-existing seeds in the Big Bang.

Of course, cosmology might not be that simple. There might be much more exotic processes happening, and many scientific papers have been written about them. For example, perhaps the voids exist because some enormous explosions occurred within them, which then blew away all the remaining matter and left no trace of themselves. Or perhaps there are cosmic strings, stretching across huge distances, that are left over from the quantum mechanical processes in the earliest times. These could have immense energies and could also cause voids and stringy structures in the modern universe. This question adds to the importance of finding the cosmic seeds.

In truth we have strong evidence that cosmology is not simple. We have three kinds of evidence that the universe is filled with some kind of dark matter, with as much as 10 or 100 times as much mass as ordinary matter such as atoms and molecules. The particles of this dark matter apparently interact so weakly with ordinary matter that they have never been detected in laboratories. They can be detected only indirectly, by the gravitational forces that they exert. First, we have known for some decades that spiral galaxies rotate strangely. The rotational velocity seems to be a constant in the outer parts of the spiral, and if we apply Newton's laws of gravity we find that the mass contained within a given radius from the center is roughly proportional to the radius. However, the mass of the luminous stars and the hydrogen gas that we can measure there does not increase in the same way. There's something missing. Second, we can measure the gravitational forces in clusters of galaxies because many of them emit X-rays from gas clouds in their centers. The temperature of the gas clouds can be measured from the X-ray brightness, and used to compute the gravitational forces that hold them in place. Again, the calculated mass of the cluster of galaxies is much larger than the measured masses of all the stars and gases. Third, we can find clusters of galaxies that act like lenses, magnifying glasses for the even more distant galaxies and quasars behind them. The lensing action comes from the gravitational field of the clusters, which bends light so much that sometimes we can find multiple images of distant quasars or galaxies. Again, we can use the bending of light to calculate the mass of the cluster, and it is much larger than the mass of ordinary stars. Still, we don't know anything about this mysterious dark matter.

The only particles detected in human laboratories that are serious candidates are neutrinos. Ever since they were first discovered, they have appeared to be massless, but if they happen to have an extremely small mass, only 1/100 000 as much as an electron, they could solve the mystery. Unfortunately, making a neutrino requires a nuclear reaction, which is so violent that it's nearly impossible to find the effects of such a small mass. So we still don't know.

## 4 Open questions and the history of the background radiation

There are thus four very important questions of cosmology. First, is the hot Big Bang really the right picture? There have been some serious alternatives, including particularly the Steady State theory. Second, what makes the large scale structures of galaxy clusters and voids? Third, how did the primeval material turn into condensed objects like stars and galaxies? Fourth, what has happened to the gaseous material that did not condense into stars? It seems unlikely that all of the primordial stuff has turned into stars already. All four of these questions can be attacked by measuring the cosmic microwave and infrared background radiation.

The history of the background radiation goes all the way back to the Big Bang itself. It's thought that in the earliest times, particles of light (photons) and particles of each kind of matter and antimatter were about equally abundant, and all of them were extremely hot and packed very tightly together. So in the beginning there was a lot more than light. Then the universe expanded and cooled, and most of the antimatter particles found matter particles and annihilated them, releasing energy that was converted into more photons. The last antimatter particles to disappear were the anti-electrons (also called positrons), because these are the least massive and therefore the most easily created again from energy. This happened around the time when the universe was about $10^{10}$ times smaller than it is now, and a fraction of a second old. The next important thing that happened to the radiation occurred a few minutes later, when many of the neutrons left over from the explosion spontaneously decayed, or reacted with protons to make helium nuclei (alpha particles). However, even the enormous energy release from these nuclear reactions had little effect on the background radiation, mostly because there are about a billion ($10^9$) photons for every proton or neutron. Also, the matter and the radiation were in thermal equilibrium, so the radiation must have had a very nearly perfect blackbody spectrum.

A blackbody is an object that absorbs all the light that falls on it. If it is warm enough, it can also emit light. In the late 19th century Planck found a simple formula for the brightness of a blackbody. It shows that the brightness depends only on the temperature of the blackbody and on the wavelength at which it is measured.

About a year later, when the universe had expanded to about a millionth of its present size, something new happened. The processes that create and destroy photons almost stopped, because the temperatures had become too low. After that time, the number of photons was fixed, but it might have happened that the energy in the radiation field could have increased anyway if some of the more exotic ideas about the early universe are true. If this happened, it would result in a particular shape of distorted spectrum of the background radiation. This shape is called a Bose–Einstein distribution with a chemical potential, and occurs because the frequent collisions of the photons with the electrons can change the wavelengths of the photons. In other words, by proper measurements of the background radiation, we can test the theory all the way back to just 1 year after the Big Bang.

Around 1000 years later, even this wavelength changing by collisions with electrons essentially stopped, or became too slow to matter. That means that some exotic objects such as black holes could emit radiation that would simply add to the cosmic background radiation. If we could find this radiation we could see (with the aid of equations and imagination) all the way back to 1000 years after the Big Bang.

About 300 000 years after that, the universe rather suddenly became transparent. This is the moment called the 'decoupling' or the 'recombination', which happened when the atomic nuclei became able to hold onto the electrons and become complete atoms. The temperature then was only about 3000 K. The collisions of photons with electrons nearly ceased, so the radiation was free to move throughout the universe. We now see it as it was then, unchanged by further interaction with matter, but with longer wavelengths because of the cosmic expansion. This is a critically important event, because it means we can make a snapshot of a piece of the universe only 300 000 years after the beginning of time.

It took a long time, maybe hundreds of millions or billions of years, for the first galaxies to appear. We don't know because we can't see them growing. That's also one of the great mysteries. However, when galaxies did turn on, they may have been very bright. The energy from this burst of activity would be redshifted by the cosmic expansion and would appear in the infrared region of the spectrum.

A cartoon by S. Harris shows two angels discussing a little sparkling point. One of them asks if that was it, was that the big bang? We would like to answer the same question by direct measurement. Was the universe constituted according to this simple picture, or not?

We made a graph to show the intensities of the cosmic microwave and infrared radiation and compare them to other more recent radiation sources (Fig. 3). At wavelengths from about 0.5 mm to 1 cm, the cosmic microwave radiation is brighter than everything else, like electrons and dust grains in interstellar space that we have measured, up to a million times brighter. This is quite a remarkable result, and it means that we can hope to measure features, hot and cold spots, in the radiation that are very faint, only a few parts per million of the total.

At shorter wavelengths, the situation is quite different. We live inside a dust cloud, the debris of the asteroid belt where there should have been a planet orbiting the Sun in the space between Mars and Jupiter. It seems that maybe there was more than one planet in that zone, and they collided with each other. Now there are only small planets and rocks there, called asteroids, but they are continually colliding with one another and making tiny dust grains. Worse yet, the beautiful comets that come to visit the inner solar system every year or so are also balls of gravel and dust, held together by frozen gases like water. They disintegrate when the Sun warms them up, and the inner solar system is again filled with dust grains.

These tiny dust grains reflect sunlight, called zodiacal light, which we can see with our eyes in a dark enough place. They also emit a great deal of infrared light, which prevents us from searching for the predicted light of the first galaxies. There are a few ways around it. We can measure the light from

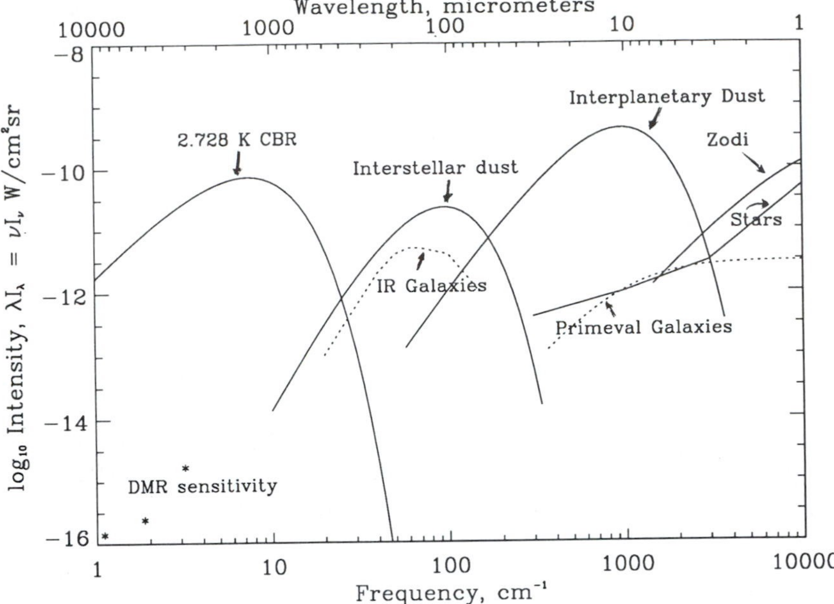


**Fig. 3** Brightness of diffuse radiation versus wavelength.

the dust grains very precisely and subtract it from the total. We can also concentrate our attention at those special wavelengths where the dust grains are not very bright, around 3 micrometers and longer than 200 micrometers. Maybe in the future we will have the ability to make a measurement from deep space, outside the asteroid belt.

Just to make life even more interesting, interstellar dust grains do the same thing at longer wavelengths. They are important in explaining how stars form, and you can see them with your own eye too. They are responsible for the dark lane running down the middle of the Milky Way, separating it into two rivers.

## 5 COBE satellite

We were therefore motivated to build a scientific satellite to measure all the sources of radiation shown on this graph. The mission has been well described by Boggess *et al.* (1992). Our first objective was to measure the spectrum of the microwave background radiation and compare it with the theoretical prediction, the perfect blackbody, with an accuracy a thousand times better than ever before. I am the Principal Investigator for this instru-

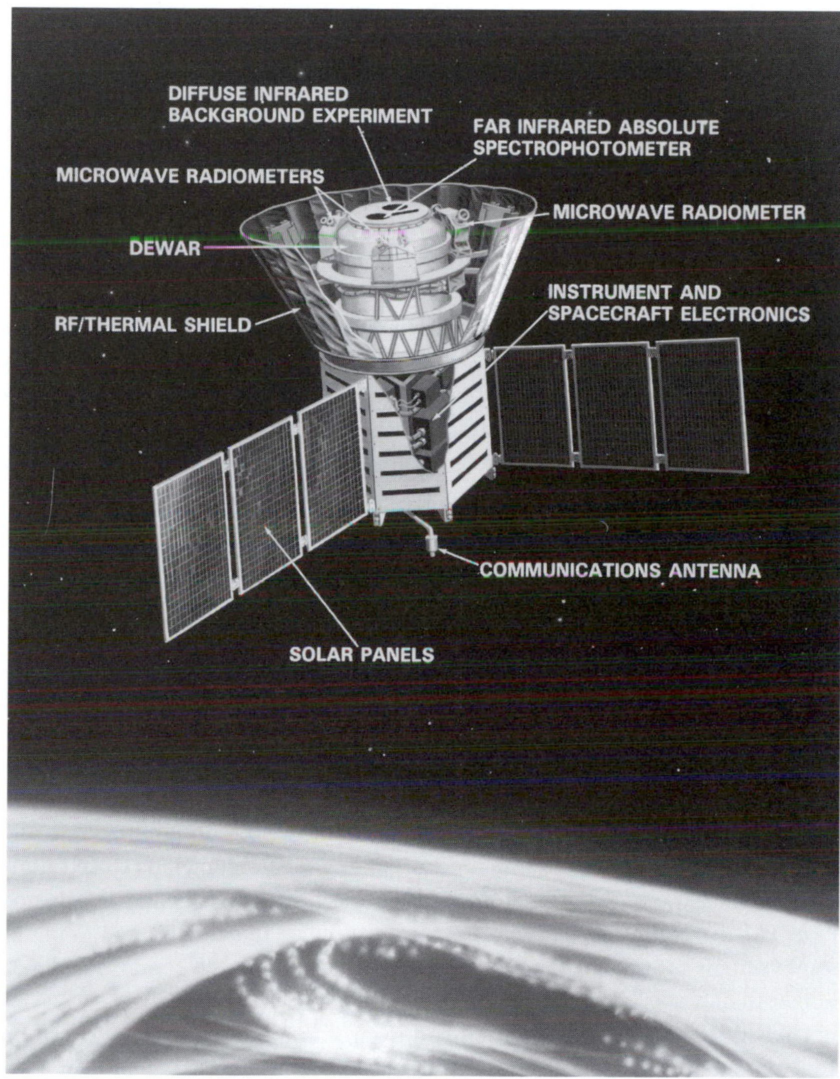


**Fig. 4** Artist's concept of COBE in space.

ment, which is called the Far Infrared Absolute Spectrophotometer (FIRAS), and Rick Shafer is my deputy PI. It is described by Mather, Fixsen, and Shafer (1993). The second objective was to measure the cosmic background anisotropy, to find out if it is equally bright in all directions, with an accuracy ten times better than before. George Smoot is the PI for this one, which is called the Differential Microwave Radiometers (DMR), and Charles Bennett is the Deputy. It is described by Smoot *et al.* (1990). The third objective was to measure the light from the first luminous objects, which would make a diffuse infrared background light at shorter wavelengths, with an accuracy a hundred times better than before. Michael Hauser is the PI for this instrument, called the Diffuse Infrared Background Experiment, and Thomas Kelsall is the Deputy. It is described by Silverberg *et al.* (1993).

The artist's concept shows the COBE in orbit, pointed away from the Earth and perpendicular to the Sun. The instruments are located in the top half, inside a conical shield that protects them from the heat of the Earth and the Sun. The FIRAS and DIRBE instruments are cooled to the temperature of liquid helium, only 1.5 degrees Kelvin (degrees above absolute zero), inside a large vacuum-insulated flask called a cryostat or Dewar. The DMR receivers occupy three boxes attached to a ring around the cryostat. The spacecraft is powered by solar cells on three large wings that unfolded after launch, and the electronic boxes are located in the lower half where they can be kept warm. The whole spacecraft spins at 0.8 revolutions per minute about its axis, so that the DMR and DIRBE instruments can map a large fraction of the sky quickly. Their lines of sight are oriented 30 degrees away from the spin axis. The orientation is sensed by Sun and Earth sensors, and the rate of spin is sensed by gyroscopes. Complicated electronics use the signals from these sensors to control the speeds of spinning reaction wheels and the currents through magnetic torquer bars, which push against the Earth's magnetic field to control the rotation of the spacecraft.

The COBE was assembled in very clean conditions at Goddard, and was cleaned over and over again, to make sure that it would not carry dust and gases with it to contaminate the measurements. The technicians wore bunny suits of special cloth that makes almost no lint, with masks and hoods and latex gloves. The basic spacecraft is about twice as tall as a human being. It took us many months to complete the construction in these conditions, much longer than it would be without all these precautions.

The COBE was finally launched southward on a Delta rocket at dawn on 18 November 1989, from the Vandenberg Air Force Base at Lompoc, California. The coastline there runs east and west, so the rocket can fly immediately over the water, minimizing the danger to human life, and the spent booster rocket stages can fall into the water. The Delta had 9 solid fuel boosters attached to its liquid fueled first stage. The second stage, also liquid fueled, put the COBE into a perfect orbit for us, 900 km above the Earth, with an orbital period of 103 minutes. With almost exactly 14 orbits per day, the COBE came overhead at the same time every morning and evening. The orbit inclination is 99 degrees, chosen to make the plane of the orbit precess at the rate of one revolution per year to follow the apparent motion of the Sun. This special feature enabled us to keep the Sun from shining on the instruments, and the limb of the Earth never rose more than a few degrees above the top of the shield.

## 6 Spectrum of the cosmic microwave background radiation

Now I'll describe the COBE instruments, starting with the FIRAS, the Far Infrared Absolute Spectrophotometer (see Fig. 5). This is the one designed to test the prediction of the Big Bang theory that the Cosmic Microwave Background Radiation (CMBR) should have the same spectrum as a blackbody. The FIRAS does this by taking its own blackbody with it, a chunk of epoxy plastic filled with iron powder that absorbs microwaves very well. Its temperature can be adjusted from 2 to 20 Kelvin to calibrate the instrument, and it can be set to match the temperature of the background radiation. We put the calibrator into the aperture of the instrument once a month at the

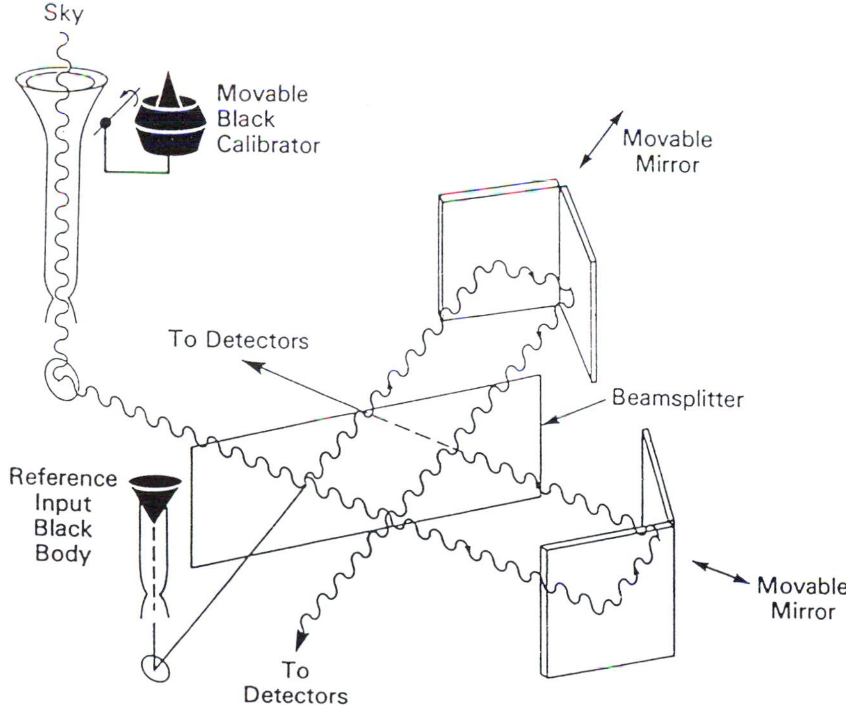


**Fig. 5** Concept of the FIRAS, the Far Infrared Absolute Spectrophotometer. The real instrument is much more complicated.

beginning of the mission, and then every week near the end. The instrument measures the wavelengths of the radiation by dividing the incoming waves into two equal parts, sending them out in two directions, and then recombining them. This design, which was invented by Michelson in the late 19th century, produces an interference pattern when the waves recombine, according to whether the peaks of the two waves line up together or oppose each other. We measure this interference pattern as we change the difference in distances that the two waves have traveled, and then we digitize it and send it back to the computers on the ground for analysis. The detectors we use to measure the intensity of the radiation are just thermometers, but they are extraordinarily sensitive, capable of detecting a few parts of a millionth of a billionth of a watt in just a second. They are so sensitive that they respond to individual cosmic rays from the Sun or the Van Allen belts that pass through them, which is a bit of a problem for us.

The whole optical portion of the instrument is cooled by liquid helium to about 1.5 Kelvin. If it were not so cold, the radiation emitted by the instrument itself would be so bright that it would confuse the measurement. The accuracy of the instrument depends entirely on the calibrator blackbody, which must get at least as cold as the CMBR to which it is compared.

When we mapped the sky with the FIRAS, we found that as expected there is interstellar dust everywhere. This is a problem for cosmology, because there is no direction to look where there is a completely clear view of the beginning of the universe. On the other hand, the dust is arranged in a layer, concentrated in the plane of the Milky Way galaxy. We made a map of the brightness of the dust at a wavelength of 700 micrometers, which shows bright spots in the Galactic Plane, above the Galactic Center, and in the constellations of Orion and Cygnus where bright stars are forming from the interstellar gas and heating up the dust. Of course, this is a very interesting map for regular astronomers, because it gives a snapshot view of the whole sky at once. We used this map to help make a mathematical model of the emission of the dust, so we could subtract it and imagine what the sky would be like if we could move the COBE far away from home, far from the Milky Way.

The calibration, analysis, and interpretation of all these data is described by Fixsen *et al.* (1996), building on Fixsen *et al.* (1994*a*,*b*), Mather *et al.* (1994), Mather *et al.* (1990), Wright *et al.* (1994), Wright *et al.* (1991), and Bennett *et al.* (1994). The result of that subtraction is that the CMBR spectrum is extraordinarily close to the prediction of the Big Bang theory. The maximum difference is only 0.01% of the peak brightness, a part in 10000, over the wavelength range from 0.5 to 5 mm where the radiation is the brightest. I have plotted this brightness versus the radiation frequency, the number of waves per centimeter (Fig. 6), and on this plot the differences

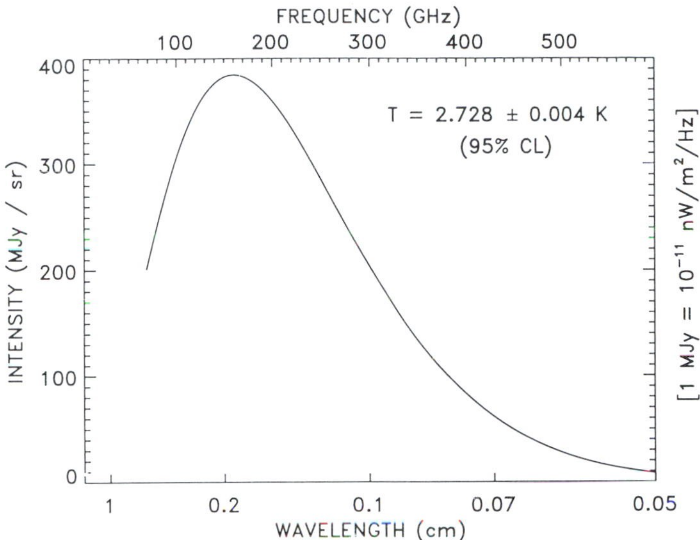


**Fig. 6** Spectrum of the cosmic microwave background radiation. Theory and observation agree within 0.01%.

between observation and theory are much too small to see, much less than the width of the line. We also determined that the temperature of the radiation is 2.728 Kelvin, with an uncertainty of only 0.004 K. Of course, the exact temperature doesn't matter very much, since it is gradually running down as the universe expands, but we do need the number for comparison to the total density of various kinds of matter. The comparison tells us much about the laws of physics in the very early universe.

One of the most immediate conclusions from this test is that almost all of the energy in the CMBR was released all at once, in the first year. This is not a surprise in the simple Big Bang picture, but the experimental scientist must test everything. According to calculations, the shape of the distortion depends on when the hypothetical energy release occurs. Cosmologists measure these by the $\mu$ parameter for a Bose–Einstein spectrum, which describes energy release between year 1 and year 1000, and the $y$ parameter for Compton scattering, which describes energy release from year 1000 to year 300 000. We have plotted the result in terms of the fraction of the energy of the CMBR that could have been added after the Big Bang, versus $1 + z$, where $z$ is the redshift, and $1 + z$ is the expansion factor of the universe, the relative size now versus then. The result is that only about 0.006% of the energy of the radiation could have been added after the first year. There are a lot of ideas about how such energy could have been added, mainly by conversion from other forms. Black holes could grow or evaporate, elementary subatomic particles could be unstable and decay, or primordial cosmic turbulent energy could be dissipated. Cosmic strings, which are hypothetical remnants of quantum mechanical instabilities in the early universe, might release energy. Gravitational waves might also be remnants of the early conditions, and could release energy. Our results don't say these processes didn't happen, but only that they didn't add more than a little bit to the CMBR.

## 7 Anisotropy of the cosmic microwave background radiation

Now I'd like to turn to the subject of the anisotropy of the CMBR, the differences in brightness from one direction to another. The effect that we expect to see was investigated theoretically by Sachs and Wolfe in 1968. According to them, a region of space that is denser than average will have a gravitational field that will cause the photons of light to lose energy when they leave. For photons, lower energy means longer wavelength, or in other words a redshift. That means that we ought to be able to map the density of matter in a particular slice of the early universe as it was at the decoupling, 300 000 years after the Big Bang, by mapping the temperature of the CMBR that we see from here.

There is another aspect too. At the moment that a photon of light last bounces off an electron before heading in our direction, the electron and the light are responding to the conditions there. In that volume of space there may also be moving matter of both ordinary and dark types, and gravitational fields arriving from distant regions. The motion of the matter is in response to the gravitational fields and the pressures of the matter and the radiation. The radiation arriving at that spot might already be anisotropic, different brightnesses in different directions. The little electron is like an astronomer for us, measuring a certain combination of things and sending a signal in our direction.

One of the amusing but important things about the CMBR is that we can tell whether we are moving relative to the rest of the universe. If the whole universe were just expanding smoothly, with no parts going slower or faster, the CMBR would seem perfectly uniform to an observer moving with the cosmic expansion (going with the flow). If we are not going along with the flow, then there is a part of the anisotropy of the CMBR which is caused by the Doppler effect of our own velocity. The direction toward which we are headed appears hotter, with a fractional temperature change of $\Delta T/T = v/c$, where $T$ is the average temperature of the CMBR, $\Delta t$ is the difference between the temperature in the forward direction and the average, $v$ is our velocity, and $c$ is the speed of light.

Our velocity is a vector sum of many parts. The COBE orbits around the Earth at 7.4 km $s^{-1}$, the Earth circles the Sun at 30 km $s^{-1}$, the Sun orbits the Milky Way Galaxy at around 200 km $s^{-1}$, the Galaxy is falling towards the middle of the Local Group of galaxies at around 40 km $s^{-1}$, and the Local Group of galaxies is falling toward distant clusters and superclusters of galaxies at around 600 km $s^{-1}$. It's an important question for cosmology to see if the measured velocity relative to the CMBR can be accounted for by a sum of all these parts. Getting the answer is pretty tricky, because the distant galaxies are moving too. From the COBE data we were able to get the velocity relative to the CMBR within a few km $s^{-1}$.

To measure the anisotropy, we built the Differential Microwave Radiometers (DMR) (Smoot *et al.* 1992; see Fig. 7). They use a classic design called a Dicke-switched microwave radiometer, which has a standard sort of microwave receiver, with a diode mixer to convert the microwaves to a lower frequency where they can be amplified by transistors. After amplification, the output of the receiver is detected by a diode, which measures the total intensity of the microwaves coming in to the input. The special feature is a switch at the input. It connects the receiver alternately to two different antennas, pointing 60 degrees apart on the sky, 100 times per second. The output of the receiver goes up and down 100 times a second too, according to the difference in intensity of the microwaves received through the two antennas. A second switch at the output is used to measure this difference, which is then digitized and sent back to the computers on the ground.

Just to be sure we got good data, we measured at three different wavelengths, 3.3, 5.7, and 9.6 mm, and we had two separate receivers at each frequency. The multiple wavelengths were important because they helped us recognize the difference between emission from our Milky Way Galaxy and the cosmic microwave background radiation. The receivers were mounted in three boxes around the outside of the helium cryostat that cools the FIRAS and DIRBE. They were not cooled by the helium, but had to be kept warm anyway because they could radiate their heat to the dark sky. The 3.3 and 5.7 mm receivers operated at a temperature of 140 Kelvin, while the 9.6 mm receivers were warmed up to room temperature, about 290 K.

To make a map of the sky, we measured hundreds of millions of differences, with the antennas pointed in all possible combinations of directions. As the COBE orbits and spins, and the orbit plane precesses throughout the year, all these combinations are measured hundreds of times. Then we fed all the measurements into a computer program that made a map that would fit all the data as well as possible. This is called a 'least squares fitting program' and it required months of computer time every time we ran it.

The result of our effort is a map of the sky, in which we have already removed as best we can the effects of the Milky Way, and the dipole pattern

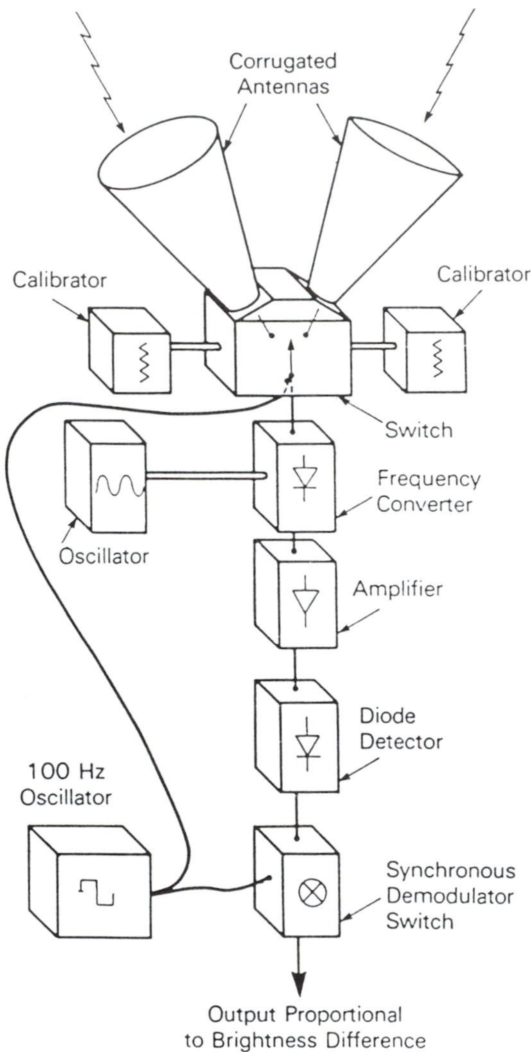


**Fig. 7** DMR conceptual design.

that results from the Doppler shift from the motion of the Earth. We take out the Milky Way by making a particular linear combination of the maps at the three different DMR wavelengths, and we take out the Doppler shift dipole by just measuring and subtracting it. The remains are a map of the cosmic anisotropy, with a bit of receiver noise added in. The process of this analysis is described in Kogut *et al.* (1992), Smoot *et al.* (1992), Bennett *et al.* (1992), Wright *et al.* (1992), and Bennett *et al.* (1996). In the map I show here (Fig. 8), based on four years of flight data, the receiver noise is significantly

COBE DMR 4−year Sky Map

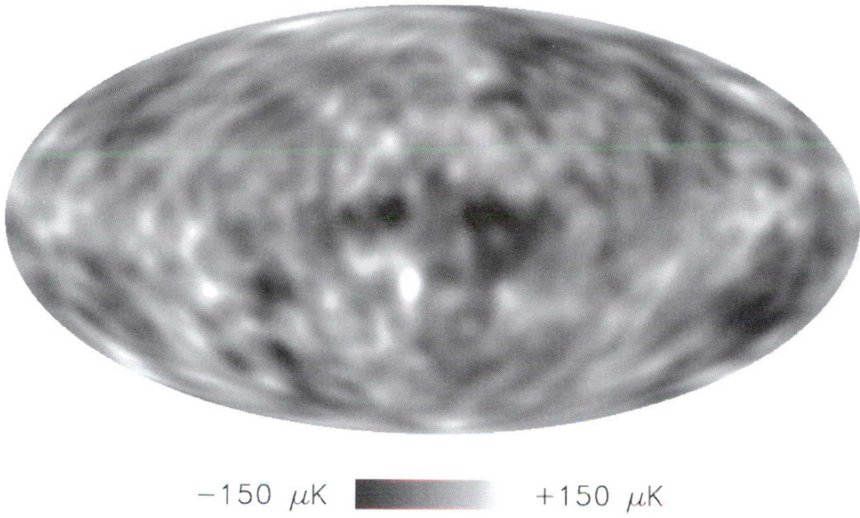


**Fig. 8** Map of the cosmic anisotropy in Galactic coordinates, with the Galactic Center in the middle.

less than the cosmic anisotropy, so most of the large blobs on this map are real, but many of the smaller ones are not. The map is plotted in Galactic coordinates, so the Milky Way runs right across the middle, and the Galactic Center is right in the middle. You can see that we have failed to remove some traces of the Milky Way, since there is a large dark blob just to the right of the middle of the map. To make sure our cosmological conclusions were not biased by that problem, we just cut out the part of the map near the Milky Way before doing the calculations.

The important conclusion is that we have mapped the Universe as it was only 300 000 years after the Big Bang. The dark regions on the map are about 0.01% more dense than the average, and will grow up to become clusters and superclusters of galaxies. The lighter regions on the map are about 0.01% less dense, and will grow up to become giant voids, regions of space with few or no galaxies. We can't identify a particular blob on this map with a nearby cluster of galaxies though, because when we account for the travel time of the light, we see the nearby material as it was only a little while ago, while we see this distant stuff as it was just after the Big Bang.

It is very important to compare the sizes and intensities of the blobs on this map with the sizes and intensities of galaxy clusters now. If we could really understand how the universe worked, we should be able to work out the process mathematically and simulate the whole thing in a computer. The people who do that have told us that the picture won't come out right unless there's a great deal of the dark matter, the invisible material whose presence can only be deduced by its gravitational effects. The dark matter is allowed to start moving even before the decoupling event at 300 000 years.

To carry this argument a lot farther, theorists have made many predictions about the amount of anisotropy for various different angular scales, which they measure with multipole moments. A multipole moment $l$ is roughly 180 degrees divided by the angular size of the structures being studied. There are several leading theories, for various amounts of cold dark matter, dark matter with ionized matter, and another version without dark matter called Primordial Baryon Isocurvature. If the matter could not affect the CMBR, the prediction would be a 'scale invariant' power law, which can be plotted as a horizontal line across the middle of a graph. The measurements by the COBE and other groups tell us that a horizontal line across the middle is not right.

There's something else that's very important. There are many bumps and wiggles on the plots, that we can call the first music of the spheres. They are literally the result of acoustic oscillations of the primordial material, as it begins to fall in to the dense regions at the decoupling, and then bounces back. Sometimes these bumps on the plot are called Doppler peaks, and sometimes they are named after Sakharov, the great Soviet physicist who worked on cosmology as well as on the hydrogen bomb. The importance of these bumps is that they can be measured, and the measurements will tell us the properties of the early universe. From the details of these curves, we should be able to know the density of matter, both ordinary and dark kinds, the total curvature of the universe, the value of the Einstein repulsive force (the lambda constant), and the Hubble constant, all within a few percent accuracy. The quest for these bumps has become the new Holy Grail of cosmology.

To do these measurements we will need a much better instrument than the one on COBE, to give a much sharper image, with finer angular resolution and better sensitivity. One can actually simulate the sky as it would appear with such an instrument. One can even recognize qualitative differences by eye, but clearly the analysis into multipoles is statistically more powerful.

There are at least 5 major projects currently proposed or under study to get these maps, three in the USA, one in Europe, and one in Russia. One mission, called the Microwave Anisotropy Probe (MAP), is led by Chuck Bennett at Goddard and David Wilkinson at Princeton. As this article was being edited, NASA approved the MAP mission, to be launched in about 2000. It is a greatly improved version of the DMR, with better receivers, and with two back-to-back telescopes instead of two little antennas. NASA has decided to get away from its concentration on large projects that take a long time and a large budget. There are also strong hopes that the European mission, entitled COBRAS/SAMBA, will be approved for an even more complete and ambitious measurement.

## 8 The search for the cosmic infrared background radiation

There's one more cosmic fossil to find, the cosmic infrared background radiation from the first generations of stars and galaxies and maybe other kinds of luminous objects in the early universe. It's really not known how they formed or when or what they looked like. One of the great hopes is that we could find their accumulated emissions, even if we could not recognize individual objects. In principle one needs only a small instrument to find a diffuse, uniform glow from those early generations, while one needs a very large telescope to magnify them enough to see individually. So far it seems we don't have such a large telescope, which would probably have to work at infrared wavelengths. Of course there's always a catch: here it is that almost everything, from stars to galaxies to interstellar and interplanetary dust, can also emit infrared light. The difficulty is in proving that the measured signal is really cosmic. In principle it's possible because we know where all of these nearby things are located.

To attack this problem we designed the Diffuse Infrared Background Experiment, the DIRBE (Silverberg *et al.* 1993, see Fig. 9). It has a small Gregorian telescope, only 19 cm in diameter, to collect light from a region 0.7 degrees square. Regular astronomers would not call this a telescope, because the view that it gives is much fuzzier than what you can see with your own eye, but the compensation is that we have designed it for exceptional accuracy and sensitivity at the infrared wavelengths that interest us. It covers 10 different wavelengths over the range from 1.25 to 240 micrometers, with four different types of detectors. It can measure polarization at short wavelengths, from 1.25 to 3.5 micrometers, because scattered sunlight from the interplanetary dust is polarized and can be recognized that way. Like the FIRAS, the DIRBE is cooled to about 1.5 Kelvin to keep its own emissions from exceeding the brightness of the cosmic signals we seek.

The DIRBE data have not yet yielded their cosmic secrets, but they do give us a new view of the universe. We have made maps of the sky at the 10 wavelengths, and combined some of them in false color images so that color indicates temperature, with blue indicating warm, yellow meaning average, and red indicating cool. In the map showing 1.25, 2.2, and 3.5 micrometers, the main feature is our own Milky Way Galaxy. The brightest part occupies the center of the picture, but of course we are a part of the Milky Way

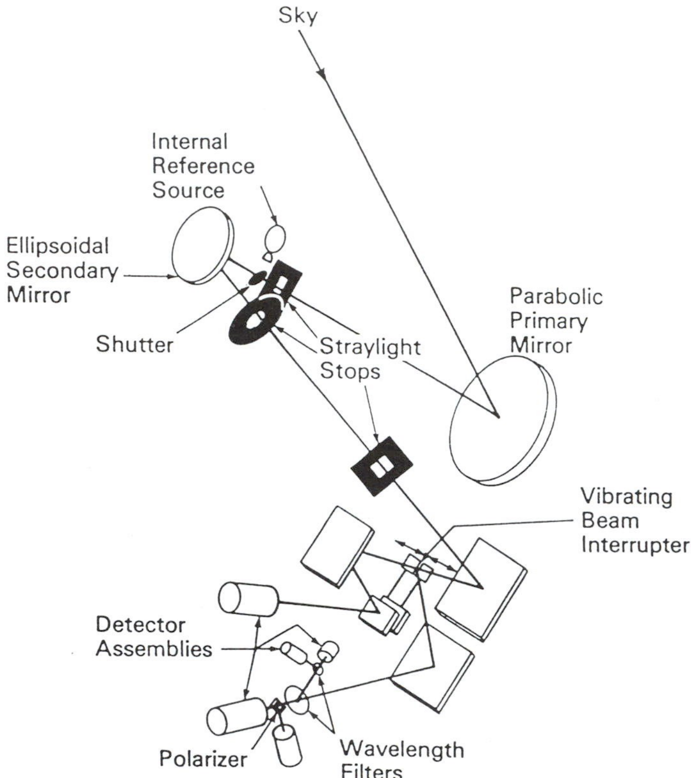


**Fig. 9** Diffuse Infrared Background Experiment, DIRBE.

Galaxy ourselves, so this map actually wraps around us and shows the whole sky except for our own vantage point. In the middle we can see that there is an oval bulge, which is made mostly of old faint stars, and a thin orange line going through the middle. We are looking at the inner part of our own spiral Galaxy edge on, because our own location is in the plane of the spiral. The plane of the Galaxy is orange because interstellar dust is absorbing the shorter wavelength light from the stars there. The dust is the same dust that makes the Milky Way divide into two great rivers. For visible light, the dust is just opaque, but the long wavelength infrared can go right through it. Far from the middle of the picture, there are a few recognizable stars, but if we had printed the map differently the whole map would be covered with stars.

The image changes dramatically when longer wavelengths are used. The map for 4.9, 12, and 25 micrometers has a bright diagonal S-shaped orange band, which is the interplanetary dust emission. The interplanetary dust, the debris of asteroid collisions and comet disintegration, actually lies close to the Ecliptic Plane, the plane in which the planets move, but it looks like a tilted S shape because the map is plotted in Galactic coordinates. With these false colors, the stars in the Galactic Plane are blue, and there are a handful of red regions in the Galactic Plane which are places where stars are forming inside dust clouds. It's very difficult to look for the cosmic infrared background light at the same wavelengths where the interplanetary dust is bright.

When the view shifts to longer wavelengths the picture changes dramatically again. In the map for 25, 60, and 100 micrometers, the interplanetary dust is now blue, and the interstellar dust is yellow and red. The dust clouds are very bright in the Cygnus region to the left of center, in the Sagittarius region in the middle, and in the Orion region at the far right. All of these are locations where stars are being formed in abundance. In the middle of the lower right quadrant one can even see the Large Magellanic Cloud, where the supernova exploded in 1987. The dust grains were not present in the Big Bang, because the elements from which they are made did not yet exist. The heavier elements were made by nuclear reactions inside stars, which then shed their atmospheres either gradually, or in great supernova explosions.

The DIRBE measurement of the cosmic infrared background light is so far an upper limit (Hauser 1996*a*,*b*); we didn't measure it yet, but we know it is somewhat fainter than the foreground objects we know about. That's already an important cosmological result, because some of the theories of the early universe imagined rather exotic processes that would liberate great quantities of energy in the infrared region of the spectrum. We have plotted the DIRBE measurements of the darkest parts of the sky on the same chart with the theoretical predictions, and some extrapolations of the measurements of individual galaxies (Fig. 10). It's clear that our best chance to test more interesting theories is at a wavelength like 3.5 micrometers, or longer than 200 micrometers, where the interplanetary dust and starlight are not too bright.

## 9 Summary and conclusions: history of the universe

Now is a good time to recap the whole history of the universe, leading up to our present existence, and to speculate a little on the long-term future. In the beginning was the Bang, and the temperature and density were inconceivably great, so great that we doubt that we know the laws of physics that should apply there. The universe expanded and cooled, the antimatter annihilated most of the matter, and around 3 minutes later the helium nuclei were formed from neutrons and protons. Around 300 000 years later the universe abruptly became transparent, when the atomic nuclei grabbed the free electrons and become complete atoms. We see the cosmic microwave back-

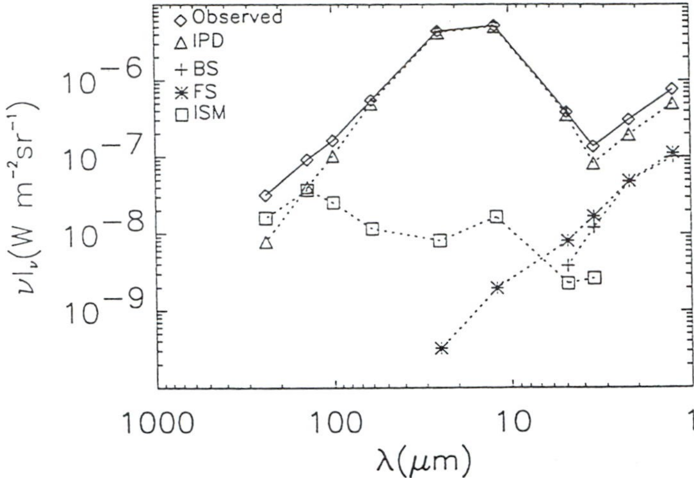


**Fig. 10** DIRBE measurement at North Galactic Pole compared with foreground estimates from IPD (interplanetary dust), bright galactic sources (BS), faint galactic sources (FS), and interstellar medium (ISM).

ground radiation as it was then, complete with hot and cold spots that trace the density distribution of matter. The denser regions eventually formed galaxies and clusters of galaxies, maybe a billion years later. Stars formed too, but we don't know which came first. The stars made the heavier elements out of the primordial hydrogen and helium by nuclear reactions, and some of the elements were expelled back into space when the first generation of stars died by explosions.

One of them blew up right near the nebula, the cloud of gas, that would become the Sun and the planets, about 4.5 billion years ago. Rich Muller calls this the second Big Bang. The Solar System formed rather quickly, but the early days were quite violent. There were many fragments left over from the formation of the planets, and the fragments bombarded the Earth and the Moon and the other planets leaving craters everywhere. It is thought that conditions were quite hostile to life here, with hazardous chemicals, frequent volcanic activity, and extreme temperatures as well as the falling planetary fragments. Eventually things settled down, the liquid water made oceans, and the atmosphere of nitrogen and carbon dioxide blanketed the land, if there was any. Somehow life got started. It used to be thought that starting life requires an extraordinary coincidence, but that is no longer obvious. Instead, it seems that some kind of chemical concentrations may have existed quite commonly, and some of them, maybe many of them, made life. There is no trace of the early life, except for the one or more forms whose descendants now cover the Earth. If there were more than one kind, then we would not know about it now because Darwinian evolution has erased all but the successful competitors. Things went along rather peacefully for a long time, with very slow changes of life forms, until suddenly in the Cambrian era life expanded to fill every available spot.

Then there was the third of Muller's Big Bangs, the asteroid that collided with the Earth at the end of the dinosaur era, that may have actually caused their extinction. There's strong evidence of the impact, from the worldwide distribution of the rare element iridium in the sedimentary rocks formed at the time, but it's much more difficult to show that the mass extinction was caused by it. At any rate, mammals survived the holocaust, as did the birds, which seem to be the modern descendants of the dinosaurs. Human beings are a very recent arrival on the scene, only a few hundred thousand years to a few million years old, depending on your definition of 'human.'

To look to the future a little, one may speculate that humans could create another form of virtual life, based on artificial intelligence and silicon chips perhaps. It we are able and dare to give it the capability of self-reproduction, then it could also evolve very rapidly. The science fiction writers have more imagination here than I, but I see no fundamental reason why a silicon-based electronic intelligent life form could not travel to the stars if it wanted to. I have strong doubts that ordinary carbon-based life could make the trip without help from the artificial kind. We already depend on computers for the most ordinary things, like running an automobile. In a practical sense the answer may already be available—if such travel is possible, why have we not already had a visit? While some would say we have, I've never believed the evidence.

A popular cartoon answers one of the questions I frequently receive: what's beyond the observable universe? The answer is, 'lots and lots of unobservable universe.'

## 10 Data products

The COBE data and images are available from the National Space Science Data Center and are fully documented online. The World Wide Web address is

http:/ /www.gsfc.nasa.gov/astro/cobe/cobe_home.html.

One can also use anonymous FTP to nssdca.gsfc.nasa.gov, with a change to the [.cobe] directory. This site provides a full bibliography of COBE publications and links to online versions of most of them.

## Acknowledgments

The COBE Project is supported by the Astrophysics Division of NASA Headquarters, and executed by Goddard Space Flight Center under the guidance of the COBE Science Working Group. My own role in the project was only possible because of the guidance that I received from my scientific mentors: Paul Richards, my thesis advisor; Pat Thaddeus, my postdoctoral advisor who suggested that I propose a satellite project; Michael Hauser, who brought me to Goddard and serves as the PI for the DIRBE; Rainer Weiss, the Chairman of the Science Working Group; and Nancy Boggess, who supported us vigorously at NASA Headquarters. Most of the work that I have reported here was done by the COBE team, not by myself. The COBE team included 1500 people who participated in varying degrees, led by the Project Manager Roger Mattson and the Deputy Project Manager Dennis McCarthy. The COBE Science Working Group includes Charles L. Bennett, Nancy W. Boggess, Edward W. Cheng, Eli Dwek, Samuel Gulkis, Michael G. Hauser, Michael Janssen, Thomas Kelsall, Phillip M. Lubin, John C. Mather, Stephan S. Meyer, S. Harvey Moseley, Thomas L. Murdock, Richard A. Shafer, Robert F. Silverberg, George F. Smoot, Rainer Weiss, David T. Wilkinson, and Edward L. Wright. David T. Leisawitz is Deputy Project Scientist and is in charge of the Guest Investigator program and public distribution of the data.

The National Aeronautics and Space Administration/Goddard Space Flight Center (NASA/GSFC) is responsible for the design, development, and operation of the Cosmic Background Explorer (COBE). GSFC is also responsible for the development of the analysis software and for the production of the mission data sets.

## References


Bennett, C. L., *et al.*, 1992, *Astrophys. J. Letters*, **396**, L7.

Bennett, C. L., *et al.*, 1992, *Astrophys. J.*, **434**, 587.

Bennett, C. L., *et al.*, 1992, *Astrophys. J. Letters*, **464**, 1

Boggess, N. W., *et al.*, 1992, *Astrophys. J.*, **397**, 420.

Fixsen, D. J., *et al.*, 1994*a*, *Astrophys J.*, **396**, 445.

Fixsen, D. J., *et al.*, 1994*b*, *Astrophys J.*, **396**, 447.

Fixsen, D. J., *et al.*, 1996, *Astrophys J.*, **473**, 576.

Hauser, M. G., 1996*a*, In *Proc. IAU Symposium 168, Examining the Big Bang and diffuse background radiations*, 99–108, Kluwer: Dordrecht, M. Kafatos and Y. Kondo (eds.).

Hauser, M. G., 1996*b*. In *Unveiling the cosmic infrared background, AIP Conf. Proc 348*, 11, E. Dwek, ed.

Kogut, A., *et al.*, 1992, *Astrophys. J.*, **401**, 1.

Mather, J. C. *et al.*, 1990, *Astrophys. J. Letters*, **354**, L37.

Mather, J. C., *et al.*, 1994, *Astrophys. J.*, **420**, 439.

Mather, J. C., Fixsen, D. J., and Shafer, R. A., 1993, *SPIE Conf. Proc.*, **2019**, 168.
Silverberg, R. F., *et al.*, 1993, *SPIE Conf. Proc.*, **2019**, 180.
Smoot, G. F., *et al.*, 1990, *Astrophys. J.*, **360**, 685.
Smoot, G. F., *et al.*, 1992, *Astrophys. J. Letters*, **396**, L1.
Wright, E. L., *et al.*, 1991, *Astrophys. J.*, **381**, 200.
Wright, E. L., *et al.*, 1992, *Astrophys. J. Letters*, **396**, L13.
Wright, E. L., *et al.*, 1994, *Astrophys. J.*, **396**, 450.

# The complexity of our singular universe

Sir Roger Penrose

## 1 Introduction

It is a pleasure and an honour to have the opportunity to pay my respects to E. Arthur Milne, who was the first holder of the Rouse Ball Chair in Oxford, which it is my privilege to occupy now. My comments will indeed have considerable relevance to some of Milne's innovative research in cosmology, as well as having a close relationship to some of the other articles in this commemorative volume.

My title has been deliberately chosen to be a pun, in more than one way. Our universe is certainly singular, in that it has been given to us once only, which makes cosmology different from other experimental sciences. But it is also singular in another way, in apparently possessing regions—called *space–time singularities*—at the big bang and inside black holes, where our presently understood physical laws break down. I shall have more to say about these later. There is, perhaps, a third sense in which our universe is singular. The mathematical laws themselves, with which our universe so closely accords, are themselves of a 'singular' character, in the sense that they are not just 'randomly' chosen, but have an exquisite and tightly controlled nature which depends upon unique propensities of particular mathematical structures. We shall be seeing something of this later as well.

The word 'complexity' is also a pun. The space–time singularities in the past—at the big bang—seem to have a particularly simple character and constitute a region having an elegant and symmetrical mathematical form. On the other hand, the space–time singularities in the future—in black holes or at the all-embracing big crunch (*if* that will eventually take place)—appear to be enormously complex, effecting a general progression from the simple to the complex that has a close association with the time-asymmetric second law of thermodynamics. Yet, there is quite a different sense in which our universe appears to be 'complex', and I shall be saying something about this at the end. The very laws which govern physical behaviour at the minutest scales seem to depend intimately upon those mathematical numbers which are technically referred to as 'complex'. We shall be seeing that the properties of complex numbers reveal themselves also on a macroscopic scale, and might well have something to say about the overall structure of our universe.

## 2 Space–time

In Fig. 1, I have indicated how the space–time picture of the physical world appears according to Newtonian theory. The planes which go across the picture horizontally are supposed to represent space. I should like to have depicted 3-dimensional space and then the whole thing would be 4-dimensional, but of course there is a difficulty in presenting that to you. So you have to imagine that the flat planes are actually 3-dimensional spaces. The one in the middle is called 'today', and the one above it is 'tomorrow', followed by 'day after'; and below 'today' we have 'yesterday' and 'day before'. Thus, you imagine all these instantaneous spaces, stacked up 'on top' of each other to give the entire space–time. Any particle in the universe will be represented by what is called a *world-line*—in terms of our diagram, a curve drawn on the space. I have also tried to represent Newtonian *mechanics*. Some of the world-lines are curved because they are being acted on by the forces between these particles (shown by the little arrows). This means that the world-line is deformed from a straight line (which it would otherwise be) in space–time. A straight world-line means the particle is moving uniformly with constant speed. The acceleration force will cause the world-lines to be

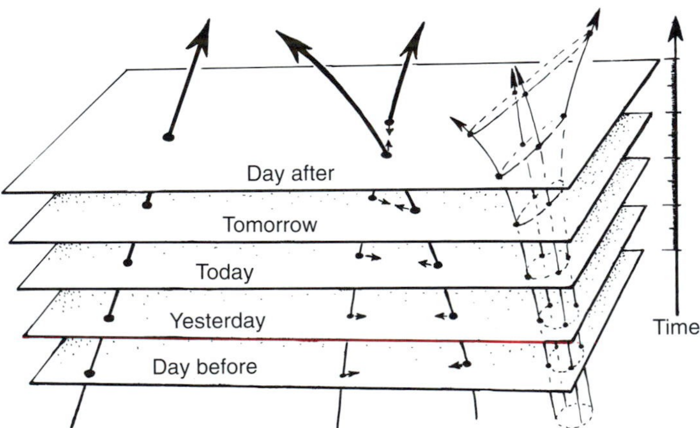


**Fig. 1** Newtonian space–time. Particle histories are represented as *world-lines* which are curved whenever forces act on the particles. On the right, the *tidal effect* is illustrated, which is a feature of gravitational fields in empty space.

bent. So that is the picture of Newtonian physics. Notice that you have a very well defined notion of *time*, which labels these different sections through the space–time.

Actually, this is not really the entire picture we have, even just within the general Newtonian framework, because there are things called *fields*, and these fields—for example, electromagnetic fields—add a lot of confusion to the picture. However, they have very precise equations governing how they behave in time, and do not actually alter the picture very much. What alters the picture a great deal more is the introduction of what is called Minkowskian geometry—this is the space–time of Minkowski shown in Fig. 2.

What Minkowski did was to put time and space on a more equal footing (although not entirely). This geometry is a bit like a Euclidean geometry of four dimensions in which you would have an expression for the distance (or 'metric') between two points involving the sum of squares. The difference is that in *Minkowskian geometry*, one puts some minus signs into the expression. (It depends, a little, on whether you are here or at Cambridge as to whether you put three minus signs and one plus or three plus signs and one minus. In Cambridge they tend to do it the other way, but I like to put three minus signs and one plus, for reasons which I shall come to shortly.)

The main features of the Minkowskian picture are what are called the *light cones*. A light cone represents the history of a flash of light emanating from (or converging on) a space–time point. Points in space–time are often referred to as *events*—they have only an instantaneous temporal existence as

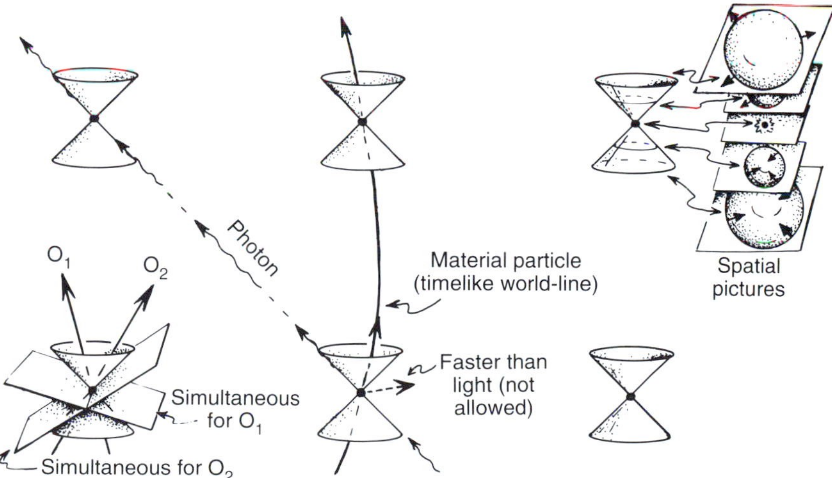


**Fig. 2** The Minkowskian space–time of special relativity. The events simultaneous for observer $O_1$ are different from those simultaneous for observer $O_2$ (lower left). The light cones represent histories of light flashes which converge on an event and then re-expand from it (upper right). A material particle's world-line must lie within the light cones whereas that of a photon lies along them.

well as zero spatial extension. At a particular event, you can imagine there is produced a flash of light and the future history of the flash is obtained by proceeding upwards in the diagram. At the top right-hand corner of Fig. 2 is a series of *spatial* pictures which represent the history of the flash of light. In the middle is the 'flash' event itself; the next minute the flash of actual light occupies a sphere surrounding that point; the next minute after that, it occupies another (larger) sphere. This gives the *future* light cone of the event. Similarly, you can define the *past* light cone—as the history of a spherical shell of light converging in on the event. All these spatial spheres, past and future, represent cross-sections through the space–time light cone, but of course you have to imagine the dimensions are all one more, so it should *really* be a sphere—but don't worry too much about that. Not *yet*—there will be a moment later on when we are supposed to worry about it!

Rather than having the sections through the space–time which are *absolute* things, which is the situation according to Newtonian physics depicted in Fig. 1, one must consider slices through the space–time which *depend on the observer.* I have drawn two of them at the bottom left of Fig. 2. The slopes of these slices (actually 3-dimensional planes) are dependent upon the velocity of the observer. So here we have one observer going off to the left who thinks that all the events on the plane sloping up on the left are simultaneous, whereas the observer going off to the right thinks that the events on the other plane are. They have different ideas about what is happening simultaneously with the event at the centre (the light flash event).

For example you might have two people walking fairly slowly past each other, but the time-difference between what each of them judges to be simultaneous with the moment they pass one another on, say, the Andromeda galaxy might be several weeks, I think. So it need not be a trivial matter. You can imagine science fiction scenarios where one person thinks something has already happened on Andromeda, say the launching of a space fleet, but another thinks the decision hasn't even been made yet whether to launch. This seems to lead you to all sorts of apparent paradoxes, but they're not really paradoxes when you think about them enough.

In my picture of Minkowski space–time in Fig. 2 (which should be 4-dimensional, but that problem is familiar by now!), the most essential features are the light cones. I have drawn both the future and the past cone at each marked event. Their significance is not so much that they represent what light *does*, but that they represent the limitations on communication. In fact any particle (any material massive particle) has to have a world-line which always lies within the light cone. On the other hand, a photon (a particle of light) can be represented as a world-line which travels along tangentally to the light cone. This expresses the fact that you cannot—in special relativity—travel faster than the speed of light. Travelling faster than the speed of light would mean having a world-line which cuts sideways across these light cones, and that is not allowed. A curve whose direction is always inside the light cone is called *timelike*. Material particles always have timelike world-lines.

## 3 The metric of relativity theory

Now, there are one or two other things which are important in relativity—well, there are *lots of things* which are important in relativity, but there are a few things which are *essential*. One of these is to do with the 'metric', which I mentioned earlier. Metric is a measure of distance between points—you can imagine putting a ruler between them—or you might have a curve and the metric defines how long that curve is. Now in space–time, what the metric tells you is basically the length of the world-line of a particle, normally a timelike world-line. What does the 'length' mean? Well, the length means the time that it would take, according to an ideal clock.

In Fig. 3(a), you can see the world-line of a particle (drawn as a little clock) and the difference in the time at event $A$ and the time at event $B$ (two points on the world-line) is simply given by the Minkowskian length of the curve between $A$ and $B$. Mathematically, this is written $\int_A^B \mathrm{d}s$. This expression means that you just add up all the little individual intervals $\mathrm{d}s$. In general spaces, this can be a horrendous expression when written explicitly, although in Minkowski space it is simple enough. What the metric $\mathrm{d}s$ is telling you is the time interval between very close events. It is physically determined by

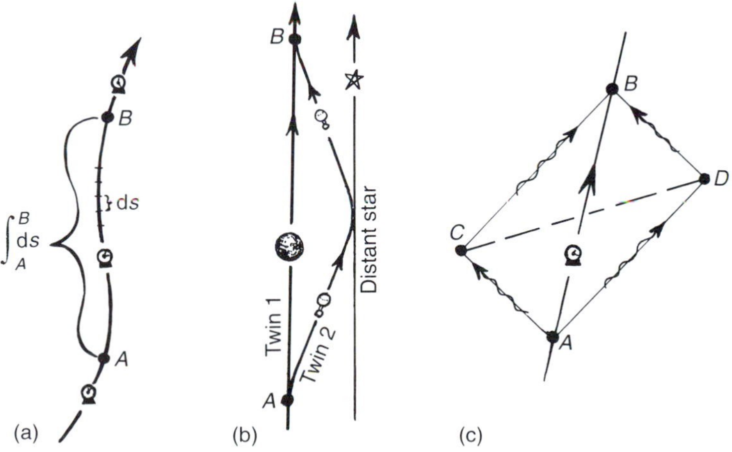


**Fig. 3** (a) The space–time metric element d*s* defines the time-interval element as measured by an ideal clock. (b) The 'clock paradox' can be understood in these terms whereby 'Twin 1', who stays on earth, ages more than 'Twin 2', who travels to a distant star and back. (c) The spatial distance *CD* can be identified with the temporal interval *AB*, where *AC*, *AD*, *CB*, and *DB* are light signals (*A*,*B*,*C*,*D* lying in a space–time plane).

how clocks behave. This assumes the existence of ideal clocks. In fact, modern atomic and nuclear clocks are very close to being 'ideal', agreeing very precisely with one another, and they provide an excellent physical interpretation of $\mathrm{d}s$ (or of $\int \mathrm{d}s$).

Some readers will have heard about the clock paradox: you have two twins, one of whom stays on the earth, while the other one goes off in a rocket ship to some distant planet and comes back. Then the time taken between departure and return, according to the stay-at-home, is the length of the straighter of the two curves, and the time taken by the other one is the length of the bent curve. Now those of you who know a little about relativity might worry that the lengths seem to come out the wrong way around, because the bent one looks longer than the stay-at-home's straight one, whereas we know that the one who goes out in the space ship is supposed to age less than the other. This is just a funny feature of Minkowski geometry. That is what the minus signs do for you. It happens that the distance measured as you slope up closer and closer to the cone gets smaller and smaller, and when you measure along the cone, it actually goes to zero. So in this funny kind of geometry, the sloping lines actually are shorter than the less sloping ones, so it works out that the bent line representing the space traveller, in this funny geometry, is actually *shorter* than the other one. You have just got to get used to it—that is how it works! (See Fig. 3(b))

Now once you know about times, you might ask: what about spatial distances? This notion applies to the two events, $C$, $D$ of Fig. 3(c), which are 'space-like separated'. This means that each one of the events lies outside the light cone of the other. The wavy lines are light signals and we want to know how far apart are the two points $C$ and $D$. The answer is, find the two events $A$ and $B$, all in a plane with C, D, so you have two light signals going out from $A$ to $C$ and $D$, respectively, reflected back to reach the event $B$, together. Then you ask: how long did it take for a clock to move uniformly from $A$ to $B$? That answer directly gives you the spatial distance between $C$ and D, in units where the speed of light is equal to one (i.e. with spatial units of light-seconds or light-years, etc). That is what relativists like doing, and it is more or less all you need to know about the space–time structure of special relativity—at least for this talk.

But what about *general* relativity? Well, the basic difference is that the expression for the time-interval can now be more complicated. Consequently, the light cones, in contrast to the situation for special relativity, are now all higgledy piggledy (Fig. 4). They are not uniformly arranged, but can be tilted at funny angles (usually not very much, in normal situations), and the fact that they are not all uniformly arranged is an expression of the gravitational field. Another important thing about the gravitational field is what is called the *equivalence principle*, which really goes back to Galileo. He

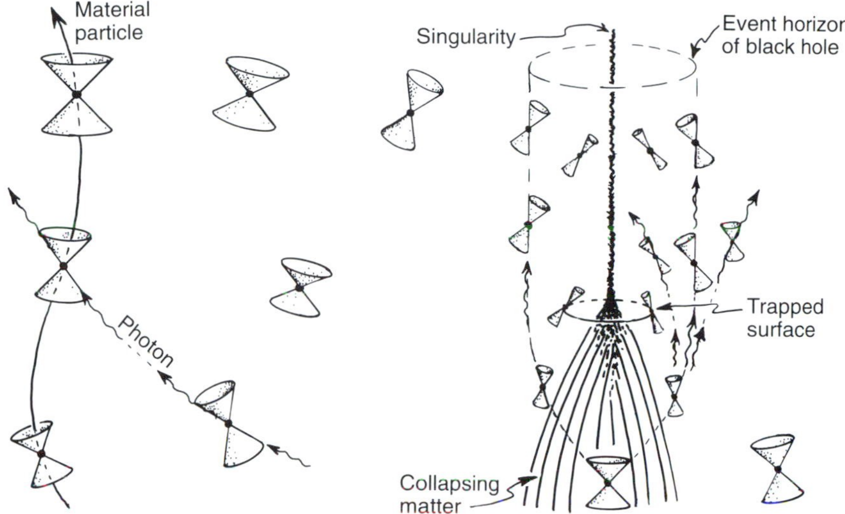


**Fig. 4** In the space–time of Einstein's general relativity, the light cones are non-uniform. On the right, the history of the collapse of a star to form a black hole is illustrated, with its singularity, event horizon, and a trapped surface, which will be discussed later in this chapter.

is supposed to have dropped a big and a little rock from the leaning tower of Pisa. Whether he actually did this or not, he certainly did things very like it. The essential point is that, ignoring air resistance, they would have dropped together. Now imagine you were sitting on one of the rocks (I don't recommend it, but imagine you did it); you would see the other one hovering in front of you. Well, people do things like that now. We know about astronauts going around the earth. Imagine one on a 'space-walk'. The spaceship appears to 'hover' just by the astronaut, and neither of them seems to experience gravity at all—despite the huge globe of the earth nearby. This is because astronaut and spaceship are all falling together. Falling doesn't necessarily mean straight down towards the centre of the earth; falling can be in orbit, as long as you are freely following the dictates of the gravitational field. Then everything will move together, and you do not seem to *feel* any gravitational field. And that is general relativity for you. Well, there is a bit more, which I shall tell you about shortly, but that is a good part of it anyway.

## 4 The Milne cosmology

I should like to say something which pertains directly to Milne. It is perhaps rather fortunate that, in more than one respect, what was probably Milne's most original contribution to cosmology was something of considerable

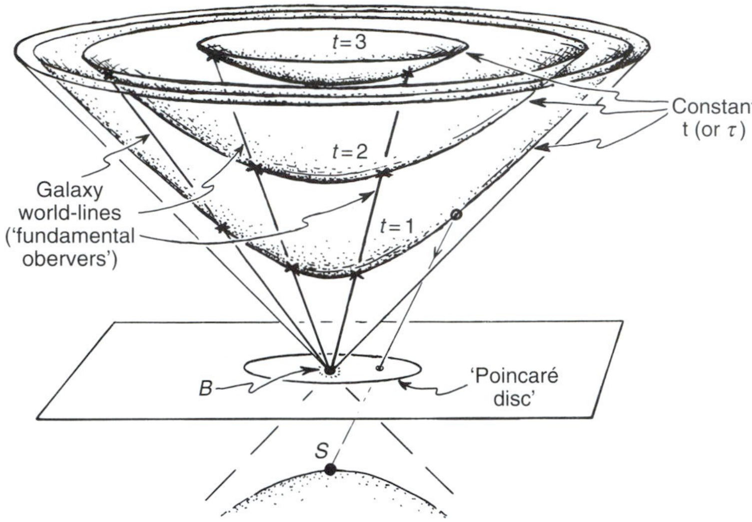


**Fig. 5** The Milne universe. The 'fundamental observers' are galaxy world-lines emanating from the big bang *B*. The bowl shapes represent surfaces of constant *t* (or $\tau = \log t$) for these observers. The Lobackevskian geometry of each of these surfaces can be represented as a 'Poincaré disc' by stereographic projection from a point *S* (the projection of $t = 1$ being illustrated).

relevance to this talk. I want to show you something that is known as the 'Milne universe'. He developed an idea that he referred to as *kinematic relativity* in a book published in 1948. I'm not sure if he was trying to find an alternative to general relativity, or simply a different way of looking at things which might be of particular value to our understanding of cosmology. You can imagine the Milne universe in the following way.

Think of Minkowski space—flat Minkowski space–time—and imagine that there is a cataclysmic explosion which takes place at some event *B*, which is to be the big bang (Fig. 5). You have particles ejected in all directions from *B*. This can be done in a way which is *uniform* with respect to relativity. Milne referred to these particles as 'fundamental observers'. We can think of them, approximately, as the galaxies in Milne's cosmology. If you make one of the transformations moving from one fundamental observer to another, according to special relativity, the distribution of particles looks just the same to each. You can actually have a distribution like that, so long as you do not go outside the future light cone of *B*. What happens at that cone is that the density gets infinite there—along the 'boundary' of the Milne universe. But you do not worry about that; the entire universe in the Milne picture is represented by the interior of this cone, where the density is finite and these particles all recede from one another with various speeds less than the speed of light. If you were sitting on any one of these particles, the others would seem to be moving away from you uniformly, in a symmetrical arrangement—the expansion of the Milne universe.

What was particularly original about Milne's way of looking at things was that he really addressed this question of *time* rather closely. In ordinary relativity, one thinks of the time measurements as being fundamental. The quantity $ds$ tells you what clocks do as you carry them along world-lines. Now it might be that different types of 'clock' behave differently, so that they have different notions of '$ds$'. As far as we know now, this is not the case. But, of course, Milne was writing some while ago (recall that 1948 was the date of his book), and a lot more is now known about the accuracy with which different types of clock will agree with each other. In fact, there was a discrepancy in those days about how old the universe seemed to be. One type of measurement gave an older universe than the other one, by a serious amount. And so Milne had the idea that perhaps there were different kinds of time that need not agree with each other. One time might govern the frequencies of atoms and molecules (he called this $t$-time) while the other might control the gravitational dynamics of orbiting bodies (his $\tau$-time). In the sort of model he produced, one of them is logarithmically related to the other, by a formula of the form

$$\tau = \log t,$$

where $t$ is the time you would get from the ordinary Minkowskian geometry measured out from the point $B$, thinking of this as applying to the inside of the future light cone of $B$ in Minkowski space. You also have the $\tau$-time related to $t$ by the above logarithmic formula. When $t$ goes to zero, which is the big bang $B$, the other time $\tau$ goes to minus infinity. So in the $\tau$-describtion, you find that the universe was there for ever! It is a *static* universe in fact, with respect to $\tau$-time. The galaxies just sit there, and the red-shift which Hubble had observed, and which seemed to indicate that the universe was expanding, is explained because oscillating atoms use a different kind of time-scale, namely that which determines the spectral frequencies that give us this red-shift. The things you are looking at in cosmology are indeed usually spectral lines which have to do with $t$-time, and they might behave according to a different kind of law from the dynamical time that bodies use as they go round each other, and so on. This was taken quite seriously by the famous biologist J. B. S. Haldane, who actually wrote an article exploring the implications of Milne's idea for biology. The essential point was that quantum mechanics was much more important in the early days of the universe, and little tiny things could produce huge effects. (The term 'quantum leap' would have been much more appropriate then than now!) And Haldane examined what effects this might have for the early stages of evolution.

Nowadays this idea does not seem to stand up very well because it does appear that all the different notions of atomic and nuclear clock satisfy very, very closely the same time. It is not quite so clear whether the gravitational constant might vary with time, but if it does, the variation must be relatively small, and not consistent with models of this nature. But, nevertheless, the idea is still of considerable interest for more than one reason.

## 5 Lobachevskian geometry

Let us return to Milne's bounding light cone again. In Fig. 5, I have drawn the different surfaces of constant time starting from $B$, where you imagine the explosion, the big bang at $B$. Here we have one unit of $t$-time after the big bang, two units of $t$-time and so on—as measured by the particles (atomic time). These surfaces of constant time do not look flat any more, they look like bowl-shaped things, and they have a rather interesting internal geometry. If you are familiar with stereographic projections, you might recognize what I have done in the lower part of Fig. 5.

The point is that if you take one of these bowl shapes, which represents one instant of time (say $t = 1$), you can project it from the point S (which is a point on $t = -1$) to the inside of what would be a circle in the reduced space–time depicted (two space dimensions, one time dimension). This is a disc sitting in a Euclidean plane, and it represents the geometry which takes place within the bowl-shaped surface ($t = 1$). Milne noticed that this geometry is what is called a Lobachevskian or hyperbolic geometry, and the particular projection that I have indicated gives you the kind of representation which is called the Poincaré disc (perhaps because it was discovered by Beltrami—it couldn't very well be called the Beltrami disc because that terminology already refers to a different representation, also due to Beltrami).

The Dutch artist M. C. Escher depicted Lobachevskian geometry very beautifully in a series of prints called 'Circle Limit'. In Fig. 6, I have reproduced one of these (Circle Limit I). Escher had the idea from H. M. S. Coxeter, who told him that it might be a fruitful thing to do, to represent hyperbolic Lobachevskian geometry in this way, since Escher was a great expert at illustrating geometrical notions in terms of repeating fish and things like that. This is a 'non-Euclidean geometry', and it requires me to give a quick explanation. First, in Euclidean geometry, if you draw a straight line and take a point not on the line, then there is a *unique* line through that point which is parallel to the first one. If you have a triangle made of three straight lines, you add the angles up and they give you 180°. A very beautiful fact. Now, there is another kind of geometry called non-Euclidean, where if you take a point not on the line, there's a *whole family* of possible lines which are parallel to it—in the sense that they do not meet it (while still lying in the

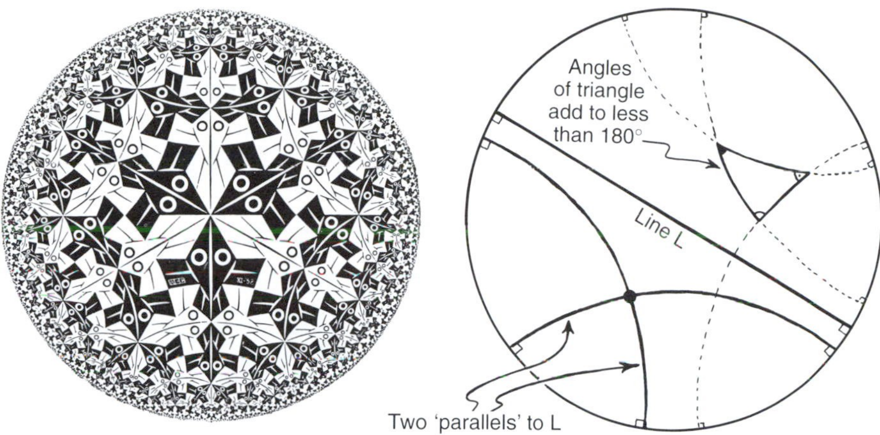


**Fig. 6** On the left is M. C. Escher's 'Circle limit 1' which provides a beautiful illustration of Lobachevskian (hyperbolic) geometry. On the right, features of this (Beltrami–Poincaré) representation are illustrated: there is more than one 'parallel' to the line *L* through a point; the angles of a triangle add to less than 180° (straight lines are represented as circles meeting the boundary at right angles).

plane containing the point and line), no matter how far they are extended. And it makes a perfectly good geometry. It took a long time for people to realise that it did make sense. In this geometry, if you draw a triangle and add up the angles, you find they don't add up to 180°. You might say, well that's ugly, but hang on for a moment; if you take 180°, and subtract the sum of the angles from it, you get the area of the triangle (in appropriate units). Now that is a remarkable fact. If you try and work out the area of a triangle in Euclidean geometry, it is much more complicated.

This Escher picture has that kind of behaviour, as we can see if we compare it with the 'straight lines' of this geometry, in the 'Poincaré disc' (Fig. 6). These are represented as arcs of circles which meet the boundary at right angles—and these lines correspond nicely with the Escher picture. Now the thing is that if you were one of these fish who lived way up near the boundary, you would think the world looked just the same to you as if you were one of these fish in the middle. The geometry has this kind of uniformity—even though it looks in this representation as if it is very crowded up against the edge. But that is just because of the way the geometry has to be drawn to represent it in the Euclidean plane. One thing that is particularly nice about this Beltrami–Poincaré representation is that the *angles* are correctly represented—it is called a *conformal representation*. In another of Escher's circle limits ('circle limit III') he used interlocking angels and devils. It is interesting that he used the same motif, 'Symmetry Work 45' and 'Heaven and Hell Sphere', to illustrate the two other kinds of uniform geometry that you can have. He used interlocking angels and devils on the ordinary Euclidean plane

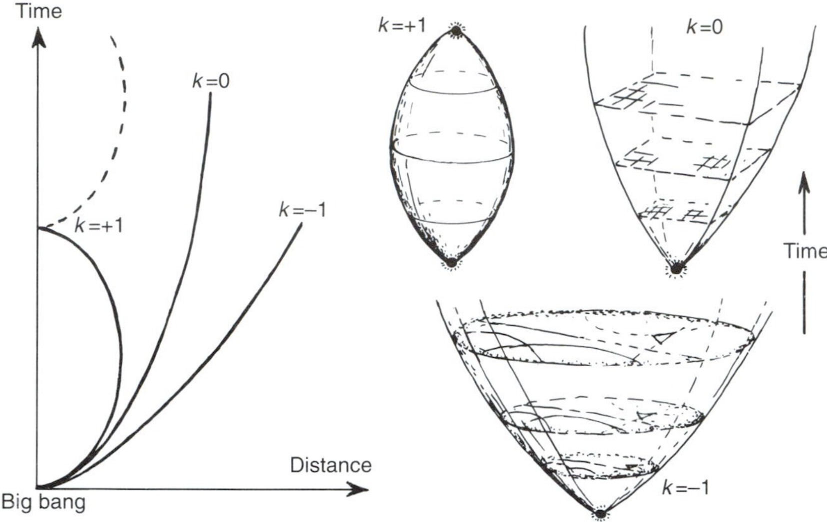


**Fig. 7** The standard (Friedmann) cosmological models. The case $k = +1$ has spatial sections which are (three-dimensional) spheres; $k = 0$ has Euclidean spatial sections; $k = -1$ has Lobachevskian spatial sections. All three models have a big bang, but only $k = +1$ encounters a big crunch.

and on a sphere. These geometries are of particular interest in cosmology, because (in three dimensions) they are the spatial geometries of the three standard models of cosmology (Fig. 7).

## 6 The standard cosmologies

These are the space–time pictures of what are sometimes called the Friedmann models, sometimes Friedmann–Robertson–Walker models, and sometimes Friedmann–Lemaître–Robertson–Walker models, just to get everybody in who deserves a mention. Now these cosmologies have spatial sections which are the 3-dimensional versions of one of these three geometries. The flat case is called $k = 0$; the positive-curvature (or spherical) case is called $k = +1$; the negative-curvature (or hyperbolic or Lobachevsky case) illustrated by the Escher picture of Fig. 6 is called $k = -1$. These different spatial geometries go along with a very particular behaviour in time (at least, so long as cosmic pressures can be ignored). These cosmological models start off with a big bang—the explosive origin of the universe itself. (The case for the actual existence of a big bang has been argued for very effectively in the articles by Rees, Sciama and Longair in this volume.)

The spatially closed universe $k = +1$ starts with a big bang, but eventually comes back to what is called the *big crunch*. Both the big bang and big crunch are space–time singularities. Next we have the spatially Euclidean universe $k = 0$. It starts with a big bang, but it goes on expanding for ever, but more and more slowly. Finally, there is the spatially Lobachevskian one which expands for ever and never really finally slows down. (In all this, I am assuming that Einstein's equations hold in their pure form, without the so-called cosmological constant.)

In the closed $k = +1$ case, if you take the curve on the left in Fig. 7 which Friedmann came up with, describing the radius of the universe as a function of time, you find that it is actually a cycloid and the universe keeps expanding and collapsing all the way up. So this closed model has been referred to as an 'oscillating universe'. I am not going to call it that because that really depends on taking the remaining arcs of the cycloid seriously. There is no reason to take them seriously because you have to get through the moments where the entire universe is concentrated in a single point—at a *space–time singularity*—where the density of matter goes to infinity.

The presence of these singularities is an embarrassment for Einstein's general relativity, but they have to be there, at least in the classical models that describe plausible cosmologies. I am going to take a leaf out of Milne's book here, by taking advantage of this idea he had of using two different time-scales. Remember that, in the Milne universe there are two different times, $t$-time (atomic time) and $\tau$-time (dynamical time). But as far as the light cones are concerned, they are not affected at all by the change. It is just the metric scaling that is altered. There can be a considerable advantage in this kind of re-scaling—called a *conformal* rescaling—not necessarily quite for the reasons that Milne had in mind, although my use of it is indeed related to his. Recall that by using his $\tau$-time he was able to get rid of the big bang singularity by pushing it right out to minus infinity. That is one of the things you can do with conformal rescalings; you can sometimes scale away singularities. Another thing you can sometimes do is bring infinity somewhere where you can have a look at it.

Roughly speaking, there are two different kinds of advantage here. You can either stretch or squash the space–time, depending upon whether it is a singularity that you want to 'remove', by stretching it out, or the infinite regions that you want to squash in and have a look at. (These are very useful mathematical devices.) In practice, it is best not to stretch a singularity so much that it disappears to infinity, but only by an intermediate amount (if you can), so that it becomes a smooth surface. In fact, in the case of the Friedmann–Lemaître–Robertson–Walker models, it is indeed possible to do this. You can take that unpleasant big bang and blow it up to obtain a very nice surface, which is absolutely smooth. This surface has the same kind of (3-dimensional) geometry as the spatial sections of the universe-model: spherical ($k = +1$), Euclidean ($k = 0$), or Lobachevskian ($k = -1$).

Now this has an advantage in that you can easily understand things like cosmological horizons. One of the puzzling features of cosmology is that when you look out at the universe, even though the whole universe at the big bang singularity was squashed into a point, you can obtain information only from certain parts of that singularity, not all of it. This makes more sense when the singularity is blown up, by a conformal rescaling, into a finite surface. There is a well-defined light cone at each point of this surface, defining the influence of that element of the singularity. Looking back at the big bang, you only see the portion of he singularity surface encompassed by your past light cone. That explains the existence of your cosmological horizon. If you could live many centuries, then eventually, as your time progresses, your light cone would encompass more of the singularity surface, and more galaxies would come into view. Thus, one way to make this concept of horizon easy to grasp is to use Milne's idea of conformal rescaling.

## 7 General relativity and space–time curvature

I shall need to come back to some of these ideas later, but before doing so I want to say something more about general relativity proper. In Fig. 4, I have tried to indicate how light cones are affected by the presence of a gravitational field, but I have not said anything detailed about how a gravitational field makes itself felt. For this we need to understand what 'curved space–time' means. You might think that this is a strange idea, but it is as well to get used to it. The idea of curved space–time and its relation to gravity is really a very beautiful idea. It is precise to an extraordinary degree. And in fact if you consider the Taylor–Hulse observations on the double neutron star system PSR 1913+16, as described in Taylor's article in this volume, you find that Einstein's theory is confirmed extremely precisely. Hulse and Taylor have observed this system now for some 20 years, and over that period of time, the precision of the signals that they obtain from the system is so great that it provides an accuracy of something like 1 part in $10^{14}$ over that period of time, agreeing with the predictions of Einstein's theory with a precision which is over two orders of magnitude better than what quantum electrodynamics (already extremely impressive) can do. This is an extraordinarily accurate series of measurements, and it tests the whole range of things gravitational, from the Newtonian end of the scale through the Einsteinian corrections right up to the contributions due to the gravitational waves which are emitted by the system. So general relativity seems to be, in one sense at least, the most precisely tested theory that we know.

Let me say a little more about general relativity. I referred earlier to the principle of equivalence: all bodies fall together in a gravitational field, so each of the bodies would appear to be unaccelerated to someone sitting on another. This asserts that in free fall—that is, in free orbit—you do not feel any gravitational force. Now you might well ask: where has gravity gone? Let me tell you where it has gone. You have to be a little bit more careful about what happens to objects in free fall, say in orbit around the earth—or dropping, if you like. Consider an astronaut, surrounded by a sphere of particles, initially set all at rest relative to the astronaut, so that they appear just to hover there. However, because of the non-uniformity in the earth's gravitational field, the particles nearest to the earth are accelerated, in Newtonian terms, by a little bit more towards the earth than the astronaut is, and the ones farthest from the earth by a little less, so the sphere gets stretched out after a little while. In addition, the particles at the side are attracted inwards, relatively speaking, because of the centre of the earth being at finite distance. Thus, the sphere becomes distorted into an elongated ellipsoid (see Fig. 8(a)).

This is called the *tidal effect*, with good reason, because replacing the earth by the moon and the astronaut by the earth, the distorted sphere of particles would be the bulging tides in the direction of the moon and also in the opposite direction. This is a well-known phenomenon, though sometimes we get puzzled by the way that the tides bulge on the opposite side, but you can see it from Fig. 8(a). It is just a question of the relative accelerations because of the non-uniformity of the earth's field.

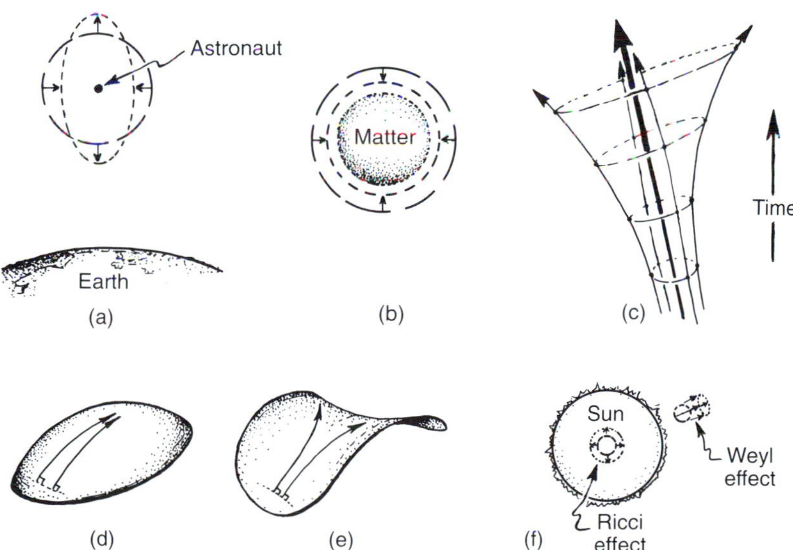


**Fig. 8** The gravitational tidal effect as space–time curvature. (a) Tidal distortion, near the earth, of the freely falling sphere of particles surrounding an astronaut in free orbit (Weyl effect). (b) If the sphere surrounds matter, there is a volume-reducing contribution due to the mass surrounded (Ricci effect). (c) The space–time description of tidal distortion (compare Fig. 1). (d) Positive curvature causes convergence. (e) Negative curvature causes divergence. (f) The focussing effect on light rays clearly distinguishes the Ricci effect (pure magnification) from the Weyl effect (pure distortion). Here the sun's disc is depicted (the sun being taken as transparent).

Now, Einstein's vacuum equations are characterised by the fact that the initial volume of this ellipsoid is equal to the volume of that sphere. So it is a volume-preserving effect. However, if you put mass inside the sphere, then you have an acceleration inwards all the way around which is roughly proportional to the total mass surrounded (Fig. 8(b)). So you have a volume-reducing effect. Einstein's equations tell you that matter indeed introduces volume-reducing effects to freely moving particles, whereas in free space, you just have the 'tidal' distortion effect.

There is a mathematical object called the *Riemann curvature tensor*, which measures the curvature of a space. In four dimensions, this can be split into two 'equal' pieces, one called the *Weyl* tensor and the other called the *Ricci* tensor. In space–time, the Ricci tensor measures the volume-reducing effect I have just been describing, and the Weyl tensor similarly contributes to the distortion. How are we to understand this? We need to think of a picture which shows what is happening to our astronaut and the surrounding sphere of particles. As time goes on, this sphere becomes distorted into an ellipsoid, its radius growing in some directions, and shrinking in others. This is indicated in Figure 8(c) (see also Fig. 1) by the circles (which represent spheres) becoming lengthened in one direction, and squashed in another. It is a combination of the two effects that you can get: with positive curvature they move closer and with negative curvature they move away (Fig. 8(d,e)). They are both mixed up together in the space–time picture. This is what space–time curvature does.

Once of the most striking instances of the distortion is what it does to light rays. One of the earliest tests of general relativity—the one which put Einstein's theory on the map—was a test of the light deflection by the sun. Arthur Eddington headed a group that went to the island of Principe in 1919 and actually observed this, and what they saw was that the star field is displaced outwards by the presence of the sun. This was during an eclipse so the sun did not actually blot things out. It is a kind of *lensing effect*, which is now extremely important in cosmology and astronomy, as nowadays people can *use* this effect to discover detailed facts about what the universe is like, such as how massive objects actually are. The sun, acting as a positive lens, distorts the starfield. The starfield appears to be pushed out, so a circular pattern in the sky just outside the sun's disc gets distorted into an ellipse (Fig. 8(f)). This will have importance for us later.

## 8 Singularity theorems

We have now seen how general relativity relates irregular matter distributions to space–time curvature. But the smoothed-out cosmological models of general relativity have these nasty singularities. You might say: perhaps they would not be there if you introduce perturbations into the space–time geometry? And people tried to do just that; Einstein in particular was keen on that possibility. However, does it work? To investigate this, let us turn to the singularities of black holes. This takes us back to another of the contributors to this volume. Subrahmanyan Chandrasekhar, in his early work, showed that if you had a white dwarf star of more than a certain mass (about 1½ times that of the sun), it cannot hold itself apart, and it will just collapse. Chandra had difficulty in getting his results accepted by some senior astronomers, but late in his life he came back to the study of black holes. In a sense, he started off the subject, because he showed that stars which have too much mass simply collapse and there is nothing you can do to stop them—although lots of people tried to get around this. The first picture which clearly represented collapse to a black hole was that produced by Oppenheimer and Snyder in 1939. On the right-hand side of Fig. 4, I have drawn a space–time diagram of the history of a collapsing star, which tried to be a white dwarf or neutron star but it was too massive, so it collapsed inwards and just went on and on and couldn't stop. The thing is that it gets to a situation where the escape velocity exceeds the velocity of light. Escape velocity is the smallest velocity where if you hurl something from a gravitating body at that speed, it will leave it altogether, never to return. At the earth's surface it is about 25 000 mph (40 000 km $h^{-1}$). If the earth were more concentrated, the speed would also be a lot greater, and if it were more massive, the speed would be greater. And because the speed of light is the limiting speed for all signals, when the escape velocity reaches the speed of light, *nothing* can escape the body. What happens is that the light cones tilt over in the picture (right-hand part of Fig. 4) and you find that there is a surface—called an *event horizon*—which prevents all signals from within it getting out. Material bodies that cross it are all dragged in towards the central region where there is the *singularity* where space–time curvature goes to infinity. Now here we have another place where there is a space–time singularity. People often used to argue, in the discussion of this black-hole singularity: OK isn't it just like the big bang but the other way around? Here we have, in effect, a (local) *end* of the universe. If you fall into a black hole, you're dragged inevitably into this singularity and squelched. (I suppose that is the right word. You get stretched in one direction and squashed in another.)

Not everybody likes singularities; I don't like them much myself. But the question is whether singularities are inevitable. An important thing is that the Oppenheimer–Snyder model assumed that the star was exactly spherically symmetrical, and they also assumed that the matter was 'dust', i.e. that there were no interactions between the particles falling in. And if you do not make these assumptions—in particular if there is no symmetry—the exact model they had cannot be relied upon. Could it be that things come into the middle, get messy and come shooting out again, as various people suggested? So this is something that I worried about in the early days of my interest in the subject. The point was: how do you characterise a black hole in such a way that you do not need to assume any symmetry? What I did was to introduce this notion of a *trapped surface*, which is a 2-dimensional surface (normally surrounding the material of the collapsing star after it has gone through the horizon) where if you imagine a flash of light emitted at that 2-surface, the area of the light emitted from it goes down, not only for the inward flash, but also the outward flash. Because the light cones are tilted in this peculiar way, the area of the outward flash also goes down. So the definition of a trapped surface is one where the area of the light flash, as you follow it into the future, goes down; moreover, it has got to be a *closed* surface with that property. Now we suppose that we just have the trapped surface. What can we deduce from that?

Well, there is a theorem. I am not even going to attempt to state it properly, let alone try and prove it to you here. Roughly speaking, it tells you that whenever there is a trapped surface, there must be a singularity somewhere. It is a very negative theorem. It does not tell you where the singularity is—just that somewhere there is a singularity. What use is that if you do not know where? Well, that is sometimes all that we mathematicians can do for you! Stephen Hawking then came in with various improvements, and finally we got together and deduced a theorem which more or less encompassed all the other ones we had thought of before. Instead of having a trapped surface, you can also use a light cone which reconverges. That is sometimes a little easier to work with. Imagine an event somewhere, where all the future light rays from that event start crumpling up in some way and re-focussing. If that ever happens, you know there must be a space–time singularity somewhere. In fact, in the 1960s and 1970s, when people were worrying about whether black holes existed or not, it was possible to say, in principle, that they have got to. You could imagine a mad dictator, for example, who inhabits some galaxy and he sends out all his troops, with orders to rope in all the stars in the galaxy and let them fall in towards the general area of the middle. You can work out that the amount of light focussing, produced by the stars, would be enough to produce the refocussing light-cone condition well before the stars even collide. So there is no problem about not knowing what the physics inside a compressed white dwarf or neutron star is—these are just ordinary stars sitting out in space, things we are supposed to know all about—and still you have this reconverging light-cone condition. In principle, you cannot avoid singularities. Moreover, in practice they will occur sometimes.

## 9 Must black holes exist?

Now, to show that there are actually black holes we need an assumption—and this is the big assumption that is still unproved in general relativity—called *cosmic censorship*. I have to confess to having invented the name. The thing is that singularities might be inside horizons, or they might be what are called *naked*. The idea of the terminology is that somehow these naked ones are not proper, and the Deity clothes them all with a horizon—and that is cosmic censorship. So, is that true? We do not know, but we strongly suspect so. (When I saw 'we' I should point out that there are a lot of people who have their doubts. I used to be a sceptic myself, but later I became more persuaded of the likelihood of cosmic censorship. It is just a question of age, I think.)

If cosmic censorship is true, that means you cannot see the singularity that results from collapse, certainly not from a long way off. Thus there is a definite region which cannot communicate with distant observers. This implies that there is what is called an event horizon, so cosmic censorship together with the existence of one of these conditions like a trapped surface or reconverging light-cone tells us that you have not only a singularity, but a singularity that is hidden inside a horizon. You have what is called a *black hole*.

Now this leads to a remarkable situation. In fact a number of theorems have been proved which show that if you have a black hole which settles down to become stationary, then it has to have a space–time geometry that we know very, very precisely. It is the geometry known as a *Kerr space–time*. This space–time depends only on two numbers—the mass and the angular momentum (spin). Chandreaskhar was so struck by this that he calls these structures the most perfect macroscopic objects in the universe, made just out of empty space. He never ceased to marvel on that fact.

The first example of what was convincingly a black hole was the X-ray source Cygnus X-1. It is part of a double star system—its partner is a blue supergiant and they are in orbit about each other. You cannot see the black hole, but X-ray signals are received from its vicinity. Because of the dynamics of the whole thing, it is concluded that this is an object which is too massive to be a self-supporting star—and therefore it should be a black hole. Unfortunately the argument is a bit negative: it would be nice to have direct evidence for the Kerr geometry. If someone could get enough information from a black hole, one could really start checking up on this geometry. That would be a really wonderful test. But in the absence of that, indirect evidence is the best we can do.

## 10 Irregularities in cosmology

In Fig. 7, I showed you the three standard models of the universe. However, they are not really quite realistic, because there ought to be irregularities such as black holes around. In Fig. 9, I have (schematically) indicated what we really think a closed or open universe would be like. The actual big bang does seem to have the kind of smooth character exhibited by the standard models, but the final singularities are quite different. We expect an awful mess at the end! The singularities in black holes gradually clump together uniting into larger and larger conglomerations. In the case of a closed universe $k = +1$, the whole thing clumps together and forms a great dreadful mess of a big crunch at the end. The open universe on the other hand, just goes on expanding, but you still get 'little' local dreadful messes. Now, as I remarked earlier, people used to say: 'Well you have got to take these singularities in black holes seriously because they are just like the big bang in reverse'. But they are not just like the big bang in reverse. In fact they are completely different. I have mentioned the Weyl curvature. This is the pure distortion part of he curvature which is present in the absence of matter (see Fig. 8(a,f)). That is what seems to go wild in these future singularities. Ricci curvature (which, we recall, is directly produced by matter) is what goes infinite at past singularities (the big bang). The structure is completely different. In the past you have diverging Ricci curvature and zero Weyl curvature, but in the future, diverging Weyl curvature (but the Ricci curvature may

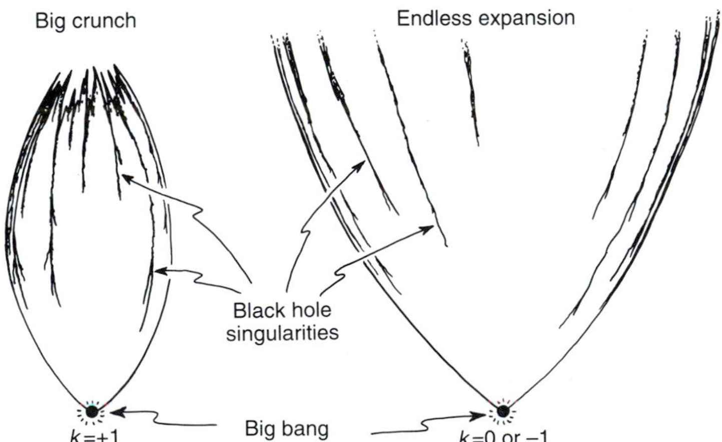


**Fig. 9** When perturbations are introduced into the standard models of Fig. 7, black holes now form. The big crunch of the $k = +1$ model is now very different from the big bang and represents the congealing of black-hole singularities. The black-hole singularities are generic (maximum entropy) while the big-bang singularity is extraordinary special (as demanded by the second law of thermodynamics). Apparently the Weyl curvature was constrained to be zero (or very small) at the big bang.

sometimes diverge too). Accordingly, I have posed the hypothesis that initial singularities have to be constrained to have vanishing Weyl curvature. This ensures that the big bang will have the nice, highly organised uniform structure of the standard models, rather than the sort of 'generic' mess which we expect at the end. Now if it were not like that, if the big bang were a mess too, then we would have a universe that is nothing like the universe we see. In fact, you can work out how improbable the universe we see would be if it were just chosen by chance. If the Creator were choosing universes by chance, how likely would we find one at all like ours as opposed to a mess like the one we expect in the big crunch? The odds against it would be at least as big as $10^{10^{123}}$:1. You couldn't even write that big number down in the normal way, even placing one digit on each of the particles in the universe!

The fundamental conundrum is: what, in our theories, do we do with these space–time singularities? These things are blemishes in our classical theory. Our beautifully confirmed general theory of relativity leads to this kind of a picture. Now general relativity is not the whole of physics, and it certainly does not include or incorporate the phenomena of quantum mechanics. It is anticipated that as one approaches these singular regions, the effects of quantum mechanics become more and more important. Remember what J. B. S. Haldane did with the Milne model? Well this time it is for real. The Milne model was in a sense a precursor of this kind of thing because the effects of quantum mechanics have to become very large near a singularity. We expect that. But now we need quantum gravity. Perhaps the effects of quantum gravity served to smooth out the initial singularity. Why do we not see this at that end? Well, all sorts of people have come up with different answers. My answer is that the correct way of combining general relativity with quantum mechanics cannot be a way which leaves the rules of quantum mechanics unchanged. You have to replace quantum mechanics by some new theory in which there is a *time asymmetry*.

## 11 The Riemann sphere

Next, I want to say something about the kind of mathematics that especially interests me in my attempts to understand how one would go beyond space–time structure in accordance with these requirements. Up to this point, I have just been giving you the descriptions that classical general relativity provides. Before going any further, I want to say something about quantum mechanics, but really what I want to tell you is a piece of geometry.

In Fig. 10, I have drawn a sphere, just an ordinary sphere. There is an ordinary plane intersecting it at its equator, and I am projecting the sphere from the south pole to the plane. Now in quantum mechanics, electrons, protons or neutrons have what is called *spin*. The points of this sphere are to

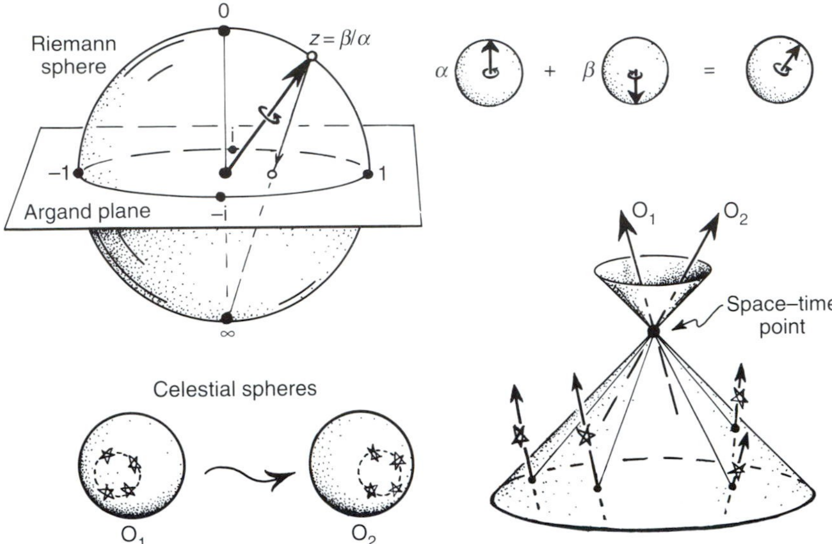


**Fig. 10** The Riemann sphere. It plays a fundamental role in quantum mechanics, clearly illustrated by the geometry of spin states for a spin ½ particle (top right). It also plays a fundamental role in relativity theory as the celestial sphere of an observer. These two facts provide the starting-point of twistor theory.

represent possible directions of the spin axis. The spin can be upwards (north pole) or downwards (south pole)—or, indeed, in any other direction, but all these other directions can be considered to be *linear combinations* of up and down. In fact, in quantum mechanics you can always form such combinations of alternatives, added together with *complex-number* weightings. It is absolutely crucial to quantum mechanics that these weightings are indeed complex numbers—numbers which involve the square root of minus 1, called 'i'. Such numbers can be plotted in a plane, sometimes called the *Argand plane*. The rules for adding and multiplying complex numbers can be made entirely geometrical. Complex analysis is a very beautiful piece of mathematics; it permeates all sorts of subjects, but only with quantum mechanics have we seen complex numbers right there at the roots of the way the world operates —at its tiniest levels. The complex number system is embedded into the structure of the world.

Now, what does this have to do with my sphere in Fig. 10? We are to think of the equatorial plane as an Argand plane (the equator being a unit circle, containing 1, i, −1, −i). Then the projection depicted—called a *stereographic projection*—assigns a complex number to each point of the sphere except for the south pole, the south pole being assigned '∞'. A sphere labelled in this way is called a *Riemann sphere*. Thus the Riemann sphere is the Argand plane, together with ∞, wrapped up into a sphere. That sphere plays a fundamental role in quantum physics. It describes the family of different quantum states that can be built up from just two quantum states, by the 'linear combinations' I referred to. Each possible ratio of the two complex weighting factors involved in these combinations gives a different point on the Riemann sphere. In the case of the spin of an electron (or proton or neutron) these combinations correspond to the different possible spin directions, and these are directly represented as points on the Riemann sphere.

Now let us consider another fundamental role of the Riemann sphere, this time in relativity theory. Suppose you look out at the sky—what do you see? You see a sphere. That is the *celestial sphere* and if you imagine two people, travelling at different velocities, looking out at that same sky, what do they see? Well, each sees a slightly transformed version of the other's celestial sphere. One person's version is a little different from the other's, owing to the aberration of starlight, first observed and interpreted by Bradley in 1725. As the Earth changes its velocity during its annual motion around the Sun, the pattern of the stars in the sky shows an annual variation. In relativity it has a very interesting property: the way you go from one celestial sphere to another is such that a circular pattern in the sky goes to a circular pattern in the sky.

The point that I want to make here is that when we think of the celestial sphere as a Riemann sphere, this kind of transformation of the field can be described by some very simple transformations. These are called *bilinear transformations*, or *Möbius* transformations, and they take the form

$$z \rightarrow \frac{az + b}{cz + d}.$$

Here, $z$ is the complex number labelling a point on the celestial sphere, thought of as a Riemann sphere, and $a, b, c, d$ are complex numbers determining the particular transformation. Circles go to circles and angles are preserved under this transformation of that plane or sphere (properties also of stereographic transformation). The thing that has always been important to me here is that you can regard that sphere as a *complex curve*. A complex curve is really a surface of some kind. With complex numbers the number of dimensions is always twice as many as it would be in real terms. Two real coordinates (the real and imaginary parts) give you one complex co-ordinate. The dimensions in the real sense are always twice as many as in the complex sense.

The simplest complex curve is this thing called the Riemann sphere and you can identify that with the celestial sphere. In Fig. 10, I have shown two observers looking at the same sky. As you pass from one to the other, the structure of the sky as a Riemann sphere is maintained.

## 12 The philosophy of twistor theory

This fact is the basis of *twistor theory*. All that I want to tell you about twistor theory is that the essential point of it is to take this complex idea seriously. This is not just for quantum mechanics, but also for relativity. For the *Lorentz group*—the system of transformations which take one observer's viewpoint into that of another, in relative motion to the first, according to relativity—can be understood as the group of transformations of the simplest of complex manifolds, namely the Riemann sphere. The idea of twistor theory is to extend that to the whole of space–time.

To do this we have to work hard, but some things come out beautifully, whereas others have doggedly resisted for years and years. Twistors describe (idealized) light rays in space–time. A space–time point is represented as a Riemann sphere in twistor space. This idea works out nicely. However, one crucial 'little' problem arises from something I hinted at earlier. Recall that in special relativity, if you look out at the sky and see a circular pattern of stars, this 'circular' property is maintained if we pass from one observer to another. Recall also what I remarked upon earlier, that in general relativity the lensing effect distorts circular patterns into elliptical ones. This is an effect of Weyl curvature, indicating the presence of a gravitational field (Fig. 8(f)). In terms of complex analysis, the circle-preserving property (for small circles) is what defines a complex function as being *holomorphic*. What is a holomorphic function? Roughly speaking, it is a function that can be built up in terms of the basic complex-number operations of adding, subtracting, multiplying and dividing, together with the process of taking limits. For a general function, which can be non-holomorphic, we allow also the operation $z \to \bar{z}$ of *complex conjugation*, where

$$\bar{z} = x - \mathrm{i}y, \text{ if } z = x + \mathrm{i}y,$$

$x$ and $y$ being real numbers.

Now, I said that the basic idea of twistor theory is that you should take these complex numbers seriously not just for quantum mechanics, but for the whole of physics. This really means that as far as you can do it, you should try to make things holomorphic. But we have seen that the lensing effect of general relativity seems to tell us that we need non-holomorphic functions. In fact this is a serious point, and it is taking more than twenty years to get everything together and make it all work.

The most important idea seems to be the following one. The thing which is spoiling the holomorphic nature of the relevant quantities—these ellipses rather than circles—is the apparent need for complex conjugation. Apparently, what you have to do is bring in what is called *twistor quantization*. In the normal process of quantization you have to replace certain variables (such as energy) by differential operators. This seems a strange idea, but it is what one does all the time in quantum mechanics. In twistor theory one adopts just the same kind of procedure, but now it is the *complex conjugate* variables that become differential operators! Explicitly, one makes the replacement

$$\bar{z} \to -\hbar \frac{\partial}{\partial z}$$

Whenever the complex conjugate of a twistor variable $z$ begins to rear its ugly head. (Here $\hbar$ is Dirac's form of Planck's constant.) The operator $\partial/\partial z$ keeps things holomorphic, so the twistor philosophy can be maintained. This tends not to be easy in practice, however.

What I find striking about this is that you are somehow *forced* into a quantum picture of what is going on. Merely from trying to do classical general relativity, you are led naturally into these quantum ideas.

## 13 Is the universe open (k = −1)?

I just want to end with a bit of twistor dogma. In Fig. 7 I showed you three universe models—these are the standard models for different values of $k$. Does twistor theory have anything to say about which is to be preferred? In a sense it does, but I suppose it is really just twistor dogma. In the absence of people *really* knowing, I am going to plump for one of these, namely $k = -1$. Why am I doing this? Recall my hypothesis of vanishing Weyl curvature at the big bang. This implies that there is a beautifully simple twistor description of the whole universe at the very beginning. If this really is in accordance with my holomorphic philosophy, there ought to be a holomorphic structure controlling the surface that represents the big bang. It has to be one of these three different geometries—either the sphere or the Euclidean plane or the 'Escher picture' Lobachevsky plane—but in 3 dimensions. Why do I plump for Lobachevsky 3-space? Because it has a symmetry group which is exactly the same as the symmetry group which is the foundation of twistor theory, namely the group of symmetries of the Riemann sphere (restricted Lorentz group). Where is the Riemann sphere here? It is the boundary of the 3-dimensional version of the Escher picture of Fig. 6. (In Escher's picture this boundary is the circle where the fish crowd in at the edge; in 3-dimensions, this boundary is a Riemann sphere.)

Finally, I should comment that all this is actually rather close to Milne. If you take the Milne model (Fig. 5) and you go right back close to the big bang, but stop just short of it (say at some small value of $t$, because quantum theory comes in), you actually get a model which, for a suitable choice of conformal scale (Milne's idea, recall), is identical with the smoothed-out cosmology that I am espousing, with Lobachevskian spatial sections. Perhaps that is an appropriate tribute to Milne's work on which to end.

## Bibliography

Bondi, H. (1961). *Cosmology*. Cambridge University Press.

Huggett, S. A. and Tod, K. P. (1985) *An introduction to twistor theory*. Cambridge University press.

Milne, E. A. (1935). *Relativity, gravitation, and world-structure*. Oxford University Press.

Milne, E. A. (1948). *Kinematic relativity*. Oxford University Press.

Penrose, R. (1989). *The emperor's new mind*. Oxford University Press.

Penrose, R. and Rindler, W. (1984, 1986). *Spinors and space–time*, Vols 1 and 2. Cambridge University Press.

Rindler, W. (1977). *Essential relativity*. Springer-Verlag, Berlin.